Zabbix 7
IT Infrastructure Monitoring
Cookbook

Explore the new features of Zabbix 7 for designing, building, and maintaining your Zabbix setup

Nathan Liefting

Brian van Baekel

Zabbix 7 IT Infrastructure Monitoring Cookbook

Group Product Manager: Pavan Ramchandani

Publishing Product Manager: Khushboo Samkaria

Book Project Manager: Ashwin Dinesh Kharwa

Senior Editor: Romy Dias and Sujata Tripathi

Technical Editor: Rajat Sharma

Copy Editor: Safis Editing

Proofreader: Sujata Tripathi

Indexer: Pratik Shirodkar

Production Designer: Gokul Raj S.T

DevRel Marketing Coordinator: Rohan Dobhal

First published: February 2021

Second edition: March 2022

Third edition: July 2024

Production reference: 1280624

Published by Packt Publishing Ltd.

Grosvenor House

11 St Paul's Square

Birmingham

B3 1RB, UK

ISBN: 978-1-80107-832-0

www.packtpub.com

To my grandparents, for supporting my education, my brother, for always being at the ready, and my mom and stepdad, for cheering me on. To my wife, for always supporting whatever new idea I get into my head. To my colleagues, throughout the years, my first mentor, Sander F., for inspiring me, and Brian, for making it all possible.

– Nathan Liefting

Foreword

Dear readers,

Brian and Nathan, the authors of this book, are well-known figures in the Zabbix community.

I met them many years ago, and I have always been impressed by their deep knowledge of the product, which would be impossible without extensive practical experience. How could it be otherwise? Brian and Nathan have been professionally working with Zabbix for many years, helping their clients with the architecture, installation, configuration, and maintenance of Zabbix. The authors of this book have earned their reputation through their vast experience and ability to continuously share this knowledge with other Zabbix users. You can find Brian and Nathan speaking at conferences and participating in various community platforms, always ready to help and answer your questions!

Zabbix allows you to quickly immerse yourself in the wonderful world of monitoring. In Zabbix development, we follow this principle: "Make simple things easy, and complex things possible." But don't let this simplicity and speed deceive you. Soon enough, you will start to realize that simple monitoring tasks are indeed easy to implement. However, the further you move forward, the quicker you will understand that you need to study and carefully read the documentation. You will discover that there are more complex monitoring tasks that require a deeper understanding of Zabbix's capabilities.

This book fills the gap between the official documentation, which describes the product's functionality, and real-life situations that require specific solutions and it does it perfectly. Especially valuable is that this book covers the new functionality of Zabbix 7.0 LTS, the latest release of Zabbix to date!

This book is a must-read for anyone who wants to learn specific techniques for solving particular monitoring and observability tasks. How great it is to have solutions to many challenges you may encounter at your fingertips!

I am grateful to the authors of the book for their work and wish everyone an enjoyable reading. You will not be disappointed!

Alexei Vladishev

Creator of Zabbix

Contributors

About the authors

Nathan Liefting, also known as **Larcorba**, is an IT consultant and trainer. He has more than 9 years of professional experience in IT. His experience ranges from managing networks running EVPN/VXLAN to Linux environments and programming. Nathan started working with Zabbix in 2016, when it was still at Zabbix 2 and Zabbix 3 was just released.

He is now working for Opensource ICT Solutions BV in The Netherlands as a Zabbix trainer and consultant. Here, he designs and builds professional Zabbix environments and Zabbix components for some of the biggest and most interesting companies around the world.

Brian van Baekel quickly discovered how powerful Zabbix is during his career as a network engineer. Ever since, he has been working with Zabbix in various (large) environments, leading to him gaining his official Zabbix Certified Trainer certification in early 2017.

In 2018, Brian founded Opensource ICT Solutions BV in the Netherlands and Opensource ICT Solutions LLC in the USA. Both companies primarily focus on building Zabbix environments all over the world. In 2021, further expansion of the business was established by opening a subsidiary in the United Kingdom with a full focus on the product Zabbix. All companies provide support, training, and consultancy services, and basically, that means Brian is working with the Zabbix product 24/7.

Fun fact: even his cat is named "Zabbix" and you also should ask Brian what's on his arm some time.

About the reviewers

Sven Putteneers got to know Zabbix more than eight years ago as part of his duties at his then-new employer, after which he swore off all other monitoring systems. He has founded a company, 7 to 7, through which he provides Zabbix consultancy and develops custom integrations with external systems. He is also an active member of the Zabbix International Community Telegram group, where he is known as @OffByOne. In his free time, he likes to swim, read science fiction, and go to conferences and meet-ups. He dislikes social media but can be found on LinkedIn.

First of all, I'd like to thank the authors of the book for providing a valuable resource to the Zabbix community. Furthermore, I'd like to give a shout-out to the Zabbix community and people who try to help each other in their free time, be it through the forums, IRC, Telegram groups, and so on …. You are awesome!

Andreas Drbal boasts a decade in tech, evolving from a self-taught frontend developer to a key figure in DevOps and CloudOps. His academic journey led him to UCLA, majoring in English with a German minor, and later acquiring a master's in information systems management and an MBA. Andreas's passion for monitoring began with Zabbix 3.x in 2015, integrating it with IoT devices. Currently, as a senior manager at Cloud Software Group, he's driven by automation, streamlining the DevOps life cycle, and ensuring team productivity.

I want to thank my wife, who encouraged me to contribute to the learning community by active participation, and also for her patience during the period of reviewing of this book.

Table of Contents

3

Setting Up Zabbix Monitoring   79

4

Working with Triggers and Alerts 149

5

Building Your Own Structured Templates 185

6

Visualizing Data, Inventory, and Reporting   217

7

Using Discovery for Automatic Creation 283

8

Setting Up Zabbix Proxies 327

9

Integrating Zabbix with External Services   361

10

Extending Zabbix Functionality with Custom Scripts and the Zabbix API   403

11

Maintaining Your Zabbix Setup   427

Preface

Welcome to *Zabbix 7 IT Infrastructure Monitoring Cookbook*. IT infrastructure ranges from Windows and Linux to networking and development, and basically anything that runs on computer hardware. In this book, we will go over various subjects useful to anyone in IT who wants to use Zabbix to monitor their IT infrastructure.

Who this book is for

Monitoring systems are often overlooked within IT organizations, but they can provide an overview that will save you time, money, and headaches. This book is for IT engineers who want to learn something about Zabbix 7 and how to use it to bring their IT environments to the next level.

What this book covers

Chapter 1, Installing Zabbix and Getting Started Using the Frontend, covers how to set up Zabbix optionally with HA and work your way through its menus.

Chapter 2, Getting Things Ready with Zabbix User Management, covers how to set up your first users, user groups, and user roles.

Chapter 3, Setting Up Zabbix Monitoring, covers how to set up almost any type of monitoring within Zabbix.

Chapter 4, Working with Triggers and Alerts, covers how to set up triggers and get alerts from them.

Chapter 5, Building Your Own Structured Templates, covers how to build templates that are structured and will work wonders for keeping your Zabbix setup organized.

Chapter 6, Visualizing Data, Inventory and Reporting, covers how to visualize data in graphs, maps, and dashboards. It also covers how to use the Zabbix inventory, reporting, and business service monitoring functionality.

Chapter 7, Using Discovery for Automatic Creation, covers how to use Zabbix discovery for automatic host creation as well as items, triggers, and more with agents, SNMP, WMI, and JMX.

Chapter 8, Setting Up Zabbix Proxies, teaches how to set up Zabbix proxies correctly for use in a production environment, with the addition of the new proxy high availability.

Chapter 9, Integrating Zabbix with External Services, teaches how to integrate Zabbix with external services for alerting.

Chapter 10, Extending Zabbix Functionality with Custom Scripts and Zabbix API, covers how to extend Zabbix functionality by using custom scripts and the Zabbix API.

Chapter 11, Maintaining Your Zabbix Setup, covers how to maintain a Zabbix setup and keep its performance up over time.

Chapter 12, Advanced Zabbix Database Management, teaches how to manage Zabbix databases for an advanced setup.

Chapter 13, Bringing Zabbix to the Cloud with Zabbix Cloud Integration, covers how to use Zabbix in the cloud with services such as AWS, Azure, Docker, and Kubernetes.

To get the most out of this book

You should have a good basis in IT to understand the terminology used in this book. This book is best for people with at least basic knowledge of monitoring systems, Linux, and network engineering.

Software/hardware covered in the book	Operating system requirements
Zabbix 7	Linux (any)
Python 3	
MariaDB (MySQL)	
PostgreSQL	
NGINX	
VIM	

Make sure you have a virtualization environment ready to create virtual machines for use with the recipes. VirtualBox, VMware, or any other type of client/hypervisor will do.

Throughout the book, we will make use of Vim to edit files, so make sure to install it. If you do not feel comfortable using Vim, you can substitute this for Nano or anything else you prefer.

If you are using the digital version of this book, we advise you to type the code yourself or access the code via the GitHub repository (link available in the next section). Doing so will help you avoid any potential errors related to the copying and pasting of code.

Download the example code files

You can download the example code files for this book from GitHub at https://github.com/PacktPublishing/Zabbix-7-IT-Infrastructure-Monitoring-Cookbook. If there's an update to the code, it will be updated on the existing GitHub repository.

We also have other code bundles from our rich catalog of books and videos available at https://github.com/PacktPublishing/. Check them out!

Conventions used

There are a number of text conventions used throughout this book.

`Code in text`: Indicates code words in text, database table names, folder names, filenames, file extensions, pathnames, dummy URLs, user input, and Twitter handles. Here is an example: "The `log_bin_trust_function_creators` function is set to 1 here to allow the initial database data to be imported.

A block of code is set as follows:

```
# listen 8080;
# server_name example.com;
```

When we wish to draw your attention to a particular part of a code block, the relevant lines or items are set in bold:

```
# MariaDB Server
# To use a different major version of the server, or to pin to a
specific minor version, change URI below. deb [arch=amd64,arm64]
https://dlm.mariadb.com/repo/mariadbserver/11.4/repo/ubuntu jammy main
```

Any command-line input or output is written as follows:

```
systemctl start mariadb
```

Bold: Indicates a new term, an important word, or words that you see onscreen. For example, words in menus or dialog boxes appear in the text like this. Here is an example: "We will also need a **virtual IP (VIP)** address for our cluster nodes."

> **Tips or important notes**
> Appear like this.

Sections

In this book, you will find several headings that appear frequently (*Getting ready*, *How to do it...*, *How it works...*, *There's more...*, and *See also*).

To give clear instructions on how to complete a recipe, use these sections as follows:

Getting ready

This section tells you what to expect in the recipe and describes how to set up any software or any preliminary settings required for the recipe.

How to do it...

This section contains the steps required to follow the recipe.

How it works...

This section usually consists of a detailed explanation of what happened in the previous section.

There's more...

This section consists of additional information about the recipe in order to make you more knowledgeable about the recipe.

See also

This section provides helpful links to other useful information for the recipe.

Get in touch

Feedback from our readers is always welcome.

General feedback: If you have questions about any aspect of this book, mention the book title in the subject of your message and email us at customercare@packtpub.com.

Errata: Although we have taken every care to ensure the accuracy of our content, mistakes do happen. If you have found a mistake in this book, we would be grateful if you would report this to us. Please visit www.packtpub.com/support/errata, select your book, click on the Errata Submission Form link, and enter the details.

Piracy: If you come across any illegal copies of our works in any form on the Internet, we would be grateful if you would provide us with the location address or website name. Please contact us at copyright@packt.com with a link to the material.

If you are interested in becoming an author: If there is a topic that you have expertise in and you are interested in either writing or contributing to a book, please visit authors.packtpub.com.

Before we get started

Whether you are a real Zabbix guru or you've just started working with Zabbix, this book will include some recipes for everyone. We will go over most of the Zabbix basics and even do some cool stuff with the Zabbix API and databases in the book.

We decided to write this book because we think it's important to be part of the community and want to supply you with the Zabbix information available online and in the official Zabbix training materials in a clear and straightforward way from a reliable source. We've all been through the process of bookmarking all those amazing community blog posts, community guides, and even official documentation. Sometimes it can be a bit much, which is where this book will help. See it as your guide with something for everyone without the need to Google until your fingers fall off.

Now even when you have gained experience, finished this and maybe other books, and you've bookmarked every useful page about Zabbix, you might still not know everything. This is where we come in. Zabbix is a free product built on an amazing open source community, but besides that, there are some real Zabbix gurus out there that decided to make a living out of it. Our company, Opensource ICT Solutions, comes from these humble beginnings, and we are there to provide our customers with everything they need when it comes to Zabbix. As a Premium Zabbix partner, we provide the following:

- Official Zabbix training
- Official Zabbix support
- Zabbix consultancy
- Turnkey solutions
- Custom integrations
- 24/7-available service desk

As providing services on Zabbix is our core business, we see different environments, different customers, and different use cases of the product every day, while working out the best solutions we can for our worldwide customer base. As an official training partner, it's not only about building environments but also about sharing knowledge and teaching others how to get the most out of the product.

So, if you've enjoyed this book, please do think about us and others in our amazing Zabbix community. Give us a follow on LinkedIn (and other social media) and whenever you need help, give us a call! We will definitely be ready to help you out with any questions you might run into.

United Kingdom	The Netherlands	United States
Opensource ICT Solutions LTD	Opensource ICT Solutions B.V.	Opensource ICT Solutions LLC
Phone: +44 20 4551 1827	Phone: +31 (0)72 743 65 83	Phone: +1-929-377-1232
https://oicts.com info@oicts.com		

Share Your Thoughts

Once you've read *Zabbix 7 IT Infrastructure Monitoring Cookbook*, we'd love to hear your thoughts! Scan the QR code below to go straight to the Amazon review page for this book and share your feedback.

https://packt.link/r/1801078327

Your review is important to us and the tech community and will help us make sure we're delivering excellent quality content.

Download a free PDF copy of this book

Thanks for purchasing this book!

Do you like to read on the go but are unable to carry your print books everywhere?

Is your eBook purchase not compatible with the device of your choice?

Don't worry, now with every Packt book you get a DRM-free PDF version of that book at no cost.

Read anywhere, any place, on any device. Search, copy, and paste code from your favorite technical books directly into your application.

The perks don't stop there, you can get exclusive access to discounts, newsletters, and great free content in your inbox daily

Follow these simple steps to get the benefits:

1. Scan the QR code or visit the link below

https://packt.link/free-ebook/978-1-80107-832-0

2. Submit your proof of purchase
3. That's it! We'll send your free PDF and other benefits to your email directly

1

Installing Zabbix and Getting Started Using the Frontend

For Zabbix 7, the developers have really outdone themselves. In this **Long Term Support** (**LTS**) release, we will find far more **quality-of-life** (**QoL**) changes along with some impressive new cutting-edge features. Coming from Zabbix 6.0, you will still find a lot of improvements made in Zabbix 6.2 and 6.4 as those have, of course, been included in Zabbix 7.0 LTS. We will detail all important changes throughout the book.

In this chapter, we will install the Zabbix server and explore the Zabbix UI to get you familiar with it. We will go over finding your hosts, triggers, dashboards, and more to make sure you feel confident diving into the deeper material later on in this book. The Zabbix UI has a lot of options to explore, so if you are just getting started, don't get overwhelmed. It's quite structurally built, and once you get the hang of it, I am confident you will find your way without issues. You will learn all about these subjects in the following recipes:

- Installing the Zabbix server
- Setting up the Zabbix frontend
- Enabling Zabbix server **high availability** (**HA**)
- Using the Zabbix frontend
- Navigating the Zabbix frontend

Technical requirements

We'll be starting this chapter with an empty Linux (virtual) machine. Feel free to choose a RHEL- or Debian-based Linux distribution (we'll be using Ubuntu in the examples). It's recommended to use a server distribution and not a desktop distribution of the Linux distribution you choose. We will then set up a Zabbix server from scratch on this host.

So before jumping in, make sure you have your Linux host at the ready. I'll be using Rocky Linux 9 and Ubuntu 22.04 in my examples.

Installing the Zabbix server

Before doing anything within Zabbix, we need to install it and get ready to start working with it. In this recipe, we are going to discover how to install Zabbix server 7.0.

Getting ready

Before we actually install the Zabbix server, we are going to need to fulfill some prerequisites. We will be using **MariaDB** mostly throughout this book. MariaDB is popular, and a lot of information is available on using it with Zabbix.

At this point, you should have a prepared Linux server in front of you running either a RHEL- or Debian-based distribution. I'll be installing Rocky Linux 9 and Ubuntu 22.04 on my server; let's call them `lar-book-rocky` and `lar-book-ubuntu`.

When you have your server ready, we can start the installation process.

How to do it...

1. Let's start by adding the Zabbix 7.0 repository to our system.

    For RHEL-based systems, run the following:

    ```
    rpm -Uvh https://repo.zabbix.com/zabbix/7.0/rocky/9/x86_64/
    zabbix-release-7.0-2.el9.noarch.rpm
    dnf clean all
    ```

    For Ubuntu systems, run the following:

    ```
    wget https://repo.zabbix.com/zabbix/7.0/ubuntu/pool/main/z/
    zabbix-release/zabbix-release_7.0-1+ubuntu22.04_all.deb
    dpkg -i zabbix-release_7.0-1+ubuntu22.04_all.deb
    apt update
    ```

2. For RHEL-based systems, we'll also remove the Zabbix **Extra Packages for Enterprise Linux (EPEL)** repository packages (if installed):

    ```
    vim /etc/yum.repos.d/epel.repo
    ```

3. Then, add the following line:

    ```
    [epel]
    ...
    excludepkgs=zabbix*
    ```

4. Now that the repository is added, let's add the MariaDB repository on our server:

    ```
    wget https://downloads.mariadb.com/MariaDB/mariadb_repo_setup
    chmod +x mariadb_repo_setup
    ./mariadb_repo_setup
    ```

5. Then, install and enable it.

    For RHEL-based systems, run the following:

    ```
    dnf install mariadb-server
    systemctl enable mariadb
    systemctl start mariadb
    ```

    For Ubuntu systems, run the following:

    ```
    apt install mariadb-server
    systemctl enable mariadb
    systemctl start mariadb
    ```

6. After installing MariaDB, make sure to secure your installation by running the following command:

    ```
    mariadb-secure-installation
    ```

7. Make sure to answer the questions with yes (Y) and configure a root password that's secure.

8. Run through the secure installation setup, and make sure to save your password somewhere. It's highly recommended to use a password vault.

9. Now, let's install our Zabbix server with MySQL support.

    For RHEL-based systems, run the following command:

    ```
    dnf install zabbix-server-mysql zabbix-sql-scripts
    zabbix-selinux-policy
    ```

    For Ubuntu systems, run the following command:

    ```
    apt install zabbix-server-mysql zabbix-sql-scripts
    ```

10. With the Zabbix server installed, we are ready to create our Zabbix database. Log in to MariaDB with the following:

    ```
    mysql -u root -p
    ```

11. Enter the password you set up during the secure installation. Next, we'll create the Zabbix database with the following commands. Do not forget to change password in the second and third commands, as this will be the password used for connecting to your Zabbix database:

    ```
    create database zabbix character set utf8mb4 collate
    utf8mb4_bin;
    create user zabbix@localhost identified by 'password';
    ```

```
grant all privileges on zabbix.* to zabbix@localhost identified
by 'password';
set global log_bin_trust_function_creators = 1;
quit
```

The log_bin_trust_function_creators function is set to 1 here to allow the initial database data to be imported. We will disable it again afterwards.

Tip

Since Zabbix 6, Zabbix uses utf8mb4 by default in all its installation documentation. We've changed utf8 to utf8mb4 in the preceding command so that everything will work. For reference, check the Zabbix support ticket here: https://support.zabbix.com/browse/ZBXNEXT-3706.

---

12. We now need to import our Zabbix database schema to our newly created Zabbix database:

```
zcat /usr/share/zabbix-sql-scripts/mysql/server.sql.gz | mariadb
--default-character-set=utf8mb4 -u zabbix -p zabbix
```

13. As mentioned, we can now disable log_bin_trust_function_creators again:

```
mysql -u root -p
set global log_bin_trust_function_creators = 0;
quit;
```

Important note

At this point, it might look like you are stuck and the system is not responding. Do not worry, though, as it will just take a while to import the SQL schema.

We are now done with the preparations for our MariaDB side and are ready to move on to the next step, which will be configuring the Zabbix server.

1. The Zabbix server is configured using the Zabbix server config file. This file is located in /etc/zabbix/. Let's open this file with our favorite editor; I'll be using Vim throughout the book (but feel free to substitute Vim with vi or nano):

```
vim /etc/zabbix/zabbix_server.conf
```

2. Now, make sure the following lines in the file match your database name, database user username, and database user password:

```
DBName=zabbix
DBUser=zabbix
DBPassword=password
```

> **Tip**
>
> Before starting the Zabbix server, you should configure SELinux or AppArmor to allow the use of the Zabbix server. If this is a test machine, you can use a permissive stance for SELinux or disable AppArmor, but it is not recommended to do this in production.

3. All done; we are now ready to start our Zabbix server:

    ```
    systemctl enable zabbix-server
    systemctl start zabbix-server
    ```

4. Check whether everything is starting up as expected with the following:

    ```
    systemctl status zabbix-server
    ```

5. Also, make sure to monitor the log file, which provides a detailed description of the Zabbix startup process:

    ```
    tail -f /var/log/zabbix/zabbix_server.log
    ```

6. Most of the messages in this file are fine and can be ignored safely, but make sure to read them well and see if there are any issues with your Zabbix server starting.

How it works...

The Zabbix server is the main process for our Zabbix setup. It is responsible for our monitoring, problem alerting, and a lot of the other tasks described in this book. A complete Zabbix stack consists of at least the following:

- A database (MySQL/MySQL fork, PostgreSQL, or Oracle)
- A Zabbix server
- Apache or NGINX running the Zabbix frontend with PHP 8.0 or higher

We can see the components and how they communicate with each other in the following diagram:

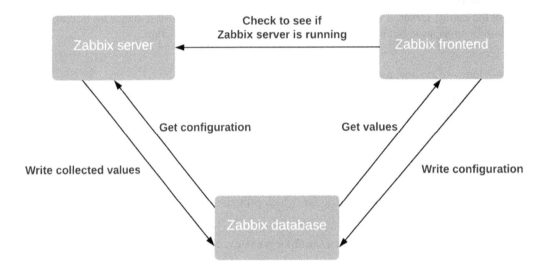

Figure 1.1 – Zabbix setup communications diagram

We've just set up the Zabbix server and database; by running these two, we are basically ready to start monitoring. The Zabbix server communicates with the Zabbix database to write collected values to it.

There is still one problem, though: we cannot configure our Zabbix server to do anything. For this, we are going to need our Zabbix frontend, which we'll set up in the next recipe.

Setting up the Zabbix frontend

The Zabbix frontend is the face of our server. It's where we will configure all of our hosts, templates, dashboards, maps, and everything else. Without it, we would be blind to what's going on on the server side. So, let's set up our Zabbix frontend in this recipe.

Getting ready

We are going to set up the Zabbix frontend using NGINX. It's also possible to use Apache, but NGINX is known to be faster and, as such, it has a slight edge over Apache. Since installation of both NGINX and Apache is quite simple, NGINX is the preferred way to go if you have a lot of frontend users. Before starting with this recipe, make sure you are running the Zabbix server on a Linux distribution of your choice. I'll be using the lar-book-rocky and lar-book-ubuntu hosts in these recipes to show the setup process on Rocky Linux 9 and Ubuntu 22.04.

How to do it...

1. Let's jump right in and install the frontend.

    For RHEL-based systems, run the following:

    ```
    dnf module switch-to php:8.3
    dnf install zabbix-web-mysql zabbix-nginx-conf
    ```

    For Ubuntu systems, run the following:

    ```
    apt install zabbix-frontend-php zabbix-nginx-conf
    ```

    > **Tip**
    >
    > Don't forget to allow ports 80 and 443 in your firewall if you are using one. Without this, you won't be able to connect to the frontend.

2. We will then have to configure our NGINX configuration.

    For RHEL-based systems, edit the following file:

    ```
    /etc/nginx/conf.d/zabbix.conf
    ```

    For Ubuntu systems, edit the following file:

    ```
    /etc/zabbix/nginx.conf
    ```

3. Then, edit the following two lines:

    ```
    # listen 8080;
    # server_name example.com;
    ```

    Make it look like this:

    ```
    listen 80;
    server_name 192.168.0.50;
    ```

    > **Important note**
    >
    > At server_name, it is important to add the IP address of where this NGINX (Zabbix) web page will be running. Add the IP address or DNS name of where you want your frontend to be available here.

4. Restart the Zabbix components and make sure they start up when the server is booted.

    For RHEL-based systems, run the following:

    ```
    systemctl enable nginx php-fpm
    systemctl restart nginx php-fpm
    ```

For Ubuntu systems, run the following:

```
systemctl enable nginx
systemctl restart nginx php-fpm
```

5. We should now be able to navigate to our Zabbix frontend without any issues and start the final steps to set up the Zabbix frontend.

6. Let's go to our browser and navigate to our server's IP. It should look like this:

```
http://<your_server_ip>/
```

> **Important note**
>
> On Ubuntu (in some cases), you might have to add port 8080, which you will probably want to change to port 80 (or 443 after adding SSL) later. In those cases, the URL should look like this:
>
> ```
> http://<your_server_ip>:8080/
> ```

7. We should now see the following web page:

Figure 1.2 – The Zabbix welcome screen

If you don't see this web page, you may have missed some steps in the setup process. Retrace your steps and double-check your configuration files; even the smallest typo could prevent the web page from serving.

8. Let's continue by clicking **Next step** on this page, which will serve you with the next page:

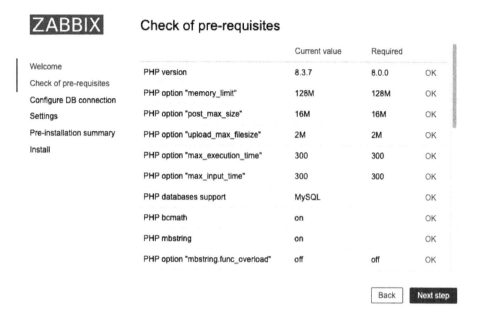

Figure 1.3 – The Zabbix installation prerequisites page

9. Every single option here should be showing **OK** now; if not, fix the mistake it's showing you. If everything is OK, you may proceed by clicking **Next step** again, which will take you to the next page:

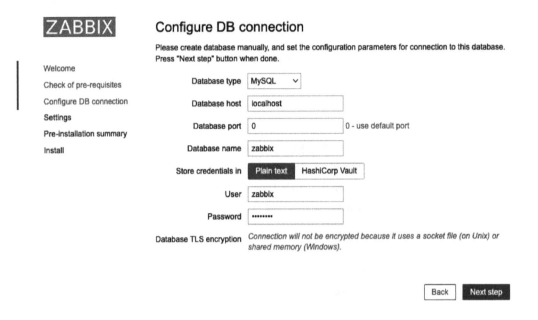

Figure 1.4 – The Zabbix installation DB connection page

10. Here, we need to tell our Zabbix frontend where our MySQL database is located. Since we installed it on `localhost`, we just need to make sure we issue the right database name, database user username, and database user password.

11. This should make the Zabbix frontend able to communicate with the database. Let's proceed by clicking **Next step** again:

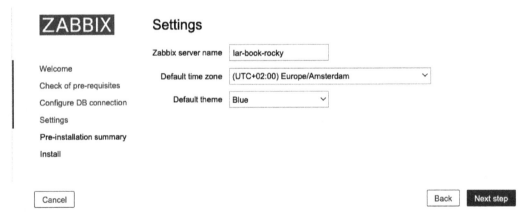

Figure 1.5 – The Zabbix installation server details page

Next up is the Zabbix server configuration. Make sure to name your server something useful or something cool. For example, I set up a production server called Meeseeks because every time we got an alert, we could make Zabbix say "*I'm Mr. Meeseeks. Look at me.*" But something such as zabbix.example.com also works.

12. Let's name our server, set up the time zone to match our own time zone, and proceed to the next step:

Figure 1.6 – The Zabbix installation summary page

13. Verify your settings and proceed to click **Next step** one more time:

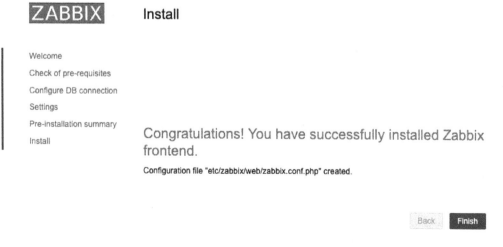

Figure 1.7 – The Zabbix installation finish page

14. You have successfully installed the Zabbix frontend. You may now click the **Finish** button, and we can start using the frontend. You'll be served with a login page where you can use the following default credentials:

- **Username**: Admin

- **Password**: zabbix

Keep in mind that the username and password fields are both case-sensitive.

How it works...

Now that we've installed our Zabbix frontend, our Zabbix setup is complete and we are ready to start working with it. Our Zabbix frontend will connect to our database to edit the configuration values of our setup, as we can see in the following diagram:

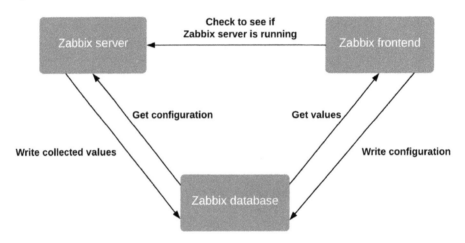

Figure 1.8 – Zabbix setup communications diagram

The Zabbix frontend will also talk to our Zabbix server, but this is just to make sure the Zabbix server is up and running and to provide some additional functionality such as a test button on items. Now that we know how to set up the Zabbix frontend, we can start using it. Let's check this out after the next recipe.

There's more...

Zabbix provides a very convenient setup guide that contains a lot of detail regarding the installation of Zabbix. I would always recommend keeping this page open during a Zabbix installation as it contains information such as a link to the latest repository. Check it out here:

https://www.zabbix.com/download

Enabling Zabbix server HA

Zabbix 6 shipped with one of the most anticipated features of all time, HA. This feature will bring your Zabbix setup to the next level by making sure that if one of your Zabbix servers is having issues, another one will take over. Of course, this feature is still available in Zabbix 7, and we will be making good use of it.

A great thing about this implementation is that it supports an easy proprietary way to put one to many Zabbix servers in a cluster: a great way of making sure your monitoring stays in the air at all times (or at least as much as possible).

The setup for the Zabbix server will be active/passive, as we cannot do anything such as load balancing on the Zabbix server side. Load balancing is, however, supported using Zabbix proxies, which we will discuss in *Chapter 8*, *Setting up Zabbix Proxies*.

Getting ready

Before getting started, please note that creating an HA setup is considered an advanced topic. It might be more difficult than other recipes in this chapter.

For this setup, we will need three new virtual machines, as we are going to create a split Zabbix setup, unlike the setup that we created in the first recipe of this chapter. Let's take a look at how I have named our three new virtual machines and what their IP addresses will be:

- `lar-book-ha1` (192.168.0.1)
- `lar-book-ha2` (192.168.0.2)
- `lar-book-ha-db` (192.168.0.10)

Two of these servers will run our Zabbix server cluster and a Zabbix frontend. The other server is just for our MySQL database. Please take note that the IP addresses used in the example may be different for you. Use the correct ones for your environment.

We will also need a **virtual IP** (**VIP**) address for our cluster nodes. We will use 192.168.0.5 in the example.

> Tip
>
> In our setup, we are using only one MySQL Zabbix database. To make sure all parts of Zabbix are set up as highly available, it might be worth looking into setting up MySQL in a primary/ primary setup. This can be a great combination with the Zabbix server's HA.

This cookbook will *not* use SELinux or AppArmor, so make sure to add the correct policies before or during the use of this guide. It's also possible to disable SELinux, but this is not recommended in production. Additionally, this guide does not detail how to set up your firewall, so make sure to do this beforehand as well.

How to do it...

For your convenience, we've split this *How to do it...* section into three parts. The first is setting up the database, the next is setting up the Zabbix server cluster, and the last is how to set up the Zabbix frontend redundantly. The *How it works...* section will then provide an explanation of the entire setup.

Setting up the database

Let's start with setting up our Zabbix database, ready to be used in a highly available Zabbix server setup.

1. Log in to `lar-book-ha-db` and install the MariaDB repository with the following command on Red Hat-based systems:

    ```
    wget https://downloads.mariadb.com/MariaDB/mariadb_repo_setup
    chmod +x mariadb_repo_setup
    ./mariadb_repo_setup
    ```

2. Then, let's install the MariaDB server application.

    For RHEL-based systems, run the following:

    ```
    dnf install mariadb-server
    systemctl enable mariadb
    systemctl start mariadb
    ```

    For Ubuntu systems, run the following:

    ```
    apt install mariadb-server
    systemctl enable mariadb
    systemctl start mariadb
    ```

3. After installing MariaDB, make sure to secure your installation with the following command:

    ```
    mariadb-secure-installation
    ```

4. Make sure to answer the questions with yes (`Y`) and configure a root password that's secure. It's highly recommended to use a password vault for storing it.

5. Now, let's create our Zabbix database for our Zabbix servers to connect to. Log in to MariaDB with the following command:

    ```
    mariadb -u root -p
    ```

6. Enter the password you set up during the secure installation. Next, we'll create a Zabbix database with the following commands. Do not forget to change `password` in the second, third, and fourth commands:

    ```
    create database zabbix character set utf8mb4 collate
    utf8mb4_bin;
    create user zabbix@'192.168.0.1' identified by 'password';
    ```

```
create user zabbix@'192.168.0.2' identified by 'password';
create user zabbix@'192.168.0.5' identified by 'password';
grant all privileges on zabbix.* to 'zabbix'@'192.168.0.1'
identified by 'password';
grant all privileges on zabbix.* to 'zabbix'@'192.168.0.2'
identified by 'password';
grant all privileges on zabbix.* to 'zabbix'@'192.168.0.5'
identified by 'password';
set global log_bin_trust_function_creators = 1;
quit
```

7. Lastly, we need to import the initial Zabbix database configuration, but for that, we need to install the Zabbix repository.

    For RHEL-based systems, run the following command:

    ```
    rpm -Uvh https://repo.zabbix.com/zabbix/7.0/rocky/9/x86_64/
    zabbix-release-7.0-2.el9.noarch.rpm
    dnf clean all
    ```

    For Ubuntu systems, run the following command:

    ```
    wget https://repo.zabbix.com/zabbix/7.0/ubuntu/pool/main/z/
    zabbix-release/zabbix-release_7.0-1+ubuntu22.04_all.deb
    dpkg -i zabbix-release_7.0-1+ubuntu22.04_all.deb
    apt update
    ```

8. Then, we need to install the SQL scripts Zabbix module.

    For RHEL-based systems, run the following command:

    ```
    dnf install zabbix-sql-scripts
    ```

    For Ubuntu systems, run the following command:

    ```
    apt install zabbix-sql-scripts
    ```

9. Then, we issue the following command, which might take a while, so be patient until it is done:

    ```
    zcat /usr/share/doc/zabbix-sql-scripts/mysql/server.sql.gz |
    mysql --default-character-set=utf8mb4 -u root -p zabbix
    ```

10. We will need to disable log_bin_trust_function_creators after this step:

    ```
    mysql -u root -p
    set global log_bin_trust_function_creators = 0;
    quit;
    ```

Setting up the Zabbix server cluster nodes

Setting up the cluster nodes works in the same way as setting up any new Zabbix server. The only difference is that we will need to specify some new configuration parameters.

1. Let's start by adding the Zabbix 7.0 repository to our `lar-book-ha1` and `lar-book-ha2` systems:

    ```
    rpm -Uvh https://repo.zabbix.com/zabbix/7.0/rocky/9/x86_64/
    zabbix-release-7.0-2.el9.noarch.rpm
    dnf clean all
    ```

    For Ubuntu systems, use the following command:

    ```
    wget https://repo.zabbix.com/zabbix/7.0/ubuntu/pool/main/z/
    zabbix-release/zabbix-release_7.0-1+ubuntu22.04_all.deb
    dpkg -i zabbix-release_7.0-1+ubuntu22.04_all.deb
    apt update
    ```

2. Now, let's install the Zabbix server application.

    For RHEL-based systems, run the following command:

    ```
    dnf install zabbix-server-mysql zabbix-selinux-policy
    ```

    For Ubuntu systems, run the following command:

    ```
    apt install zabbix-server-mysql
    ```

3. We will now edit the Zabbix server configuration files, starting with `lar-book-ha1`. Issue the following command:

    ```
    vim /etc/zabbix/zabbix_server.conf
    ```

4. Then, add the following lines to allow a database connection:

    ```
    DBHost=192.168.0.10
    DBPassword=password
    ```

    `DBName` and `DBUser` are both set to `zabbix` by default. If you used a different database and/or username, you will have to change those as well.

5. To enable HA on this host, add the following lines in the same file:

    ```
    HANodeName=lar-book-ha1
    ```

6. To make sure our Zabbix frontend knows where to connect to if there is a node failover, fill in the following:

    ```
    NodeAddress=192.168.0.1
    ```

7. Save the file, and let's do the same for our `lar-book-ha2` host by editing its file:

    ```
    vim /etc/zabbix/zabbix_server.conf
    ```

8. Then, add the following lines to allow a database connection:

    ```
    DBHost=192.168.0.10
    DBPassword=password
    ```

    DBName and DBUser are both set to zabbix by default. If you used a different database and/
    or username, you will have to change those as well.

9. To enable HA on this host, add the following lines in the same file:

    ```
    HANodeName=lar-book-ha2
    ```

10. To make sure our Zabbix frontend knows where to connect to if there is a node failover, fill
    in the following:

    ```
    NodeAddress=192.168.0.2
    ```

11. Save the file, and let's start our Zabbix server:

    ```
    systemctl enable zabbix-server
    systemctl start zabbix-server
    ```

Setting up NGINX with HA

To make sure our frontend is also set up in such a way that if one Zabbix server has issues, it fails over,
we will set them up with keepalived. Let's see how we can do this.

1. Let's start by logging in to both lar-book-ha1 and lar-book-ha2 and
    installing keepalived.

    For RHEL-based systems, run the following:

    ```
    dnf install -y keepalived
    ```

    For Ubuntu systems, run the following:

    ```
    apt install keepalived
    ```

2. Then, on lar-book-ha1, edit the keepalived configuration with the following command:

    ```
    vim /etc/keepalived/keepalived.conf
    ```

3. Delete everything from this file (if it's not empty already) and add the following text to the file:

    ```
    vrrp_track_process chk_nginx {
            process nginx
            weight 10
    }
    vrrp_instance ZBX_1 {
            state MASTER
            interface ens192
    ```

```
virtual_router_id 51
priority 244
advert_int 1
authentication {
            auth_type PASS
            auth_pass password
}
track_process {
            chk_nginx
}
virtual_ipaddress {
            192.168.0.5/24
}
}
```

4. Do not forget to update `password` to something secure and edit the `ens192` interface to your own interface name/number.

> **Important note**
>
> In the previous file, we specified `virtual_router_id 51`; make sure the virtual router ID `51` isn't used anywhere in the network yet. If it is, simply change the virtual router ID throughout this recipe.

5. On `lar-book-ha2`, edit the same file with the following command:

    `vim /etc/keepalived/keepalived.conf`

6. Delete everything from the file with dG (if you are using vim), and this time, we will add the following information:

```
vrrp_track_process chk_nginx {
        process nginx
        weight 10
}
vrrp_instance ZBX_1 {
        state BACKUP
        interface ens192
        virtual_router_id 51
        priority 243
        advert_int 1
        authentication {
                auth_type PASS
```

```
                    auth_pass password
        }
        track_process {
                chk_nginx
        }
        virtual_ipaddress {
                192.168.0.5/24
        }
    }
```

7. Once again, do not forget to update `password` to something secure and edit the `ens192` interface to your own interface name/number.

8. Now, let's install the Zabbix frontend.

    For RHEL-based systems, run the following:

    ```
    dnf install nginx zabbix-web-mysql zabbix-nginx-conf zabbix-
    selinux-policy
    ```

    For Ubuntu systems, run the following:

    ```
    apt install nginx zabbix-frontend-php zabbix-nginx-conf
    ```

9. We will then have to configure our NGINX configuration.

    For RHEL-based systems, edit the following file:

    ```
    /etc/nginx/conf.d/zabbix.conf
    ```

    For Ubuntu systems, edit the following file:

    ```
    /etc/zabbix/nginx.conf
    ```

10. Then, edit the following two lines:

    ```
    #   listen          8080;
    #   server_name     example.com;
    ```

    Make it look like this:

    ```
    listen          80;
    server_name     192.168.0.5;
    ```

Important note

At `server_name`, it is important to add the IP address of where this NGINX (Zabbix) web page will be running. Add the IP address or DNS name of where you want your frontend to be available here. In this case, we have added the VIP address that `keepalived` will be managing.

11. Start the web server and `keepalived` to make your Zabbix frontend available with the following command:

```
systemctl enable nginx keepalived
systemctl start nginx keepalived
```

12. Then, we are ready to configure our Zabbix frontend. Navigate to your VIP address (in the example IP case, `http://192.168.0.5/`), and you will see the following page:

Figure 1.9 – The Zabbix initial configuration window

13. Click on **Next step** twice until you see the following page:

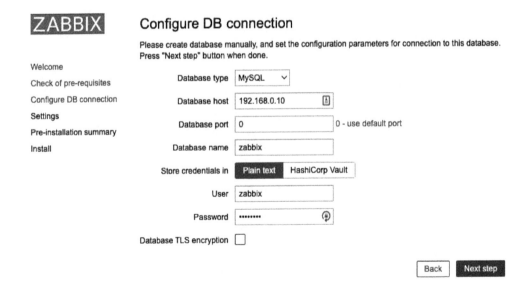

Figure 1.10 – The Zabbix database configuration window for lar-book-ha1

14. Make sure to fill in **Database host** with the IP address of our Zabbix MariaDB database (192.168.0.10). Then, fill in the database password for our zabbix database user.

15. Then, for the last step, for our first node, set up **Zabbix server name** as lar-book-ha1 and select your time zone, as seen in the following screenshot:

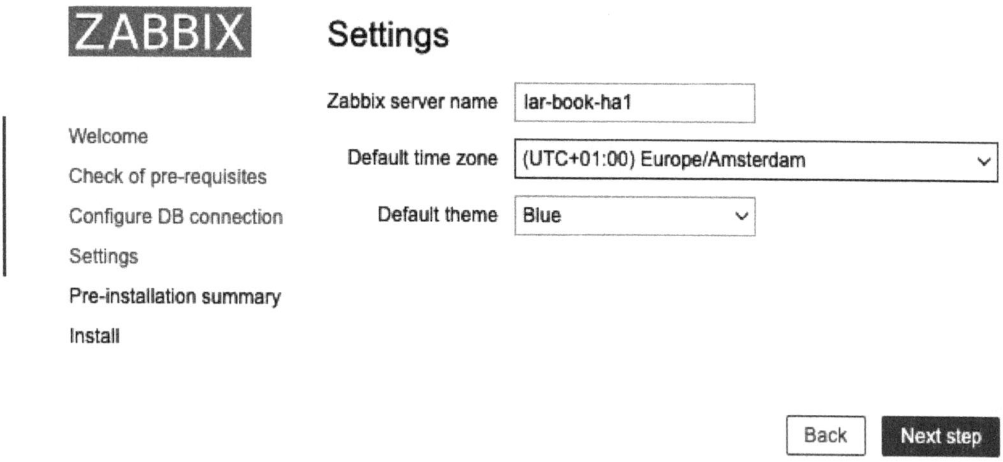

ZABBIX   Settings

Zabbix server name | lar-book-ha1

Welcome

Check of pre-requisites

Configure DB connection

Settings

Pre-installation summary

Install

Default time zone | (UTC+01:00) Europe/Amsterdam

Default theme | Blue

Back   Next step

Figure 1.11 – The Zabbix server settings window for lar-book-ha1

16. Then, click **Next step** and **Finish**.

17. Now, we need to do the same thing to our second frontend. Log in to lar-book-ha1.

On RHEL-based systems, issue the following:

```
systemctl stop nginx
```

For Ubuntu systems, issue the following:

```
systemctl stop nginx
```

18. When navigating to your VIP (in the example IP case, http://192.168.0.5/zabbix), you will see the same configuration wizard again.

19. Fill out the database details again:

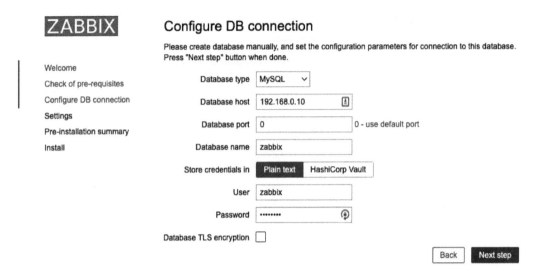

Figure 1.12 – The Zabbix database configuration window for lar-book-ha2

20. Then, make sure to set up **Zabbix server name** as `lar-book-ha2`, as seen in the following screenshot:

Figure 1.13 – The Zabbix server settings window for lar-book-ha2

21. Now, we need to enable the `lar-book-ha1` frontend again by issuing the following:

```
systemctl start nginx
```

That should be our last step. Everything should now be working as expected. Make sure to check your Zabbix server log file to see if the HA nodes are running as expected.

How it works...

Now that we have done it, how does the Zabbix server actually work in an HA mode? Let's start by checking out the **Reports | System information** page on our Zabbix frontend:

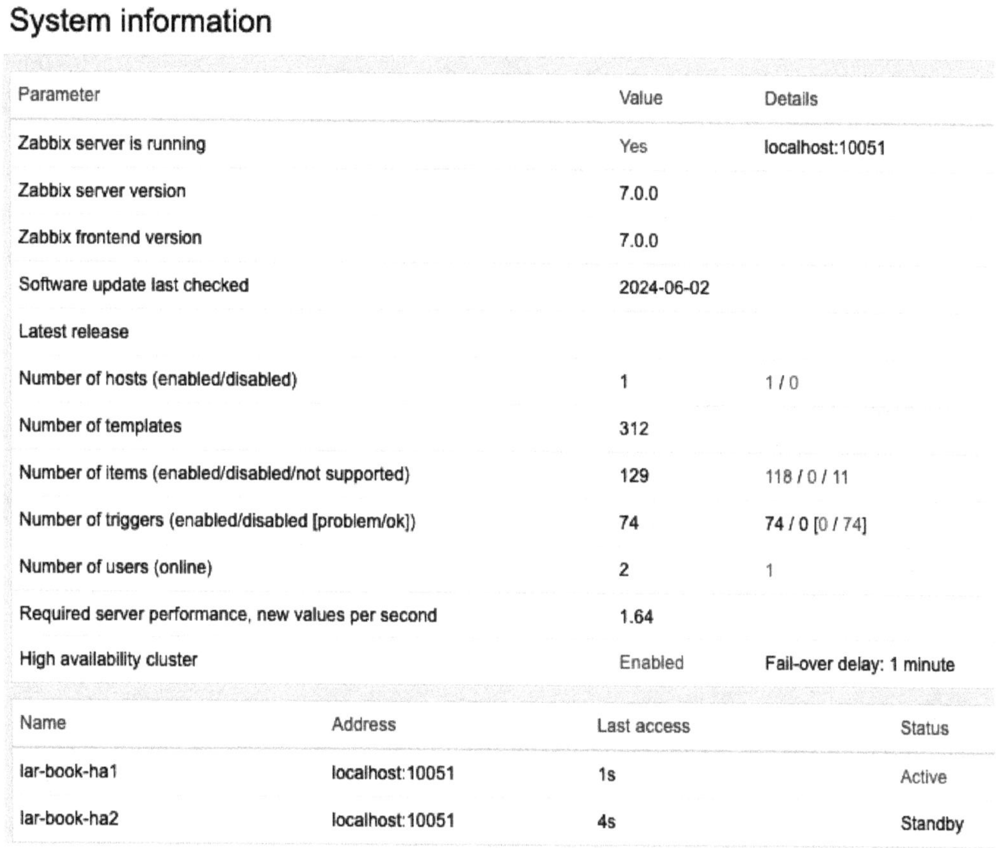

System information

Parameter	Value	Details
Zabbix server is running	Yes	localhost:10051
Zabbix server version	7.0.0	
Zabbix frontend version	7.0.0	
Software update last checked	2024-06-02	
Latest release		
Number of hosts (enabled/disabled)	1	1 / 0
Number of templates	312	
Number of items (enabled/disabled/not supported)	129	118 / 0 / 11
Number of triggers (enabled/disabled [problem/ok])	74	74 / 0 [0 / 74]
Number of users (online)	2	1
Required server performance, new values per second	1.64	
High availability cluster	Enabled	Fail-over delay: 1 minute

Name	Address	Last access	Status
lar-book-ha1	localhost:10051	1s	Active
lar-book-ha2	localhost:10051	4s	Standby

Figure 1.14 – The Zabbix server system information with HA information

We can now see that we have some new information available; for example, the **High availability cluster** parameter. This parameter now tells us if HA is enabled or not and what the failover delay is. In our case, this is 1 minute, meaning that it could take up to 1 minute before failover is initiated.

Furthermore, we can see every single node in our cluster. As Zabbix now supports one to many nodes in a cluster, we can see every single one taking part in our cluster right here. Let's take a look at the setup we have built:

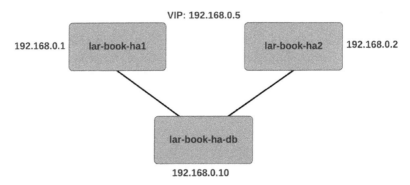

Figure 1.15 – The Zabbix server HA setup

As you can see in the setup, we have connected our two Zabbix server nodes, `lar-book-ha1` and `lar-book-ha2`, to our single Zabbix database, `lar-book-ha-db`. Because our Zabbix database is our **single source of truth** (**SSOT**), it can be used to keep our cluster configuration as well. In the end, everything Zabbix does is always kept in the database, from host configuration to history data to HA information. That's why building a Zabbix cluster is as simple as putting the `HANodeName` value in the Zabbix server configuration file.

We also included the `NodeAddress` parameter in the configuration file. This parameter is used by the Zabbix frontend to make sure that our system information (widget) and Zabbix server are not running frontend notification work. The `NodeAddress` parameter will tell the frontend which IP address to connect to for each respective server once it becomes the active Zabbix server.

To take things a bit further, I have added a simple `keepalived` setup to this installation as well. A `keepalived` configuration is a way to build simple VRRP failover setups between Linux servers. In our case, we have entered the VIP as `192.168.0.5` and added the `chk_nginx` process monitoring to determine when to fail over. Our failover works as follows:

```
lar-book-ha1 has priority 244
lar-book-ha2 has priority 243
```

If NGINX is running on our node, that adds a weight of 10 to our priority, leading to the total priority of `254` and `253`, respectively. Now, let's imagine that `lar-book-ha1` no longer has the web server process running. That means its priority drops to `244`, which is lower than `253` on `lar-book-ha2`, which does have the web server process running.

Whichever host has the highest priority is the host that will have the `192.168.0.5` VIP, meaning that host is running the Zabbix frontend, which will be served.

Combining these two ways of setting up HA, we have just created redundancy for two of the parts that make up our Zabbix setup, making sure we can keep outages to a minimum.

There's more...

Now, you may wonder, what if I wanted to go further in terms of setting up HA? First, the Zabbix HA feature is built to be simple and understandable to the entire Zabbix user base, meaning that as of now, you might not see the same amount of features you would get with a third-party implementation.

Nevertheless, the new Zabbix server HA feature has proved itself to be a long-awaited feature that really adds something to the table. If you want to run an HA setup such as this, the best way to add one more level of complexity to HA is a MySQL master/master setup. Setting up the Zabbix database with HA, which is the main **source of truth** (**SOT**), will make sure that your Zabbix setup really is reliable in as many ways as possible. For more information regarding MariaDB replication, check out the documentation here: `https://mariadb.com/kb/en/standard-replication/`.

Using the Zabbix frontend

If this is your first time using Zabbix, congratulations on getting to the UI. If you are a returning Zabbix user, there have been some changes to the Zabbix 7 UI that you might notice. We'll be going over some of the different elements that we can find in the Zabbix frontend so that during this book, you'll feel confident in finding everything you need.

Getting ready

To get started with the Zabbix UI, all we need to do is log in to the frontend. You will be served with the following page at the IP on which your server is running the Zabbix frontend:

Figure 1.16 – The Zabbix login screen

Make sure you log in to the Zabbix frontend with the default credentials:

- **Username**: Admin

- **Password**: zabbix

> **Tip**
> Just like in Linux, Zabbix is case-sensitive in most places. When entering your username, make sure to include the right cases; otherwise, you won't be able to log in!

How to do it...

After you log in, you'll be served with the default page, which is the default dashboard. This is what Zabbix has called **Global view**, and it provides us with a nice overview of what's going on. We can completely customize this and all the other dashboards that Zabbix supplies, but it's a good idea to familiarize yourself with the default setup before building something new:

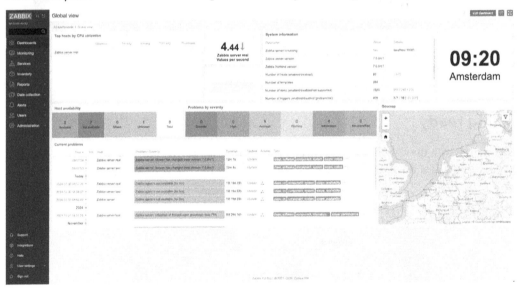

Figure 1.17 – The Global view dashboard

So, let's get started on getting to know this Zabbix 7 frontend by looking at the default dashboard. Please follow along in the frontend by clicking and checking out the content mentioned.

Zabbix uses dashboards, and they are filled with widgets to show you information. Let's go over the different widgets in the default dashboard and detail their information.

Let's start with the **System information** widget:

Parameter	Value	Details
Zabbix server is running	Yes	localhost:10051
Zabbix server version	7.0.0	
Zabbix frontend version	7.0.0	
Software update last checked	2024-05-19	
Latest release		
Number of hosts (enabled/disabled)	20	15 / 5
Number of templates	284	
Number of items (enabled/disabled/not supported)	1020	564 / 240 / 216
Number of triggers (enabled/disabled [problem/ok])	409	371 / 38 [13 / 358]
Number of users (online)	10	1
Required server performance, new values per second	8.63	
High availability cluster	Disabled	

Figure 1.18 – The System information widget

The **System information** widget, as you might have guessed, details all system information for you. This way, we can keep an eye on what's going on with our Zabbix server and see whether our Zabbix server is even running.

Let's go over the parameters:

- **Zabbix server is running**: Informs us whether the Zabbix server backend is actually running and where it is running. In this case, it's running, and it's running on localhost:10051.

- **Zabbix server version/Zabbix frontend version**: We now have separate indicators that will detail what the versions of our Zabbix server and frontend are.

- **Software update last checked**: This is another new addition. Zabbix will now check for new releases for you, indicating in the column below what the latest release is.

- **Latest release**: This will detail the latest available version of Zabbix.

- **Number of hosts (enabled/disabled)**: This will detail the number of hosts enabled (15) and the number of hosts disabled (5).

- **Number of templates**: This details the total number of templates we have (284).

- **Number of items (enabled/disabled/not supported)**: Here, we can see details of our Zabbix server's items—in this case, enabled (564), disabled (240), and not supported (216).

- **Number of triggers (enabled/disabled [problem/ok]**: This details the number of triggers. We can see how many are enabled (371) and disabled (38), but also how many are in a **problem** state (13) and how many are in an **ok** state (358).

- **Number of users (online)**: The first value details the total number of users. The second value details the number of users currently logged in to the Zabbix frontend.

- **Required server performance, new values per second**: Perhaps I'm introducing you to a completely new concept here, which is **New Values Per Second**, or **NVPS**. A server receives or requests values through items and writes these to our MariaDB database (or another database). The NVPS information detailed here shows the estimated number of NVPS received by the Zabbix server. Keep a close eye on this as your Zabbix server grows, as it's a good indicator to see how quickly you should scale up.

- **High availability cluster**: If you are running a Zabbix server HA cluster, you will see if it is enabled here and what the failover delay is. Additionally, the **System information** page will display extra HA information.

You might also see two additional values here depending on your setup:

- **Database history tables upgraded**: If you see this, it indicates that one of your database history tables hasn't been upgraded yet. Numeric (float) tables have been expanded to allow for more characters to be saved per data point. This table isn't upgraded automatically coming from Zabbix 4 to 5 or higher, as not everyone needs it, and it might take a long time to upgrade.

- **Database name**: If you see the name of your database with the value of your version, it might indicate you are running a non-supported database version. You could see a message such as Warning! Unsupported <DATABASE NAME> database server version. Should be at least <DATABASE VERSION>.

Now, that's one of the most important widgets when it comes to your Zabbix server, and it's a great one to keep on your main dashboard if you ask me.

Let's move on to the next widget, **Host availability**:

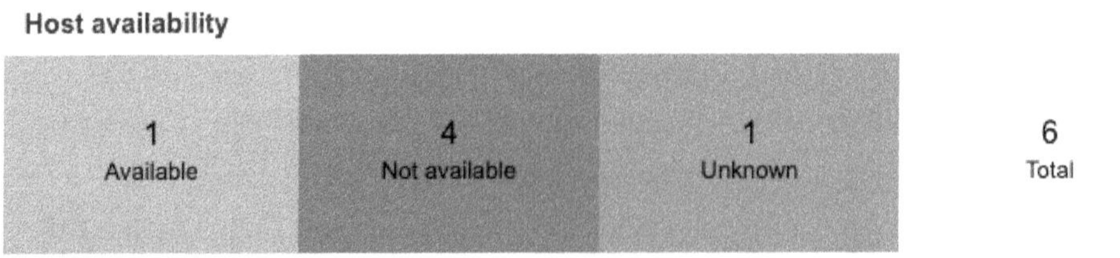

Host availability

| 1 | 4 | 1 | 6 |
| Available | Not available | Unknown | Total |

Figure 1.19 – The Host availability widget

The **Host availability** widget is a quick overview widget showing you everything you want to know about your monitored host's availability status. In this widget, it shows whether the host is available, not available, or unknown. This way, you get a good overview of the availability of all hosts you could be monitoring with your Zabbix server in a single widget.

Next to it is the **Problems by severity** widget:

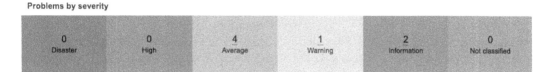

Figure 1.20 – The Problems by severity widget

It shows you how many hosts currently have a trigger in a certain state. There are several default severities in Zabbix:

- **Disaster**
- **High**
- **Average**
- **Warning**
- **Information**
- **Not classified**

We can fully customize the severity levels and colors; for example, what severity levels we want to put on which triggers. So, if you are worried about the severities right now, don't be. We'll get to that later.

> **Tip**
> Customizing the severity levels and colors can be very useful to your organization. We can customize the severity levels to match levels used throughout our company or even to match some of our other monitoring systems.

The next widget is **Clock**:

09:55
Amsterdam

Figure 1.21 – The Clock widget, indicating a time

It's a clock with the local Linux system time, finally in a digital time format. Need I say more? Let's move on to the **Problems** widget:

Figure 1.22 – One of the Problems widgets available

Now, this is an interesting widget that I use a lot. We see our current problems on this screen, so if we have our triggers set up correctly, we get valuable information here. A quick overview of how many hosts are having problems is one thing, but the **Problems** page also gives us more details about the problem:

- **Time**: At what time this problem was first noticed by the Zabbix server

- **Info**: Information about the event, with **Manual close** and **Suppressed** statuses being represented here.

- **Host**: What host this problem occurred on.

- **Problem/Severity**: What the problem is and how severe it is. The severity is shown in a color; in this case, orange, meaning **Average**.

- **Duration**: How long this has been a problem.

- **Update**: A button to allow us to make changes to this existing problem, such as acknowledging it or adding messages.

- **Actions**: What actions have been taken after this problem occurred; for example, the problem being acknowledged or a custom script that executes on problem creation? If you hover over any action, it will show you detailed information about all actions that have been taken for this problem.

- **Tags**: What tags are assigned to this problem?

The **Problems** widget is very useful. We have different types of this widget available, and as mentioned before, it is completely customizable, based on how this widget shows our problems to us. Take a quick look at some of the options, which we'll detail further in a later chapter:

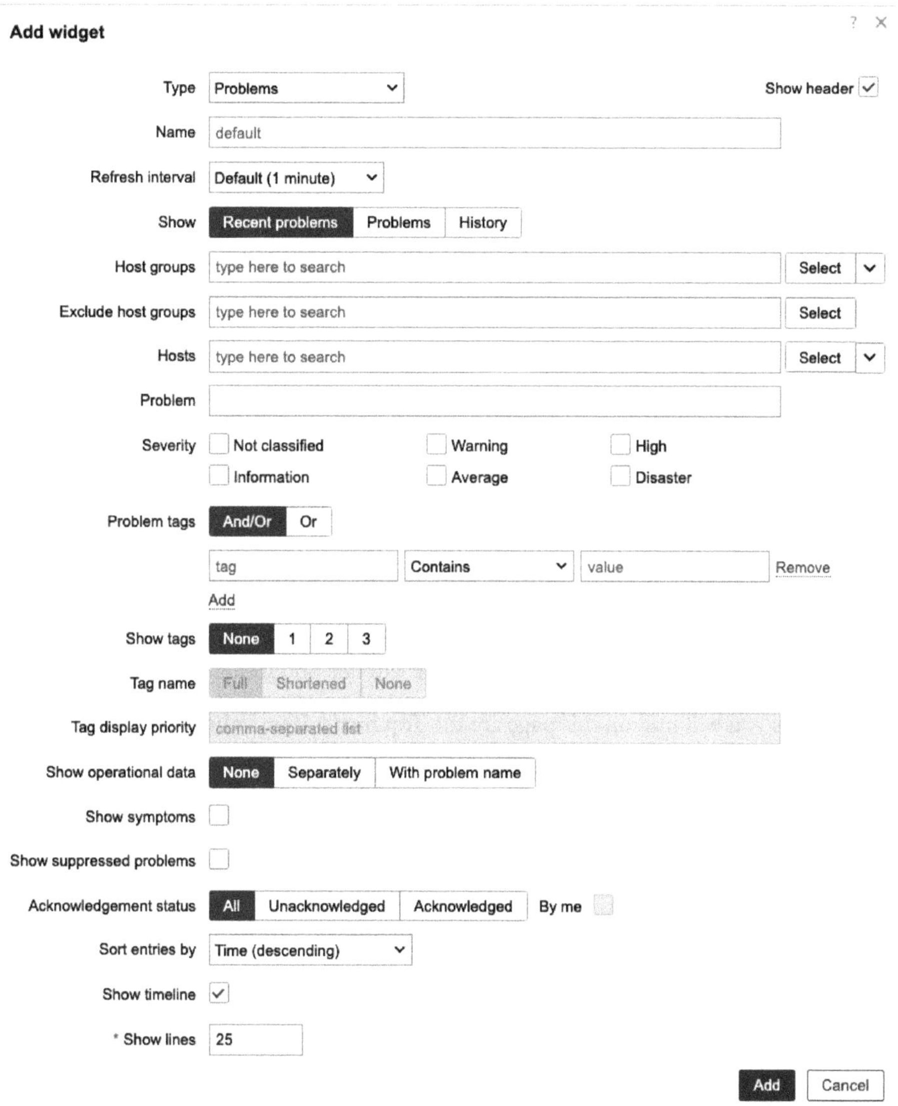

Add widget   ? ×

Type	Problems ⌄
Name	default
Refresh interval	Default (1 minute) ⌄
Show	**Recent problems** Problems History
Host groups	type here to search Select ⌄
Exclude host groups	type here to search Select
Hosts	type here to search Select ⌄
Problem	
Severity	☐ Not classified ☐ Warning ☐ High ☐ Information ☐ Average ☐ Disaster
Problem tags	**And/Or** Or
	tag Contains ⌄ value Remove
	Add
Show tags	**None** 1 2 3
Tag name	Full Shortened None
Tag display priority	comma-separated list
Show operational data	**None** Separately With problem name
Show symptoms	☐
Show suppressed problems	☐
Acknowledgement status	**All** Unacknowledged Acknowledged By me ☐
Sort entries by	Time (descending) ⌄
Show timeline	☑
* Show lines	25

Show header ☑

Add Cancel

Figure 1.23 – The Add widget screen

Tip

We can hide severity levels from these widgets to make sure we only see important ones. Sometimes, we don't want to see informational severity problems on our dashboards; it can distract us from a more important problem. Keep your dashboards clean by customizing widgets to their full extent.

There's also a very basic **Graph** widget included on the default dashboard. As you can see, it details the number of processed values per second. As we said, **System information** gives us an estimate on this value. The **Graph** widget gives us a more accurate and detailed look:

Graph

2024-06-02 08:44:27

Zabbix server: Number of processed values per second: 6.2094

Figure 1.24 – The Graph widget

These graph widgets are also fully customizable, and we will talk about them more later in the book.

Some newer widgets you will find on this page are the **Top hosts**, **Item value**, and **Geomap** widgets. Starting with the **Top hosts** widget, let's have a closer look at these:

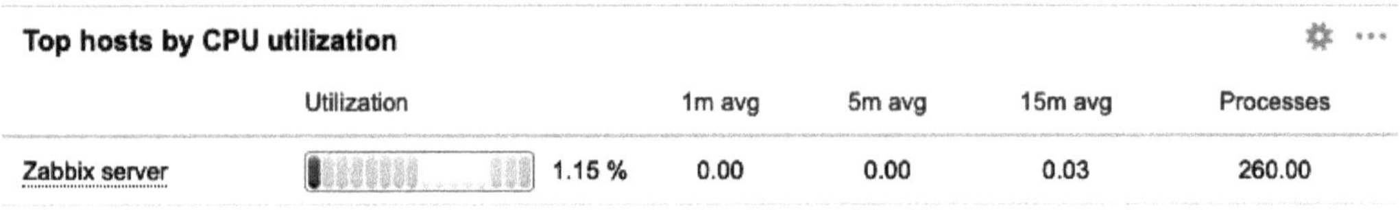

Top hosts by CPU utilization					
	Utilization	1m avg	5m avg	15m avg	Processes
Zabbix server	1.15 %	0.00	0.00	0.03	260.00

Figure 1.25 – The Top hosts widget

The **Top hosts** widget is fully customizable to show us a list of any item we want. We can then sort that list to show us the item with the highest (**Top N**) or Lowest (**Bottom N**) values, giving us a great overview of—in this case—hosts with the highest CPU load.

The **Item value** widget is also super useful, showing us a single item value:

1.00 ⬆

Zabbix server
Values per second

Figure 1.26 – The Item value widget

Then, last, we have the **Geomap** widget: a geographical map with a representation of all the hosts we would like to add. As you can see, by default, our Zabbix server is already included. In our case, I've made sure to update the location of our Zabbix server to our main office in the Netherlands:

Figure 1.27 – The Geomap widget

It also shows us that the Zabbix server currently has no issues, as the icon is green. If there is a problem the icon will show in the colour of the severity of the problem.

We now know how to work with the Zabbix frontend and can continue further on with how to navigate our instance.

Navigating the Zabbix frontend

Navigating the Zabbix frontend is easier than it looks at first glance, especially with some of the amazing changes made to the UI starting from Zabbix 5.0 and continuing into Zabbix 7.0. Let's explore the Zabbix navigational UI some more in this recipe by looking at the navigation bar and what it has to offer.

Getting ready

Now that we've seen the first page after logging in with the default dashboard, it is time to start navigating through the Zabbix UI and see some of the other pages available. We'll move through the sidebar and explore the pages available in our Zabbix installation so that when we start monitoring our networks and applications, we know where we can find everything.

So, before continuing, make sure you have the Zabbix server ready as set up in the previous recipes.

How to do it...

The Zabbix navigation bar is the gateway to all of our powerful tools and configuration settings. Zabbix uses a left-side navigation bar to keep our UI as clean as possible. On top of that, they have made the sidebar disappear so that we can keep a close look at all of our content without the sidebar blocking our vision.

> **Tip**
>
> We cannot change the Zabbix navigation menu location, but it is possible to hide it to a smaller form or completely hide it. If you want the navigation bar to be hidden (or not), click the first icon on the right side of the Zabbix logo. If you want to fully hide the navigation bar, click the second icon on the right side of the Zabbix logo.

Let's take a look at the Zabbix sidebar as we see it from our default page and get to know it. Please follow along in the frontend by clicking and checking out the content mentioned:

Figure 1.28 – The default Zabbix page as seen in your own web browser

We've got some categories here to choose from, and one level below the categories, we've got our different pages. First, let's start by detailing the categories:

- **Monitoring**: The **Monitoring** category is where we can find all information about our collected data. It's basically the category you want to use when you're working with Zabbix to read any collected information you've worked hard to acquire.

- **Services**: The **Services** category is new to Zabbix 7 and comes as part of the improved **Business Service Monitoring** (**BSM**) features. We can find all the information regarding service and **service-level agreement** (**SLA**) monitoring here.

- **Inventory**: The **Inventory** category is a cool extra feature in Zabbix that we can use to look at our host-related inventory information. You can add stuff such as software versions or serial numbers to hosts and look at them here.

- **Reports**: The **Reports** category contains a variety of predefined and user-customizable reports focused on displaying an overview of parameters such as system information, triggers, and gathered data.

- **Data collection**: The **Data collection** category is where we build everything that has to do with monitoring our devices. We will later use this data in **Monitoring**, **Inventory**, and **Reports**. We can edit our settings to suit our every need so that Zabbix can show us that data in a useful way.

- **Alerts**: The **Alerts** section of the UI is all about showing the right data.

- **Users**: The **Users** section is all about displaying user- and user-group-related data. It has everything you need to manage accounts and permissions.

- **Administration**: The **Administration** category is where we administer the Zabbix server. You'll find all your settings from the server here to enable you and your colleagues to have a good working Zabbix experience.

You'll go over all of these quite a lot while using this book, so remember them well. Let's dive a little deeper into the categories by looking at them one by one. Let's start with the **Monitoring** category:

Figure 1.29 – The Monitoring section of the sidebar

The **Monitoring** tab contains the following pages:

- **Problems**: We can look at our current problems in detail here. We are provided with a bunch of filter options to narrow down our problem search if needed.

- **Hosts**: This page will provide a quick overview of what's going on with hosts. It also provides links to navigate to pages showing data for our hosts.

- **Latest data**: Here is a page we're going to use quite a lot throughout our professional Zabbix lifetime. The **Latest data** page is where we can find collected values for every single host, which we can, of course, filter on.

- **Maps**: Maps are a very helpful tool in Zabbix to get an overview of your infrastructure. We can use them for network overviews and such.

- **Discovery**: This page provides us with an overview of discovered devices. We'll work more on this later.

Next, we have the **Services** category:

Figure 1.30 – The Services section of the sidebar

This part of the sidebar contains the following pages:

- **Services**: This is where we configure all of the services we want to monitor

- **SLA**: We can configure any SLAs here that we can then use in our services

- **SLA report**: A detailed overview of configured services with their SLAs and whether they are being met or not

Then, we have the **Inventory** category:

Figure 1.31 – The Inventory section of the sidebar

The **Inventory** tab contains the following pages:

- **Overview**: A quick overview page for your inventory information

- **Hosts**: A more detailed look into inventory values on a per-host basis

Next, we have the **Reports** category:

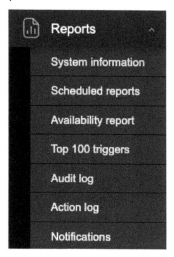

Figure 1.32 – The Reports section of the sidebar

The **Reports** tab contains the following pages:

- **System information**: You can look at system information here; it contains the same information as the **System information** widget we discussed earlier.

- **Scheduled reports**: This is where we configure any automatic PDF reporting that we might want to send out.

- **Availability report**: On this page, we can see the percentage of time a trigger has been in a **problem** state compared to an **ok** state. This is a helpful way of seeing how long certain items are actually healthy.

- **Top 100 triggers**: The top 100 triggers that have changed their state most often within a period of time.

- **Audit log**: We can see who changed what on our Zabbix server here. This is a great way to see which colleague locked you out by accident or whether it was on purpose.

- **Action log**: We can see a list of actions that have been taken; for example, due to triggers going to a **problem** state or an **ok** state.

- **Notifications**: On this page, we can see the number of notifications sent to our users.

Next, we have the **Data collection** category, which is a new entry:

Figure 1.33 – The Data collection section of the sidebar

The **Data collection** tab is almost the same as what the **Configuration** tab used to be in older versions. It contains the following pages:

- **Template groups**: We configure our template groups here; for instance, a group for all templates that will be used within our company (*Templates/Open source ICT Solutions*) or all network devices (*Templates/Networking*).

- **Host groups**: We configure our host groups here; for instance, a group for all Linux servers.

- **Templates**: This is where we configure templates that we can use to monitor hosts from the Zabbix server.

- **Hosts**: Another **Hosts** tab, but this time it is not for checking data. This is where we add and configure host settings.

- **Maintenance**: In Zabbix, we have the availability to set maintenance periods; this way, triggers or notifications won't disturb you while you take something offline for maintenance, for example.

- **Event correlation**: We can correlate problems here to reduce noise or prevent event storms. This is achieved by closing new or old problems when they correlate to other problems.

- **Discovery**: This is where we configure Zabbix discovery for automatic host creation.

New in Zabbix is the **Alerts** category:

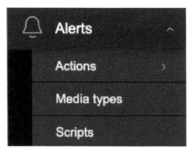

Figure 1.34 – The Alerts section of the sidebar

It consists of the following three pages:

- **Actions**: All different kinds of actions can be configured in this part of the frontend. We can set up actions for sending out alerts, creating hosts, and more.

- **Media types**: There are several media types preconfigured in Zabbix, which you'll find here already. We can also add custom media types.

- **Scripts**: This is where we can add custom scripts for extending Zabbix functionality in the frontend.

Second to last, and also new, we have **Users**:

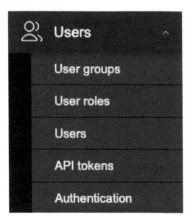

Figure 1.35 – The Users section of the sidebar

We can find five different pages here:

- **User groups**: This is where we configure user groups and the permissions for these user groups.

- **User roles**: It's possible to configure different users' roles here to limit or extend certain frontend functionality to certain users.

- **Users**: Add users to this page.

- **API tokens**: This page was kind of hidden before, but it's now easier to find. We can manage all API tokens we have permission to edit here: super useful for super admins to create and manage tokens of different users.

- **Authentication**: We can find our authentication settings here, such as **Lightweight Directory Access Protocol (LDAP)**, **Security Assertion Markup Language (SAML)**, and HTTP. It also contains settings for the new **just-in-time (JIT)** user provisioning.

Finally, we have the **Administration** category:

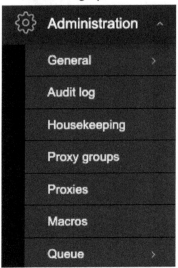

Figure 1.36 – The Administration section of the sidebar

The **Administration** tab contains the following pages:

- **General**: The general page contains our Zabbix server configuration. Settings ranging from **Housekeeping** to **Frontend theme** are found here.

- **Audit log**: The audit log settings are located here. We can enable or disable audit logging as well as change how long logs should be stored.

- **Housekeeping**: General housekeeping settings are found here. We can edit what we want to enable or disable for housekeeping, as well as change how long we store things such as history, trends, triggers, and more.

- **Proxy groups**: At this new entry, we will define proxy groups that we can use for failover and load balancing in combination with proxies.

- **Proxies**: This is where we configure proxies that should be connected to this Zabbix server.

- **Macros**: A bit more accessible now is global macros, since they have been relocated here. Define new and manage existing global macros here.

- **Queue**: View your Zabbix server queue here. Items might be stuck in a queue due to data collection or performance issues.

> Tip
>
> When using Zabbix authentication such as HTTP, LDAP, or SAML, we still need to create our users internally with the right permissions. Configure your users to match your authentication method's username in Zabbix and use the authentication method for password management. With Zabbix 7.0, however, it is possible to use JIT user provisioning to automatically create users with the correct permissions, which we will also talk about in this book.

2
Getting Things Ready with Zabbix User Management

In this chapter, we will work on creating our first user groups, users, and user roles. It's very important to set these up in the correct manner, as they will give people access to your Zabbix environment with the correct permissions. By going over these things step by step, we will make sure we have a structured Zabbix setup before continuing on with this book.

As a bonus, we will also set up some advanced user authentication using SAML and LDAP to make things easier for your Zabbix users and provide them with a way to use the login credentials they might already be using throughout your company. We will go over all these steps in the order of the following recipes:

- Creating user groups
- Using Zabbix user roles
- Creating your first users
- Azure AD SAML user authentication and JIT user provisioning
- OpenLDAP user authentication and JIT user provisioning

Technical requirements

We can do all of the work in this chapter with any installed Zabbix setup. If you haven't installed Zabbix yet, check out the previous chapter to learn how to do so. We will go through our Zabbix setup to get everything ready for our users to start logging in and using the Zabbix frontend.

Creating user groups

To log in to the Zabbix frontend, we are going to need users. Right now, we are logged in with the default user, which is logical because we need a user to create users. This isn't a safe setup though, because we don't want to keep using `zabbix` as a password. So, we are going to learn how to create new users and group them accordingly.

It's important to choose how you want to manage users in Zabbix before setting up user accounts. If you want to use something such as LDAP or SAML, it's a smart idea to make the choice to use one of those authentication methods right away, so you won't have any migration trouble.

Getting ready

Now that we know how the Zabbix UI is structured and how to navigate it, we can start doing some actual configuration. We'll start out by creating some user groups to get familiar with the process and start using them. This way, our Zabbix setup gets not only more structured but also more secure.

To get started with this, we'll need a Zabbix server like the one we used in the previous recipes and the knowledge we've acquired there to navigate to the correct frontend sections.

Looking at the following figure, we can see how our example company, **Cloud Hoster**, is set up. We will create the users seen in the diagram to create a structured and solid user setup:

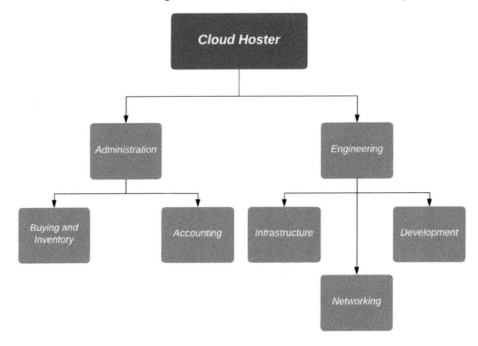

Figure 2.1 – Cloud Hoster department diagram

So, **Cloud Hoster** has some departments that need access to the Zabbix frontend and others that don't need it at all. Let's say we want to give the following departments access to the Zabbix frontend:

- **Networking**: To configure and monitor their network devices

- **Infrastructure**: To configure and monitor their Linux servers

- **Buying and Inventory**: To look at inventory information and compare it with other internal tools

How to do it...

Let's get started with creating these three groups in our Zabbix UI:

1. To do this, navigate to **Users | User groups**, which will show you the following page:

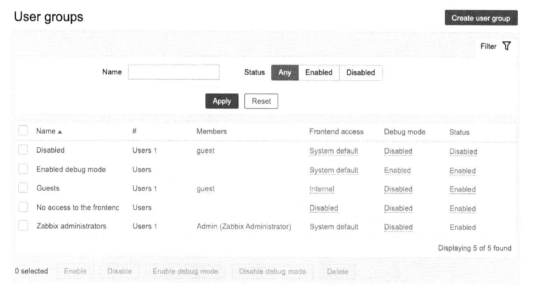

Figure 2.2 – The Zabbix User groups window

2. Now, let's start by creating the **Networking** group by clicking **Create user group** in the top-right corner. This will bring you to the following screen:

User group   Template permissions   Host permissions   Problem tag filter

* Group name	
Users	type here to search Select
Frontend access	System default ⌄
LDAP Server	Default ⌄
Multi-factor authentication	Disabled ⌄
Enabled	✓
Debug mode	☐

Add   Cancel

Figure 2.3 – The Zabbix User groups configuration window

We will need to fill in the information, starting with **Group name**, which will of course be
Networking. There are no users for this group yet, so we'll skip that one. **Frontend access**
gives us the option to provide authentication; if you select **LDAP** here, LDAP authentication
will be used for authenticating. We will keep it as **System default**, which uses the internal
Zabbix authentication system.

Multi-factor authentication

New to Zabbix 7.0 is the ability to use multi-factor authentication. If we want users to be forced
to use this, we can set that up in the user group here. Before doing that, however, make sure to
set up multi-factor authentication under **Users | Authentication**.

---

3.   Now, let's navigate to the next tab on this page, which is **Template permissions**:

Figure 2.4 – The Zabbix User groups Template permissions configuration window

Here, we can specify what host groups our group will have access to. There's a default host
group for **Network devices** already, which we will use in this example.

4. Click **Select** to take you to a pop-up window with host groups available. Select **Templates/Network devices** here and it'll take you back to the previous window, with the group filled in.

5. Select **Read-write** permissions.

6. We won't be adding anything else, so click the big blue **Add** button to finish creating this host group.

Tip

When using Zabbix authentication such as HTTP, LDAP, or SAML, we still need to create our users internally with the right permissions if we do not use JIT user provisioning. To do so, configure your users to match your authentication method's username in Zabbix and use the authentication method for password management. When using JIT user provisioning, this is not something we have to worry about.

Now we will have a new user group called **Networking** that is only allowed to read and write to the **Templates/Network devices** template group:

	Name ▲	#	Members
☐	Enabled debug mode	Users	
☐	Guests	Users 1	guest
☐	Internal	Users 2	Admin (Zabbix Administrator), guest
☐	Networking	Users 2	
☐	No access to the frontend	Users	
☐	Zabbix administrators	Users 1	Admin (Zabbix Administrator)

Figure 2.5 – The Zabbix User groups window

7. Let's repeat this process to create a new **Infrastructure** user group, except instead of adding the **Templates/Network devices** template group, we'll add the **Linux servers** host group, like this:

Figure 2.6 – The Zabbix User groups Permissions configuration window with one host group

8. Click **Add** to save this host group.

9. Repeat the steps again and to add **Buying and Inventory** user group, we'll do something differently. We'll repeat the process we've just done except for the part with the permissions. We want **Buying and Inventory** to be able to read our inventory data, but we don't want them to actually change our host configuration. Add both the **Templates/Network devices** template group and **Linux servers** host group to the user group, but with only **Read** permissions like this:

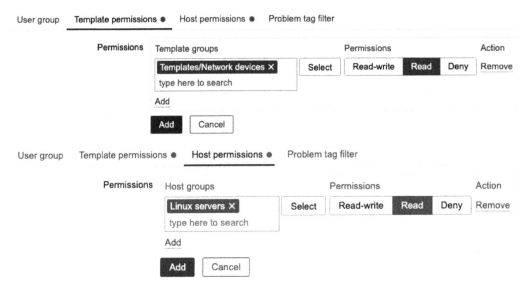

Figure 2.7 – The Zabbix User groups Permissions configuration window with two groups

Congratulations! Finishing this means you've ended up with three different user groups and we can continue to create our first new users! Let's get to it.

There's more...

Zabbix user groups are quite extensive and there is a lot more to it than there seems at first. As the entire permission system is based on what user group(s) and user role you are part of, it is always a good idea to read the Zabbix documentation first: https://www.zabbix.com/documentation/current/en/manual/config/users_and_usergroups/usergroup.

Using Zabbix user roles

Since Zabbix 6.0, we can create user roles within our Zabbix system. By creating our own user roles in Zabbix, it's possible to provide additional permission settings. In older Zabbix versions, we had the ability to assign one of three user types:

- **Users**
- **Admin**
- **Super admin**

What these user types did in earlier releases was restrict what Zabbix users could see in the frontend. This was always pre-defined though.

Although these user groups are still in Zabbix, with the addition of user roles that we can create ourselves, we can set up our own frontend-related restrictions, making it possible to only show certain parts of the UI to certain Zabbix users. This is done by limiting access to certain permissions a user group has by default as well as respecting the user group-related permissions.

Getting ready

For this recipe, we will need a Zabbix server, preferably the one set up in the previous recipe. In the previous recipe, we set up different user groups to provide for different permissions on host groups. Completely separate from the user group, we will apply certain user roles to our users to determine what they can see in the UI. Let's check out how to set up our user roles.

How to do it...

1. First, navigate to the Zabbix frontend and go to **Users | User roles**. This will show us the default user roles as you know them from older Zabbix versions:

Name ▲	#	Users
Admin role	Users	
Guest role	Users 1	guest
Super admin role	Users 1	Admin (Zabbix Administrator)
User role	Users	

Figure 2.8 – The default Zabbix User roles configuration window

2. Here, we can click on the blue **Create user role** button in the top-right corner.

3. We'll set up a new user role called User+ role. This role will be for Zabbix users with only read permissions, but who need more access than just the **Monitoring**, **Inventory**, and **Reports** navigational elements.

* Name			
User type	User ⌄		
	Access to UI elements		
Dashboards	✔		
Monitoring	✔ Problems	✔ Latest data	☐ Discovery
	✔ Hosts	✔ Maps	
Services	✔ Services	☐ SLA	✔ SLA report
Inventory	✔ Overview	✔ Hosts	
Reports	☐ System information	✔ Top 100 triggers	☐ Notifications
	☐ Scheduled reports	☐ Audit log	
	✔ Availability report	☐ Action log	
Data collection	☐ Template groups	☐ Hosts	☐ Discovery
	☐ Host groups	☐ Maintenance	
	☐ Templates	☐ Event correlation	
Alerts	☐ Trigger actions	☐ Autoregistration actions	☐ Scripts
	☐ Service actions	☐ Internal actions	
	☐ Discovery actions	☐ Media types	
Users	☐ User groups	☐ Users	☐ Authentication
	☐ User roles	☐ API tokens	
Administration	☐ General	☐ Proxy groups	☐ Queue
	☐ Audit log	☐ Proxies	
	☐ Housekeeping	☐ Macros	

* At least one UI element must be checked.

Figure 2.9 – The top part of a new Zabbix User role configuration window

4. First things first, make sure to enter User+ role into the **Name** field.

5. Let's focus on the part where it states **Access to UI elements** first. When **User** is selected for **User type**, we are not able to add access rights to the user role. So, let's change the **User type** by selecting **Admin** in the dropdown.

6. I specifically want this user role named `User+ role` to have the ability to access the maintenance page. Setting this up will look like this:

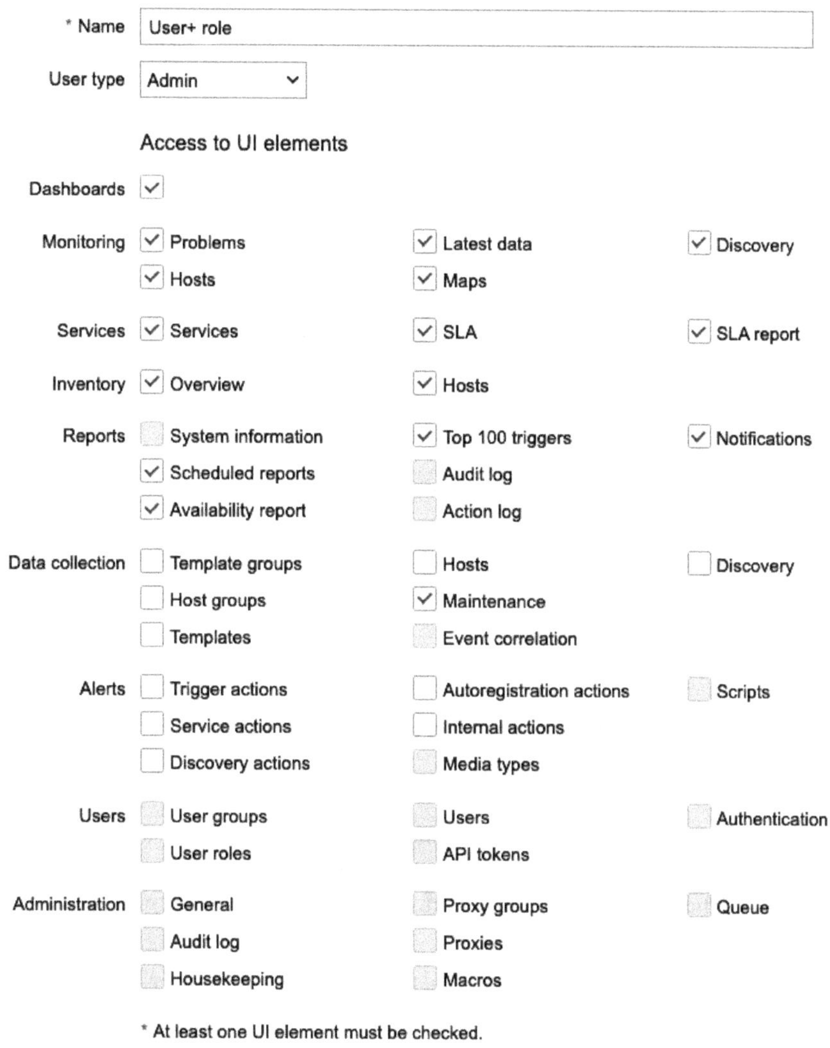

* Name	User+ role

User type	Admin ⌄

Access to UI elements

Dashboards ✔

Monitoring	✔ Problems	✔ Latest data	✔ Discovery
	✔ Hosts	✔ Maps	

Services	✔ Services	✔ SLA	✔ SLA report

Inventory	✔ Overview	✔ Hosts	

Reports	System information	✔ Top 100 triggers	✔ Notifications
	✔ Scheduled reports	Audit log	
	✔ Availability report	Action log	

Data collection	Template groups	Hosts	Discovery
	Host groups	✔ Maintenance	
	Templates	Event correlation	

Alerts	Trigger actions	Autoregistration actions	Scripts
	Service actions	Internal actions	
	Discovery actions	Media types	

Users	User groups	Users	Authentication
	User roles	API tokens	

Administration	General	Proxy groups	Queue
	Audit log	Proxies	
	Housekeeping	Macros	

* At least one UI element must be checked.

Figure 2.10 – A new Zabbix User+ role with access to Maintenance

7. Make sure to also change the **Access to actions** section of the form by deselecting **Manage scheduled reports** as follows:

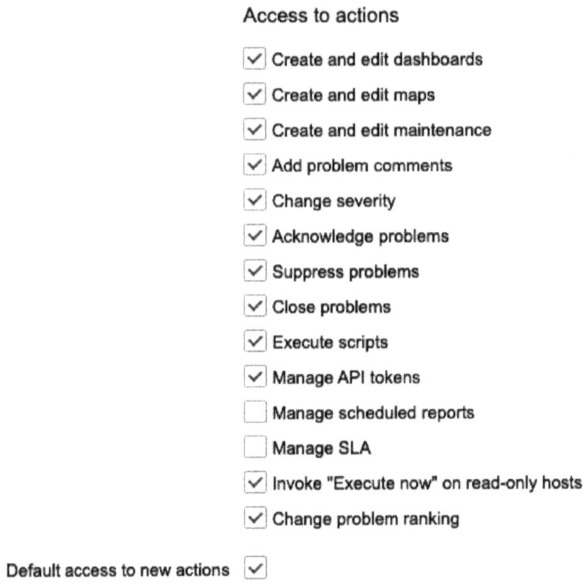

Access to actions

☑ Create and edit dashboards

☑ Create and edit maps

☑ Create and edit maintenance

☑ Add problem comments

☑ Change severity

☑ Acknowledge problems

☑ Suppress problems

☑ Close problems

☑ Execute scripts

☑ Manage API tokens

☐ Manage scheduled reports

☐ Manage SLA

☑ Invoke "Execute now" on read-only hosts

☑ Change problem ranking

Default access to new actions ☑

Figure 2.11 – A new Zabbix User+ role with correct Access to actions settings

8. Last, but not least, click on the blue **Add** button at the bottom of the form to add this new user role.

How it works...

First, let's break down the options we have when creating user roles in Zabbix:

- **Name**: We can set a custom name for our user role here.

- **User type**: User types still exist in Zabbix 6, although they are now assigned through user roles. There's still a limit to what can be seen by a certain user type, for example the User type will never have more than read access and the Super admin type is still unrestricted when it comes to permissions.

- **Access to UI elements**: Here, we can restrict what a user can see on the Zabbix UI when they are assigned to this user role.

- **Access to services**: Service or SLA monitoring can be restricted here, as we might not want all users to have access to it.

- **Access to modules**: Custom Zabbix frontend modules are fully integrated into the user role system, meaning we can select what frontend modules a Zabbix user can see.

- **Access to API**: The Zabbix API can be restricted to certain user roles. For example, you might only want a specific API user role, limiting the rest of the users' access to the Zabbix API.

- **Access to actions**: In Zabbix user roles, certain actions can be limited, including the ability to edit dashboards, maintenance API tokens, and more.

Now, let's look at what we've changed between the user role called User role and the user role called User+ role. The default user role called User role has the following access to UI elements:

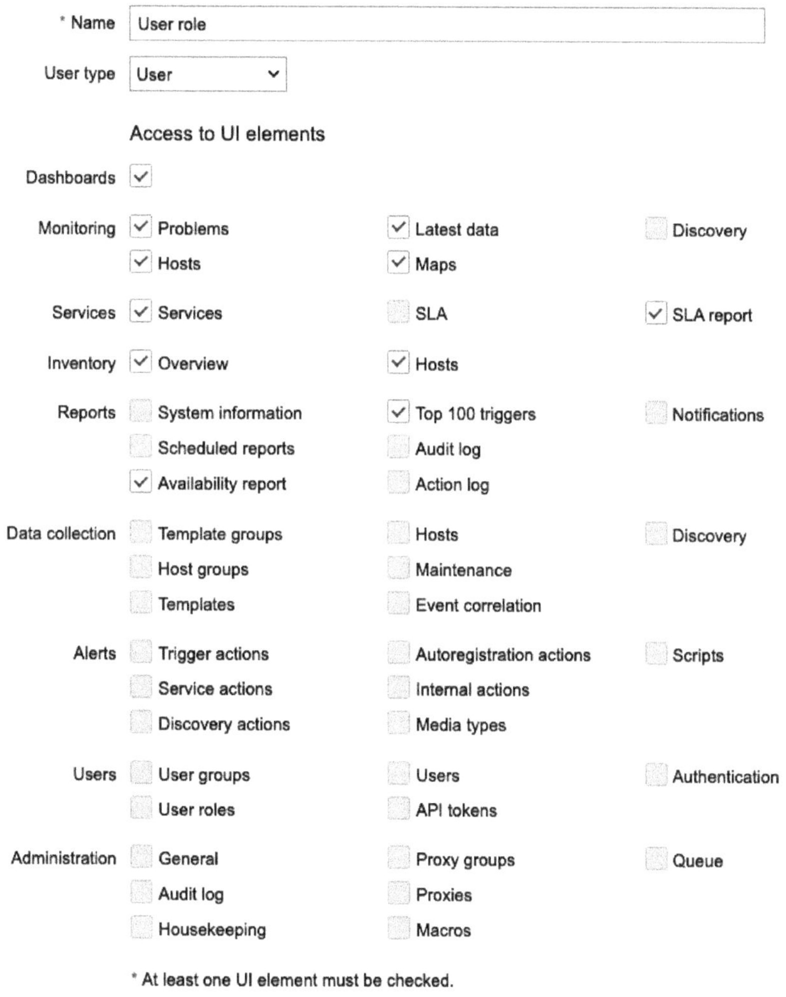

* Name	User role	
User type	User ⌄	

Access to UI elements

Dashboards	✓		
Monitoring	✓ Problems	✓ Latest data	☐ Discovery
	✓ Hosts	✓ Maps	
Services	✓ Services	☐ SLA	✓ SLA report
Inventory	✓ Overview	✓ Hosts	
Reports	☐ System information	✓ Top 100 triggers	☐ Notifications
	☐ Scheduled reports	☐ Audit log	
	✓ Availability report	☐ Action log	
Data collection	☐ Template groups	☐ Hosts	☐ Discovery
	☐ Host groups	☐ Maintenance	
	☐ Templates	☐ Event correlation	
Alerts	☐ Trigger actions	☐ Autoregistration actions	☐ Scripts
	☐ Service actions	☐ Internal actions	
	☐ Discovery actions	☐ Media types	
Users	☐ User groups	☐ Users	☐ Authentication
	☐ User roles	☐ API tokens	
Administration	☐ General	☐ Proxy groups	☐ Queue
	☐ Audit log	☐ Proxies	
	☐ Housekeeping	☐ Macros	

* At least one UI element must be checked.

Figure 2.12 – Default Zabbix user role called User role Access to UI elements

By default, we have three user roles in Zabbix 6, which mirror the available user types. The user role we see here in **Name** mirrors the user type we have called User. It gives us access to the UI elements seen above, restricting the user role called User role to only be able to see certain things and make no configuration changes.

For example, it's considered an impactful permission to be able to set **Maintenance**. Because of course, you could restrict important notifications by setting **Maintenance**. But here comes the catch, what if you explicitly want a Zabbix user to only be able to read information but still not have access to configuration pages? In Zabbix 5.0, this wasn't possible because you could only select the **User**, **Admin**, or **Super admin** types, immediately giving access to the entire configuration section when using the **Admin** and **Super admin** user types.

Now, let's see what we did by creating a new user role called User+ role:

* Name	User+ role		
User type	Admin ∨		

Access to UI elements

Dashboards	✓		
Monitoring	✓ Problems	✓ Latest data	✓ Discovery
	✓ Hosts	✓ Maps	
Services	✓ Services	✓ SLA	✓ SLA report
Inventory	✓ Overview	✓ Hosts	
Reports	System information	✓ Top 100 triggers	✓ Notifications
	✓ Scheduled reports	Audit log	
	✓ Availability report	Action log	
Data collection	Template groups	Hosts	Discovery
	Host groups	✓ Maintenance	
	Templates	Event correlation	
Alerts	Trigger actions	Autoregistration actions	Scripts
	Service actions	Internal actions	
	Discovery actions	Media types	
Users	User groups	Users	Authentication
	User roles	API tokens	
Administration	General	Proxy groups	Queue
	Audit log	Proxies	
	Housekeeping	Macros	

* At least one UI element must be checked.

Figure 2.13 – New Zabbix user role called User+ role Access to UI elements

Here, we can see what happens if we change the user type to **Admin** but do not select all the available **Access to UI elements**. We now have a user role with no access to important configuration pages but with access to **Maintenance**.

Combining that with the settings for **Access to actions**, where we added the **Create and edit maintenance** setting as seen in *Figure 2.11*, we would have full access to the maintenance settings.

When we assign this role to a user in the next recipe and log in as that user, we will be able to see the following in our Zabbix sidebar:

Figure 2.14 – Custom User role Zabbix sidebar

This, of course, is just one of the many types of configurations we can use. We also have the ability to allow Zabbix users access to menus and options through a number of parameters under a bunch of custom user roles. We are free to set this up however we please, adding a lot of user flexibility within Zabbix.

There's more...

Zabbix is currently in the process of working out user roles further, meaning that some parts might still be missing or you might see issues with them. As it is a new feature, it is constantly being improved and extended. Check out the Zabbix documentation for more information regarding this feature: https://www.zabbix.com/documentation/6.4/en/manual/web_interface/frontend_sections/users/user_roles

Creating your first users

With our newly created user groups and user roles, we've taken our first step toward a more structured and secure Zabbix setup. The next step is to actually assign some users to the newly created user groups to make sure they are assigned our new user permissions from the group, as well as making them part of a user role to provide the correct access to UI elements.

Getting ready

To get started, we'll need the server and the newly created user groups from the last recipe. So, let's start with the configuration.

We know there are three departments in the **Cloud Hoster** company that are going to use our Zabbix installation. We've created user groups for them, but there are also users in those departments that actually want to use our installation. Let's meet them:

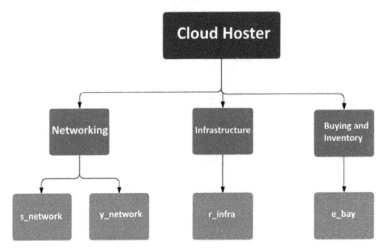

Figure 2.15 – Cloud Hoster users diagram

These are the users we need to configure for **Cloud Hoster** to use.

How to do it...

Let's start creating the users. We will start with our **Networking** department:

1. Navigate to **Users | Users**, which will bring us to this page:

Figure 2.16 – The Zabbix Users window

2. This is where all the user creation magic is happening, as we will be managing all of our users from this page. To create our first **Networking** department user named s_network, click the **Create user** button in the top-right corner, bringing us to the following screen:

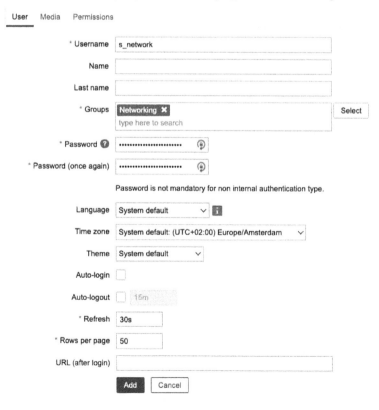

Figure 2.17 – The Zabbix Users configuration window

3. Fill out the **Username** field to provide us with the username this user will have, which will be `s_network`.

4. Also, it's important to add this user to the group we have just created to give our user the right permissions. Click **Select** and pick our group called **Networking**.

5. Last but not least, set a secure password in the **Password** fields; don't forget it because we will be using it later.

6. After this, move on to the **Permissions** tab as we won't be configuring **Media** just yet:

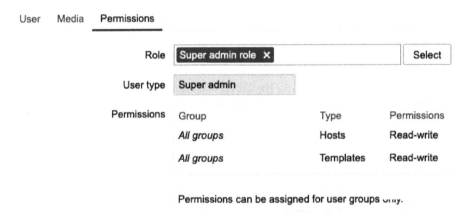

Figure 2.18 – The Zabbix user Permissions configuration window

7. Select the **Role** option named **Super admin role** here. This will enable our user to access all UI elements and see and edit information about all host groups in the Zabbix server.

The following user roles are available in Zabbix by default:

Default roles	Description
User role	The Zabbix User role has access to the visualization aspects of our Zabbix environment. Specifically, the Monitoring, Services, and Inventory and Reporting menus are available. The user will only ever have READ access to templates and hosts and they must be explicitly assigned.
Admin role	The Zabbix Admin role can additionally manage the configuration of our Zabbix monitoring. Specifically, all the menus that the Zabbix User has access to are available, with the addition of Data collection and Alerts. The user can be assigned READ-WRITE access to templates and hosts and they must be explicitly assigned.

Default roles	Description
Super admin role	The Zabbix Super admin role has access to the administrative aspects of our Zabbix environment. Specifically, all the menus that the Zabbix Admin has access to are available, with the addition of Users and Administration. The user will always have READ-WRITE access to all templates and hosts.

8. Let's repeat the previous steps for the user named `y_network`, but in the **Permissions** tab, select the **Admin role** option as follows:

Figure 2.19 – The Zabbix user Permissions configuration window

After creating these two users, let's move on to create the infrastructure user, `r_ infra`. Repeat the steps we took for `s_network`, changing the **Username**, of course. Then, add this user to the group and give our user the right permissions. Click **Select** and pick our group called **Infrastructure**. It will look as follows:

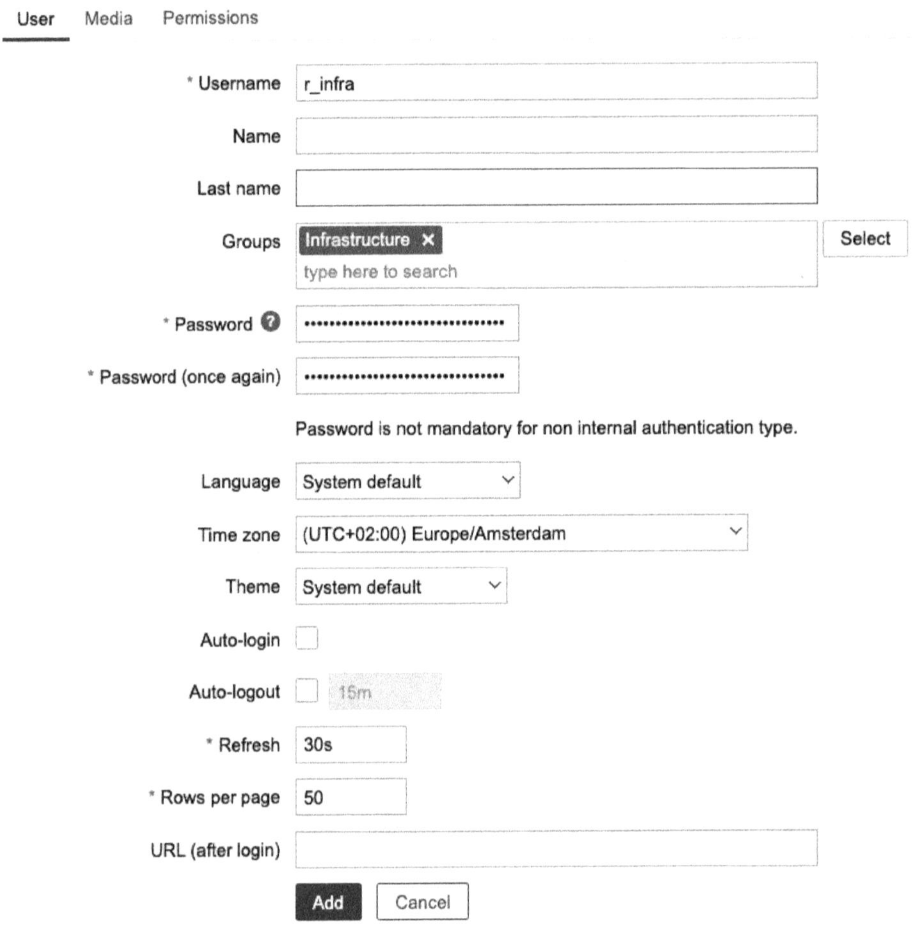

User   Media   Permissions

* Username	r_infra
Name	
Last name	
Groups	Infrastructure ✕
	type here to search
* Password ❓	••••••••••••••••••••••••••••
* Password (once again)	••••••••••••••••••••••••••••

Password is not mandatory for non internal authentication type.

Language	System default
Time zone	(UTC+02:00) Europe/Amsterdam
Theme	System default
Auto-login	☐
Auto-logout	☐ 15m
* Refresh	30s
* Rows per page	50
URL (after login)	

Add   Cancel

Figure 2.20 – The Zabbix user configuration window for r_infra

Lastly, make this user another **Super admin** on the **Permissions** page.

9. Now, for our last user, let's repeat our steps, changing the **Username** and the group in the **User** tab as follows:

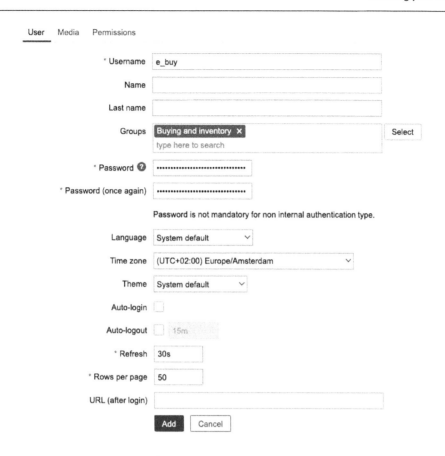

Figure 2.21 – The Zabbix User configuration window for e_buy

10. If you didn't follow the previous recipe, you can change this user's **Role** to User role at the **Permissions** tab. But if you did follow the previous recipe, we can use the User+ role we created as follows:

Figure 2.22 – The Zabbix user configuration window for e_buy

Setting the user up with the `User+` role will also let the `e_buy` user create maintenance periods.

When you're done, you'll end up with the following:

- `s_network`: A user with access to the **Networking** user group permissions with the **Super admin** role
- `y_network`: A user with access to the **Networking** user group permissions with the **Admin** role
- `r_infra`: A user with access to the **Infrastructure** user group permissions with the **Super admin** role
- `e_buy`: A user with access to the **Buying and Inventory** user group permissions with either the `User` role or the `User+ role`

Azure AD SAML user authentication and JIT user provisioning

In this recipe, we will use **Security Assertion Markup Language** (**SAML**) authentication, a widely used form of authentication in the IT world. The SAML standard allows us to exchange authorization data between applications, so we can authenticate between our Zabbix application and an authentication provider. We'll be using this as a form of managing passwords for our Zabbix users. Please note that if you only set up user authentication with passwords with SAML or LDAP, you still have to create users with their permissions manually within Zabbix. To circumvent this, we can also set up **Just In Time** (**JIT**) user provisioning since Zabbix 6.4.

Getting ready

To get started with SAML authentication, we will need our configured Zabbix server from the previous recipe. It's important that we have all the configured users from the previous recipe. We will also need something to authenticate with SAML. We will be using Microsoft Azure **Active Directory** (**AD**) SAML.

Make sure to set up users in your Azure AD before continuing with this recipe. You can use your existing AD users for authentication, so you can use this recipe with your existing AD setup.

We will be using the `s_network` user as an example as well as a new `JIT_Admin` user group in our Zabbix environment with no permissions set up. The Azure user looks as follows:

Figure 2.23 – The Azure Users and groups window

For JIT user provisioning, we also made sure to make this user part of a new `zbx_admin` group:

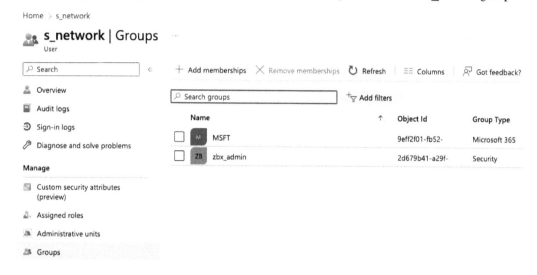

Figure 2.24 – The Azure users group details window

This group is just going to be an empty security group that we will use to assign permissions in Zabbix later:

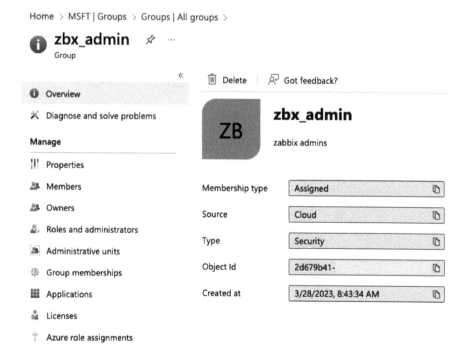

Figure 2.25 – The Azure group details window

To set up SAML, retrieve your SAML settings from your AD or another SAML provider. To work with Zabbix, we will need the following:

- IdP entity ID
- SSO service URL
- SLO service URL
- Username attribute
- SP entity ID
- SP name ID format

For the JIT user provisioning, we will need the following:

- Group name attribute
- User name attribute
- User last name attribute
- User group mapping

How to do it...

We start with the assumption that you have your Azure AD ready. Let's see how we can configure SAML using our setup:

1. Let's navigate to the following URL: `https://portal.azure.com/`.
2. After logging in, navigate to Azure AD and click on **Enterprise Applications**.

3. Now click on + **New Application** to create our new application. At the next window, click on **Create your own application**:

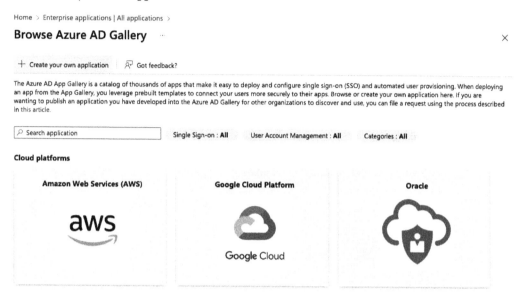

Figure 2.26 – The Azure enterprise application creation page

4. In the next window, name our new application Zabbix and click on the blue **Create** button:

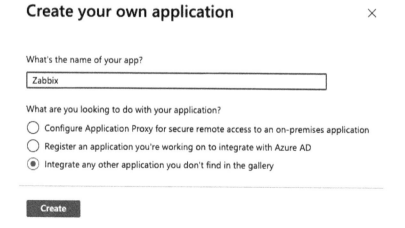

Figure 2.27 – The Azure enterprise new application page

5.   Select our new application from the list and click on **Users and Groups** to add the correct users. In our case, this will be s_network:

Selected items

Figure 2.28 – The Azure enterprise application User addition

6.   If we are setting up JIT user provisioning, make sure to also add the zbx_admin group:

Selected items

Figure 2.29 – The Azure enterprise application Group addition

With JIT user provisioning, adding the group should be enough.

7.   You will also have to assign a role. Click on **Select a role** and add the role you want to use. When using JIT you can use the **zbx_admin** group, otherwise just add the user as **User**.

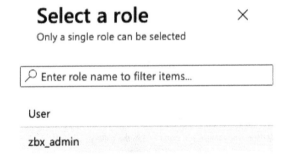

Figure 2.30 – Azure enterprise application role assignment

8.   Click on **Select** and then **Assign**.

9.   Now let's move on to the SAML settings by clicking on **Single sign-on** in the sidebar.

10. Now click on **SAML** on the page shown in the following screenshot and continue:

Figure 2.31 – The Azure enterprise application SAML option

11. Now at **1**, we can add the following information, where the black marks are our Zabbix server URL:

Figure 2.32 – The Azure SAML setting 1

12. At **2**, fill out the following:

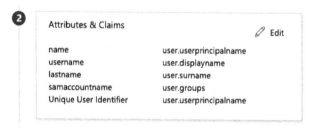

Figure 2.33 – The Azure SAML setting 2

13. **3** will be automatically filled. Click on **Download** for **Certificate (Base64)**:

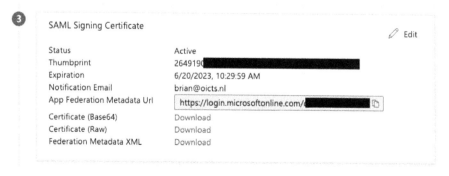

Figure 2.34 – The Azure SAML setting 3

14. Log in to the Zabbix server CLI and create a new file with the following command:

```
vim /usr/share/zabbix/conf/certs/idp.cert
```

15. Paste the contents from the file downloaded in *step 11* here and save the file.

16. Now back at Azure for **4**, we will get the following information:

Figure 2.35 – The Azure SAML setting 4

17. At the Zabbix frontend, go to the **Users** | **Authentication** | **SAML settings** page and fill in the following information:

Figure 2.36 – The Zabbix SAML settings

18. If you also want to use JIT user provisioning, enable it as seen in the previous screenshot as well as fill in the following information:

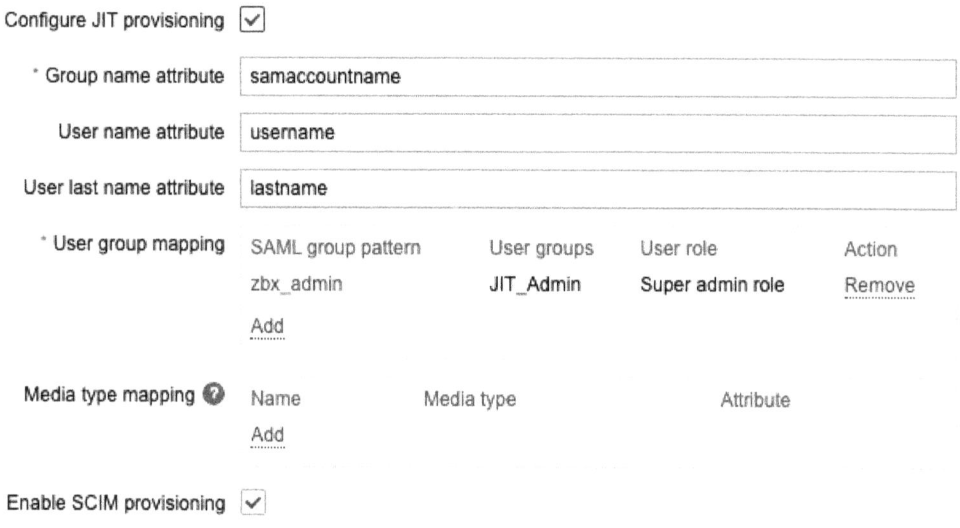

Configure JIT provisioning	☑			
* Group name attribute	samaccountname			
User name attribute	username			
User last name attribute	lastname			
* User group mapping	SAML group pattern	User groups	User role	Action
	zbx_admin	JIT_Admin	Super admin role	Remove
	Add			
Media type mapping ❓	Name	Media type	Attribute	
	Add			
Enable SCIM provisioning	☑			

Figure 2.37 – The Zabbix SAML JIT settings

> **Important note**
> I have used the **JIT_Admin** user group as suggested in the *Getting ready* part of this recipe. Please use any user group and roles you see fit and make sure to integrate the JIT user provisioning into your own groups and permissions.

19. If you have already created the s_network user and you aren't going to use JIT user provisioning, go to **Users | Users** and change the s_network user to include the used Azure domain, for example:

Figure 2.38 – The Zabbix edit user screen for our SAML setup

If you are using JIT user provisioning, you can simply log in with the new user credentials using SAML authentication and it should create the user with the correct credentials.

20. After following these steps, it should now be possible to log in with your user configured in Zabbix and use the password set in Azure AD for this:

Figure 2.39 – The Zabbix login window

How it works...

Zabbix SAML user authentication is by default used to centralize password management. In the past, we were not able to actually assign user groups and permissions to users via this setup. If we set it up without JIT user provisioning we can use it for simple password management.

This way, we can make sure it is easier for users to keep their passwords centralized:

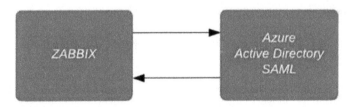

Figure 2.40 – Zabbix SAML authentication diagram

Zabbix communicates with our Azure AD SAML component when we click the **Sign in** button. The user is then authenticated against your Azure AD user and a confirmation is sent back to the Zabbix server. Congratulations, you are now logged in to your Zabbix server.

However, since Zabbix 6.4 it is also possible to enable JIT user provisioning. This new feature allows us to also assign Zabbix User groups and roles according to user groups on our SAML server. As such, the whole process with JIT user provisioning included looks something like this:

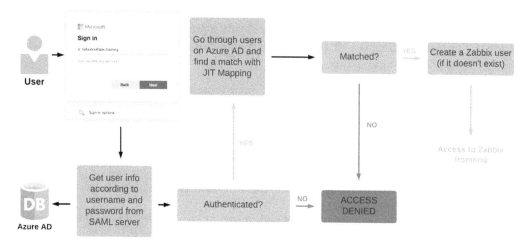

Figure 2.41 – Zabbix SAML JIT authentication diagram

There's more...

We can do this kind of authentication not only with SAML but also with HTTP and LDAP. This way, you can choose the right form of advanced authentication for your organization.

Check out the Zabbix documentation for more information on the different forms of authentication: `https://www.zabbix.com/documentation/current/en/manual/web_interface/frontend_sections/users/authentication`

It's also possible to work with an identity provider such as Okta or OneLogin, among others. This means your options aren't limited to Azure AD: as long as it supports SAML, you can use it to authenticate against your Zabbix server.

OpenLDAP user authentication and JIT user provisioning

Although a lot of people use SAML in combination with Azure Active Directory, that isn't always the case. There are loads of different methods of running your user authentication.

One of those methods is using LDAP instead of SAML with, for example, an OpenLDAP server. OpenLDAP provides us with a solid open source implementation to set up a user database with LDAP. The great thing about this is that **JIT** user provisioning doesn't just work with SAML, but also with LDAP, meaning we can apply JIT user provisioning here as well.

Getting ready

To get things going, we are going to need an OpenLDAP server set up and ready to go. It is recommended to use your own OpenLDAP environment. There are loads of guides available online to do a solid OpenLDAP implementation as well as a quick start guide for the latest version on the official website: `https://www.openldap.org/`

Another way to go is spin up a test OpenLDAP environment with Docker. We can use the following command:

```
docker run -p 389:389 -p 636:636 --name openldap-server --detach
oicts/openldap:1.0.0
docker run -p 8081:80 -p 4443:443 --name phpldapadmin --hostname
phpldapadmin --link openldap-server:ldap-host --env
PHPLDAPADMIN_LDAP_HOSTS=ldap-host --detach osixia/phpldapadmin:0.9.0
```

Please use this for testing only, since the preceding code might not be using the latest versions anymore.

How to do it...

Once OpenLDAP is set up, we can start to create some users and groups in our new OpenLDAP environment. Let's get started on that first:

1. We will open the OpenLDAP GUI by navigating to the URL in our browser:

    ```
    https://<ip_address_of_server>:4443
    ```

2. After logging in, let's create some new users. First, click on **Login** on the left-hand side of the window. The default username and password are as follows:

    ```
    Login DN: cn=admin,dc=example,dc=org
    Password: admin
    ```

3. You should see that we have already created some groups and users for you if you are using our Docker images, as you can see in the following screenshot:

Figure 2.42 – OpenLDAP server groups and users

If you are using your own OpenLDAP environment, make sure to have at least one group and one user for testing.

4.   Let's use these usernames and groups to set up LDAP authentication with JIT user provisioning.

5.   Move on to the Zabbix frontend and navigate to **Users | Users**. First, we'll give ourselves access at any time even if the default authentication method will be switched to LDAP. Switch the **Admin** user's default authentication method to internal by adding them to the **Internal** group.

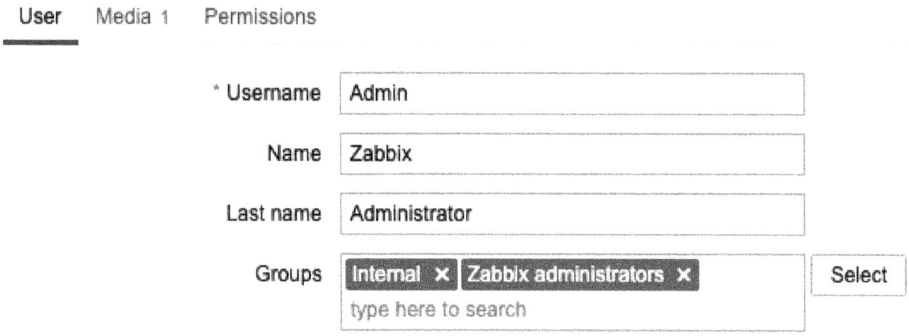

Figure 2.43 – Zabbix Admin user settings

6. Click on **Update** and it should then look like the following screenshot.

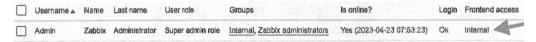

Figure 2.44 – Zabbix Admin user with Internal Frontend access

7. Then we will go to **Users | Authentication** and then **LDAP settings**. Set up the default authentication method to LDAP and the deprovisioned users group as follows.

Figure 2.45 – Zabbix Default authentication method

8. Next, we'll click on the **LDAP settings** tab. This is where we can configure our LDAP server and JIT user provisioning. Let's start by enabling the ones we would like to use.

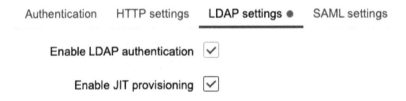

Figure 2.46 – Zabbix Default authentication method

When using just LDAP, we will have to create our users manually. By enabling JIT, users will be created and granted the correct permissions automatically.

9. It's also possible to add multiple LDAP servers in Zabbix now. Let's add our OpenLDAP server by clicking on **Add** at **Servers**.

10. Then, fill in the following.

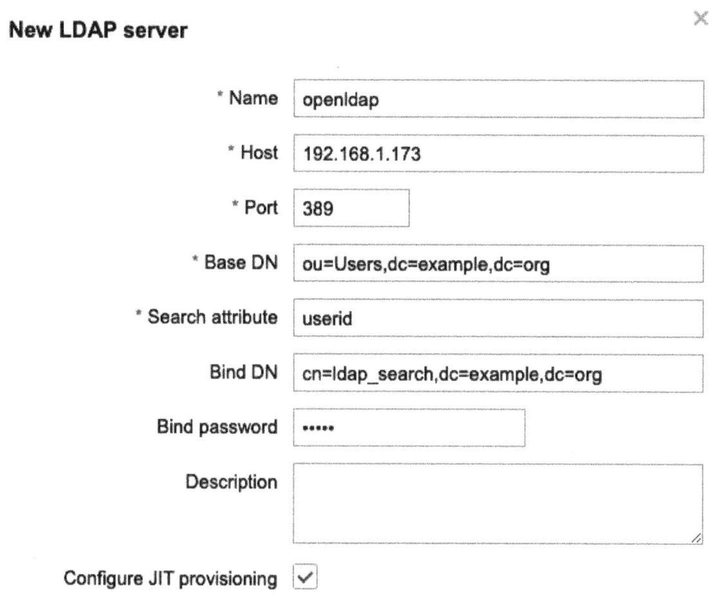

New LDAP server
* Name
openldap
* Host
192.168.1.173
* Port
389
* Base DN
ou=Users,dc=example,dc=org
* Search attribute
userid
Bind DN
cn=ldap_search,dc=example,dc=org
Bind password
Description
Configure JIT provisioning

Figure 2.47 – Zabbix LDAP authentication setup

The default **Bind password** value is `password`.

11. If we want, we can also enable JIT provisioning. Enable it and fill in the following.

Configure JIT provisioning
Group configuration
memberOf
groupOfNames
Group name attribute
cn
User group membership attribute
memberOf
User name attribute
uid
User last name attribute
* User group mapping
LDAP group pattern
User groups
User role
Action
*
Zabbix administrators
User role
Remove
Add
Media type mapping
Name
Media type
Attribute
Action
Add
Advanced configuration

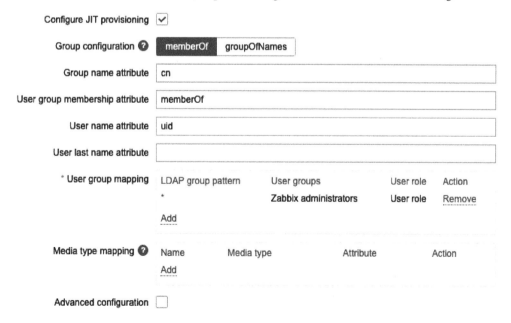

Figure 2.48 – Zabbix LDAP authentication setup with JIT

12. Now, sign out of the currently logged-in account by clicking the **Sign out** button in the lower left corner of the sidebar.

13. We should now be able to log in with the `user1` LDAP user. The password is `password`.

Figure 2.49 – Zabbix login window for user1

14. When we log in for the first time, the user will be created with the correct permissions as defined in the JIT user provisioning step. If logged in as a Zabbix super admin, we can see this under **Users | Users**.

	Username ▲	Name	Last name	User role	Groups	Is online?	Login	Frontend access	API access	Debug mode	Status	Provisioned	Info
☐	user1	user1		User role	Zabbix administrators	No (2023-03-27 09:01:28)	Ok	LDAP	Enabled	Disabled	Enabled	2023-03-27 09:01	
												Displaying 1 of 1 found	

0 selected   Provision now   Unblock   Delete

Figure 2.50 – Zabbix LDAP provisioned user

How it works...

As you can see, we can use Zabbix in combination with an LDAP server to make password management easier as a whole. There are two options for us to choose from: using LDAP with or without JIT user provisioning.

When we use Zabbix in combination with an LDAP server, but choose to not use JIT user provisioning, Zabbix will communicate with the LDAP server just to do the password authentication upon pressing the **Sign in** button.

Figure 2.51 – Zabbix LDAP authentication diagram

However, since Zabbix 6.4 it is also possible to enable JIT user provisioning. This new feature allows us to also assign Zabbix user groups and roles in line with the user groups on our LDAP server. As such, the whole process with JIT user provisioning included looks something like the following:

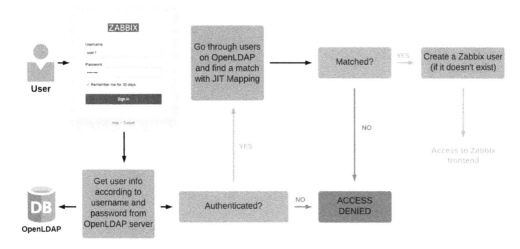

Figure 2.52 – Zabbix LDAP JIT authentication diagram

3
Setting Up Zabbix Monitoring

Zabbix is built to be flexible and should be able to monitor just about anything you could ever require. In this chapter, we will learn more about working with Zabbix to build a lot of different options for monitoring. We'll go over them recipe by recipe so that you end up with a solid understanding of how they work.

We'll cover the following recipes on the different monitoring types:

- Setting up Zabbix agent monitoring
- Working with SNMP monitoring the old way
- Setting up SNMP monitoring the new way
- Creating Zabbix simple checks and the Zabbix trapper
- Working with calculated and dependent items
- Creating external checks
- Setting up JMX monitoring
- Setting up database monitoring
- Setting up HTTP agent monitoring
- Using Zabbix browser items to simulate a web user
- Using Zabbix preprocessing to alter item values

Technical requirements

We will need a Zabbix server capable of performing monitoring, with the following requirements:

- A server with Zabbix server installed on a Linux distribution of your choice, such as Rocky Linux or Ubuntu. However, a distribution such as Debian, Alma Linux, or anything else will suit you just as well.
- A MariaDB (MySQL) server to monitor – for example, the Zabbix server database we set up in *Chapter 1*.

I'll be using the same server that we used in the previous chapter, but any Zabbix server should do.

Setting up Zabbix agent monitoring

Starting from the release of Zabbix 5, Zabbix also officially started support for the new Zabbix Agent 2. Zabbix Agent 2 brings some major improvements and is even written in another coding language – Golang instead of C. In this recipe, we will explore how to work with Zabbix Agent 2 and explore some of the new features introduced by it.

Getting ready

To get started with Zabbix Agent 2, all we need to do is install it on a host that we want to monitor. Make sure you have an empty **Red Hat Enterprise Linux** (**RHEL**)-based or Ubuntu Linux host ready to monitor.

How to do it...

Let's learn how to install Zabbix Agent 2 and then move on to working with it.

Installing Zabbix Agent 2

Let's start by installing Zabbix Agent 2 on the Linux host we want to monitor. I'll show you how to do this on both RHEL and Ubuntu systems:

1. Issue the following command to add the repository.

    For RHEL-based systems, this is as follows:

    ```
    rpm -Uvh https://repo.zabbix.com/zabbix/7.0/rocky/9/x86_64/
    zabbix-release-7.0-2.el9.noarch.rpm
    ```

    For Ubuntu systems, this is as follows:

    ```
    wget https://repo.zabbix.com/zabbix/7.0/ubuntu/pool/main/z/
    zabbix-release/zabbix-release_7.0-1+ubuntu22.04_all.deb
    dpkg -i zabbix-release_7.0-1+ubuntu22.04_all.deb
    ```

2. Then, issue the following command to install Zabbix Agent 2.

    Here's the command for RHEL-based systems:

    ```
    dnf -y install zabbix-agent2
    ```

    Here's the command for Ubuntu systems:

    ```
    apt install zabbix-agent2
    ```

Congratulations – Zabbix Agent 2 is now installed and ready to use!

> **Important note**
>
> When adding new repositories to your system, always check out the Zabbix download page. You can find the right up-to-date repository for your system here: `https://www.zabbix.com/download`.

Using a Zabbix agent in passive mode

Let's start by building a Zabbix agent with passive checks:

1. After installing Zabbix Agent 2, let's open the Zabbix agent configuration file for editing:

    ```
    vim /etc/zabbix/zabbix_agent2.conf
    ```

    In this file, we can edit all the Zabbix agent configuration values we could need from the server side.

2. Let's start by editing the following parameters:

    ```
    Server=127.0.0.1
    Hostname=Zabbix server
    ```

3. Change the value of `Server` to the IP of the Zabbix server that will monitor this passive agent. Change the value for `Hostname` to the hostname of the monitored server. We can get the IP address of our server with the following command:

    ```
    ip addr
    ```

4. Now, restart the Zabbix Agent 2 process:

    ```
    systemctl enable zabbix-agent2
    systemctl restart zabbix-agent2
    ```

5. Next, move to the frontend of your Zabbix server and add this host for monitoring.

6. Go to **Data collection** | **Hosts** in your Zabbix frontend and click **Create host** in the top-right corner.

7. To create this host in our Zabbix server, we need to fill in the values shown in the following screenshot:

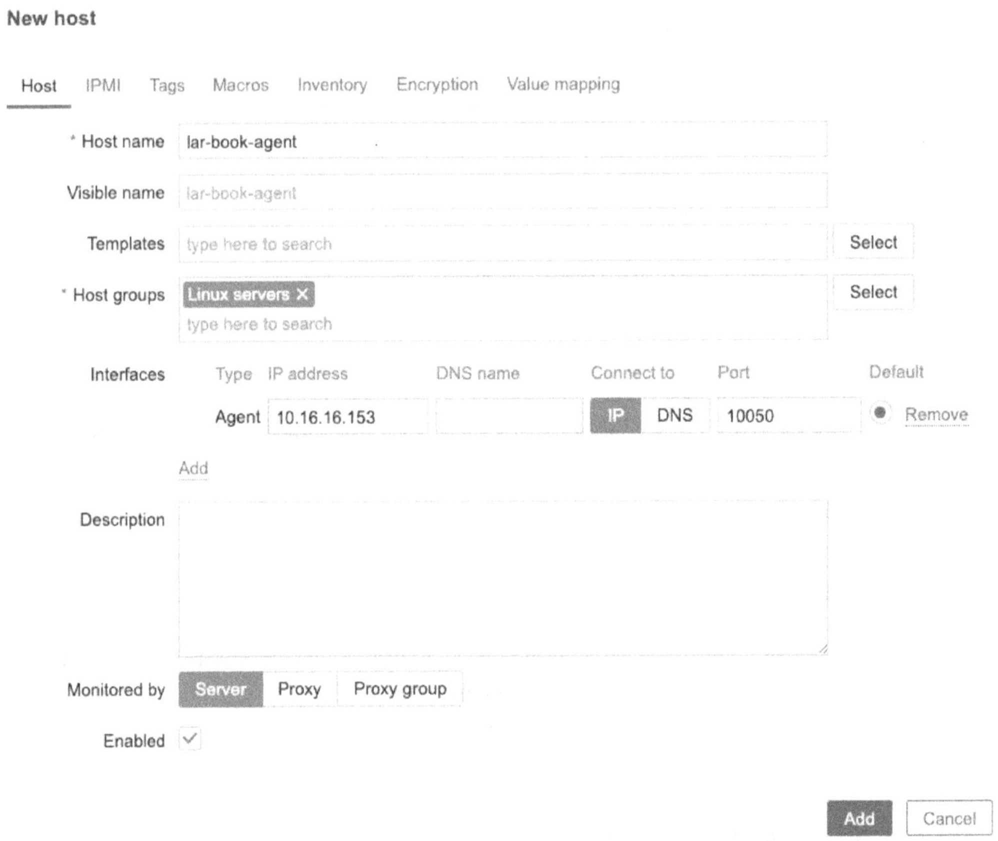

New host

Host   IPMI   Tags   Macros   Inventory   Encryption   Value mapping

* Host name	lar-book-agent
Visible name	lar-book-agent
Templates	type here to search

Select

* Host groups	Linux servers ✕
	type here to search

Select

Interfaces	Type	IP address	DNS name	Connect to	Port	Default
	Agent	10.16.16.153		IP DNS	10050	● Remove

Add

Description

Monitored by   Server   Proxy   Proxy group

Enabled   ✓

Add   Cancel

Figure 3.1 – The Zabbix host creation page for lar-book-agent

It's important to add the following:

- **Host name**: To identify the host (has to be unique).

- **Host groups**: To logically group hosts.

- **Interfaces**: To monitor this host on a specific interface. No interface means no communication. It's possible to have a host without an interface in Zabbix 7 if we don't need it. In the case of a Zabbix-agent-monitored host, an agent interface is required.

8. Make sure you add the correct IP address to the **Agent** interface configuration.

9. It is also important to add a template to this host. With Zabbix 7, this can be done on the same tab. As this is a Linux server monitored by a Zabbix agent, let's add the correct out-of-the-box template, as shown in the following screenshot:

Figure 3.2 – The Zabbix host template page for lar-book-agent

10. Click the blue **Add** button to finish creating this agent host. Now that you've got this host, make sure the **ZBX** icon turns green, indicating that this host is up and being monitored by the passive Zabbix agent:

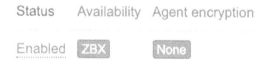

Figure 3.3 – The Zabbix configuration hosts page for lar-book-agent

11. Because we configured our host and added a template with items, we can now see the values that were received on items for this host by going to **Monitoring | Hosts** and checking the **Latest data** button. Please note that the values could take around 1 minute to show up:

Latest data

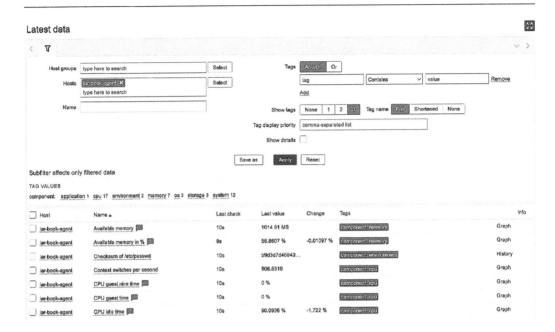

Figure 3.4 – The Zabbix Latest data page for lar-book-agent

Using a Zabbix agent in active mode

Now, let's learn how to configure the Zabbix agent with active checks. We need to change some values on the monitored Linux server host side:

1. Start by executing the following command:

    ```
    vim /etc/zabbix/zabbix_agent2.conf
    ```

2. Now, let's edit the following value to change this host to an active agent:

    ```
    ServerActive=127.0.0.1
    ```

3. Change the value for ServerActive to the IP of the Zabbix server that will monitor this passive agent. Then, change the value of Hostname to your hostname. In my case, this is lar-book-agent:

    ```
    Hostname=lar-book-agent
    ```

> **Important note**
>
> Keep in mind that if you're working with multiple Zabbix servers or Zabbix proxies, such as when you're running a Zabbix server in high availability, you need to fill in all the Zabbix servers or Zabbix proxies IP addresses when using the `ServerActive` parameter. **High availability (HA)** nodes are delimited by a semicolon (;), while different Zabbix environment IPs are delimited by a comma (,).

4. Now, restart the Zabbix Agent 2 process:

```
systemctl restart zabbix-agent2
```

5. Next, move to the frontend of your Zabbix server and add another host with a template to do active checks instead of passive ones.

6. First, let's rename our passive host. To do that, go to **Data collection | Hosts** in your Zabbix frontend and click the host we just created. Change **Host name** as follows:

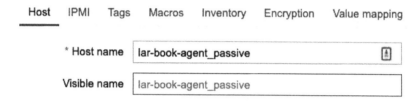

Figure 3.5 – The Zabbix host configuration page for lar-book-agent_passive

We are doing this because, for an active Zabbix agent, the hostname in the Zabbix agent configuration file needs to match the configuration of our host, as seen on the Zabbix frontend. For passive agents, this isn't the case.

7. Click on the blue **Update** button to save the changes.

8. Go to **Data collection | Hosts** in your Zabbix frontend and click **Create host** in the top-right corner.

9. Now, let's create the host, as follows:

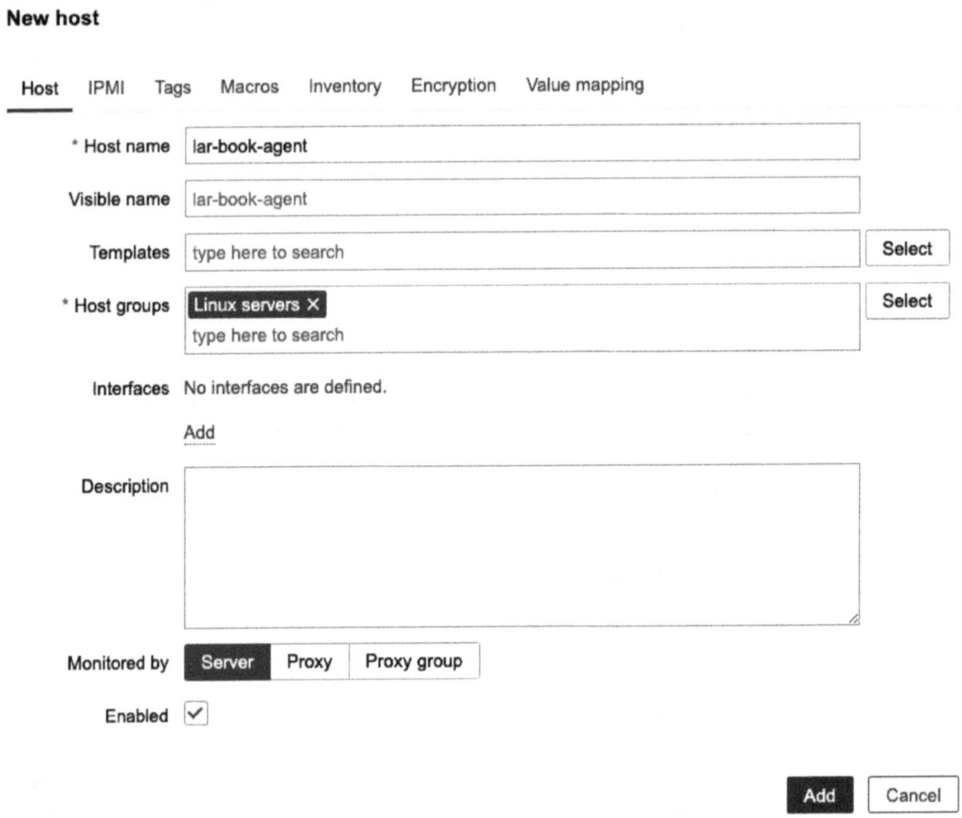

New host

| Host | IPMI | Tags | Macros | Inventory | Encryption | Value mapping |

* Host name   lar-book-agent

Visible name   lar-book-agent

Templates   type here to search   Select

* Host groups   Linux servers ✕
type here to search   Select

Interfaces   No interfaces are defined.
Add

Description

Monitored by   Server   Proxy   Proxy group

Enabled   ✓

Add   Cancel

Figure 3.6 – The Zabbix host configuration page for lar-book-agent

10. Also, make sure you add the correct template, named `Linux by Zabbix agent active`:

| Host | IPMI | Tags | Macros | Inventory | Encryption | Value mapping |

* Host name   lar-book-agent

Visible name   lar-book-agent

Templates   Linux by Zabbix agent active ✖
type here to search   Select

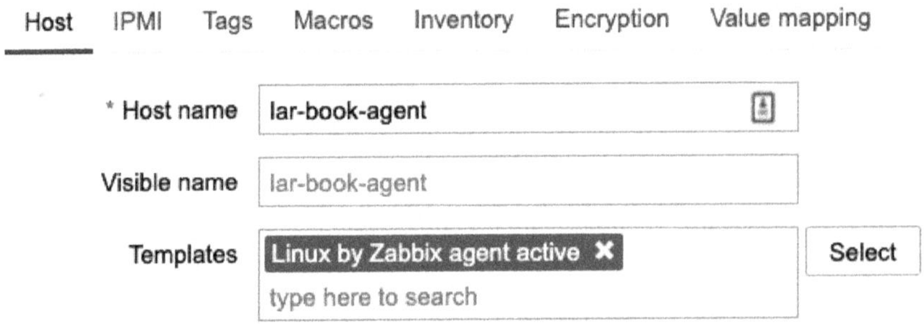

Figure 3.7 – The Zabbix host template page for lar-book-agent

Please note that as of Zabbix 6.2, the **ZBX** icon should turn green for an active agent. Note that when we navigate to **Monitoring | Hosts** and check **Latest data**, we can see our active data coming in.

> **Tip**
>
> As you might have noticed, a Zabbix agent can run in both passive and active mode at the same time. Keep this in mind when creating your own Zabbix agent templates as you might want to combine the check types. In the end, the `Item` type will determine how the checks are executed toward the agent.

How it works...

Now that we've configured our Zabbix agents and know how they should be set up, let's see how the different modes work.

Passive agent

The **passive agent** works by collecting data from our host with the Zabbix agent. Every time an item on our host reaches its *update interval*, the Zabbix server asks the Zabbix agent what the value is now:

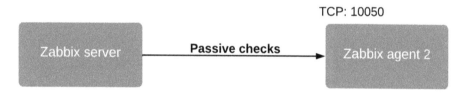

Figure 3.8 – Communication diagram between the server and the passive agent

Passive agents are great when you're working in environments where you want to keep communication initiated from the Zabbix server or Zabbix proxy side. An example of this is when there is a firewall that is only allowing outgoing traffic, as seen from the Zabbix server or proxy side.

Active agent

The **active agent** works by sending data from the Zabbix agent to a Zabbix server or Zabbix proxy. Every time an item on our agent reaches its update interval, the agent will collect the value to send to our server:

Figure 3.9 – Communication diagram between the server and the active agent

The active agent is great when you're working in an environment where there is a firewall that is only accepting outgoing connections, as seen from the Zabbix agent side. This is used in a lot of environments as it can mitigate one of the main security concerns that is mostly associated with monitoring hosts. Instead of allowing the Zabbix server to access all the different subnets (which is a bigger risk), we allow the hosts to send data to Zabbix instead – many to one instead of one to many.

On the other hand, having the Zabbix agent working in active mode can also be a lot more efficient. Most of the load that comes from getting data to your Zabbix server is now on the Zabbix agent side. Because there are more Zabbix agents out there than you have Zabbix servers or proxies, offloading a load like this is a great idea.

As mentioned previously, we can use both types of checks at the same time, giving us the freedom to configure every type of check we need. In this case, our setup would look like this:

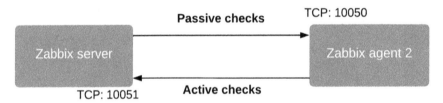

Figure 3.10 – Communication diagram between the server and both agent types

This might be the case in situations where we want to mainly monitor passively, but, for example, log file monitoring with the Zabbix agent must be done with an active Zabbix agent. In this case, we can combine our modes and make sure we use the full scale of our features provided in the Zabbix agent.

See also

There's a lot of stuff going on under the hood of Zabbix Agent 2. If you're interested in learning more about the core of Zabbix Agent 2, check out this cool blog post by Alexey Petrov: `https://blog.zabbix.com/magic-of-new-zabbix-agent/8460/`.

Working with SNMP monitoring the old way

Now, let's do something I enjoy most when working with Zabbix: build SNMP monitoring. My professional roots lie in network engineering, and I have worked with SNMP monitoring a lot to monitor all these different network devices.

Please do keep in mind that although this recipe will cover how to work with SNMP monitoring the old way, it is still a valid option. Zabbix 6.4 introduced an entirely new way of setting up SNMP monitoring. The new way utilizes bulk metric collection and is more efficient for the SNMP device and number of network sessions, so it might be a good idea to check out that recipe after this one.

Getting ready

To get started, we need the two Linux hosts we used in the previous recipes:

- Our Zabbix server host
- The host we used in the previous recipe to monitor via the Zabbix active agent

How to do it...

Monitoring via SNMP polling is easy and very powerful. We will start by configuring SNMPv3 on our monitored Linux host:

1. Let's start by issuing the following commands to install SNMP on the host we would like to be monitored by SNMP.

    For RHEL-based systems:

    ```
    dnf install net-snmp net-snmp-utils
    ```

    For Ubuntu systems:

    ```
    apt install snmp snmpd libsnmp-dev
    ```

2. Now, let's create the new SNMPv3 user that we will use to monitor our host. Please note that we'll be using insecure passwords, so make sure you use secure passwords for your production environments. Issue the following command:

    ```
    net-snmp-create-v3-user -ro -a my_authpass -x my_privpass -A SHA
    -X AES snmpv3user
    ```

    Please note that on some installations, you might have to stop snmpd before executing this command. You can start it again after.

    This will create an SNMPv3 user with a username of snmpv3user, an authentication password of my_authpass, and a privilege password of my_ privpass.

3. Make sure you edit the SNMP configuration file so that you can read all SNMP objects:

    ```
    vim /etc/snmp/snmpd.conf
    ```

4. Add the following line to the existing `view systemview` lines. If there are none, simply create this new line:

    ```
    view  systemview  included  .1
    ```

5. Now, enable and start the `snmpd` daemon so that you can start monitoring this server:

    ```
    systemctl enable snmpd
    systemctl start snmpd
    ```

    This is all we need to do on the Linux host side; we can now go to the Zabbix frontend to configure our host. Go to **Data collection** | **Hosts** in your Zabbix frontend and click **Create host** in the top-right corner.

6. Fill in the host configuration page:

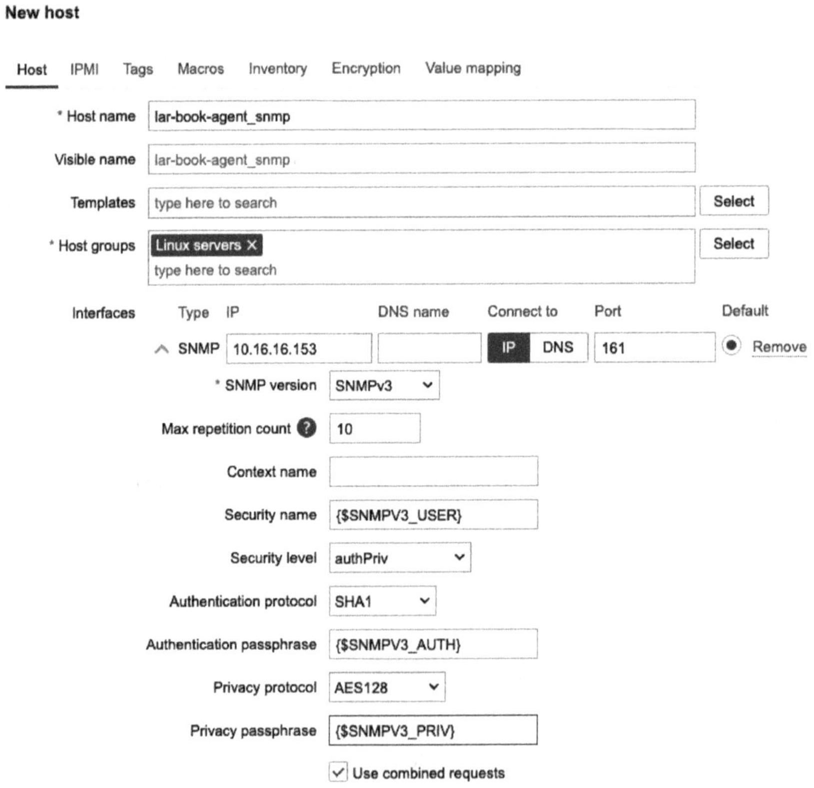

Figure 3.11 – Zabbix host configuration page for lar-book-agent_snmp

7. Don't forget to change the IP address of the SNMP interface to your own value.

8. Make sure you add the right out-of-the-box template, as shown in the following screenshot:

Figure 3.12 – Adding the Linux by SNMP template to the host

> **Tip**
> While upgrading from an earlier Zabbix version to Zabbix 6, you won't get all the new out-of-the-box templates. If you feel like you are missing some templates, you can download them from the Zabbix GitHub repository: `https://git.zabbix.com/projects/ZBX/repos/zabbix/browse/templates`.

9. We are using some macros in our configuration here for the username and password. We can use these macros to add a bunch of hosts with the same credentials. This is very useful, for instance, if you have a bunch of switches with the same SNMPv3 credentials.

   Let's fill in the macros under **Administration | Macros**, like so:

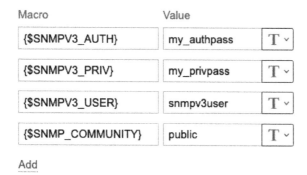

Figure 3.13 – Zabbix global macro page with SNMP macros

> **Tip**
>
> A cool feature in Zabbix 6 is the ability to hide macros in the frontend by using the **Secret text** macro type. Do keep in mind that macros of the **Secret text** type are still unencrypted in the Zabbix database, So, for fully encrypted macros, we would need something such as HashiCorp or CyberArk Vault. Check out the documentation for more information: `https://www.zabbix.com/documentation/current/en/manual/config/secrets`.

10. Use the dropdown to change **{$SNMPV3_AUTH}** and **{$SNMPV3_PRIV}** to **Secret text**:

Figure 3.14 – Zabbix Secret text used to hide sensitive (authentication) data

11. Now, after applying these changes by clicking **Update**, we should be able to monitor our Linux server via SNMPv3. Let's go to **Monitoring | Hosts** and check the **Latest data** page for our new host:

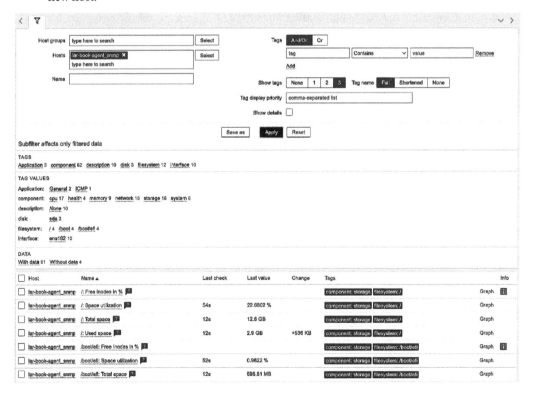

Figure 3.15 – SNMP – the Latest data page for lar-book-agent_snmp

Note that it might take around 1 minute for your data to show up here.

> **Tip**
>
> When working with macros, there are three levels in cascading order: global, template, and host-level macros. When working with global-level macros, keep in mind that they are not exported with templates or hosts. You want to use template-level and host-level macros in most cases to keep your exports independent of Zabbix global settings.

How it works...

When we create a host, as we did in *Step 4*, Zabbix polls the host using SNMP. Polling SNMP like this uses SNMP OIDs. For instance, when we poll the **Free memory** item, we ask the SNMP agent running on our Linux host to provide us the value for OID 1.3.6.1.4.1.2021.4.6.0. That value is then returned to us on the Zabbix server:

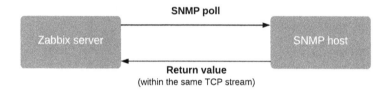

Figure 3.16 – Diagram showing communication between Zabbix server and SNMP host

The OID is like the address (or path) of where our metric is located. By requesting the OID, the metric is requested.

SNMPv3 adds authentication and encryption to this process, making sure that when our Zabbix server requests information, that request is first encrypted and the data is sent back encrypted as well.

We also included the option to use **Combined requests** when configuring our host. Combined requests request several OIDs in the same stream, making this the preferred method of doing SNMP requests as it is more efficient. Only disable it for hosts that do not support **Combined requests**. Even better is to use an SNMP bulk request, which we will discuss in the next recipe.

Lastly, let's take a look at SNMP OIDs, the most important part of our SNMP request. OIDs work in a tree-like structure, meaning that every number behind the dot can contain another value. For example, let's look at this OID for our host:

```
1.3.6.1.4.1.2021.4 = UCD-SNMP-MIB::memory
```

If we poll that OID with either the SNMPwalk CLI tool or our Zabbix server, we will get several OIDs back:

```
.1.3.6.1.4.1.2021.4.1.0 = INTEGER: 0
.1.3.6.1.4.1.2021.4.2.0 = STRING: swap
.1.3.6.1.4.1.2021.4.3.0 = INTEGER: 1679356 kB
.1.3.6.1.4.1.2021.4.4.0 = INTEGER: 1674464 kB
.1.3.6.1.4.1.2021.4.5.0 = INTEGER: 1872872 kB
.1.3.6.1.4.1.2021.4.6.0 = INTEGER: 184068 kB
```

This includes our 1.3.6.1.4.1.2021.4.6.0 OID with the value that contains our free memory. This is how SNMP is built, like a tree.

Setting up SNMP monitoring the new way

SNMP monitoring has had an entire overhaul starting with Zabbix 6.4, introducing a new way to build SNMP monitoring. The old way is still available and works, but all out-of-the-box monitoring will be overhauled to work with the new way.

The new way will utilize SNMP bulk queries, making it a lot more efficient. As such, in this recipe, we will look at how to build SNMP monitoring the new way.

Getting ready

To get started, we need the two Linux hosts:

- Our Zabbix server environment
- Any Linux host running the SNMP server

How to do it...

Let's be efficient and start building some bulk SNMP queries. First things first, get your hosts ready:

1. First, log in to your Zabbix server CLI. We will start by installing some additional tools to make building SNMP monitoring easier.

    For RHEL-based systems:

    ```
    dnf install net-snmp-utils
    ```

    For Ubuntu systems:

    ```
    apt install libsnmp-dev
    ```

2. Then, on the Linux host we would like to monitor, we must install the SNMP server.

    For RHEL-based systems:

    ```
    dnf install net-snmp net-snmp-utils
    ```

    For Ubuntu systems:

    ```
    apt install snmp snmpd libsnmp-dev
    ```

3. Now, let's configure a new SNMPv3 user on the host we want to monitor and set up the server so that we can query information:

    ```
    net-snmp-create-v3-user -ro -a my_authpass -x my_privpass -A SHA
    -X AES snmpv3user
    ```

    This will create an SNMPv3 user with a username of snmpv3user, an authentication password of my_authpass, and a privilege password of my_ privpass. Please make sure you use secure passwords in your production environments!

4. Make sure you edit the SNMP configuration file so that you can read all SNMP objects:

    ```
    vim /etc/snmp/snmpd.conf
    ```

5. Add the following line to the rest of the view systemview lines:

    ```
    view   systemview   included   .1
    ```

6. Now, enable and start the snmpd daemon so that you can start monitoring this server:

    ```
    systemctl enable snmpd
    systemctl start snmpd
    ```

    This is all we need to do on the Linux host side; we can now go to the Zabbix frontend to configure our host.

7. Go to **Data collection** | **Hosts** in your Zabbix frontend and click **Create host** in the top-right corner. We will create a new host with the following information:

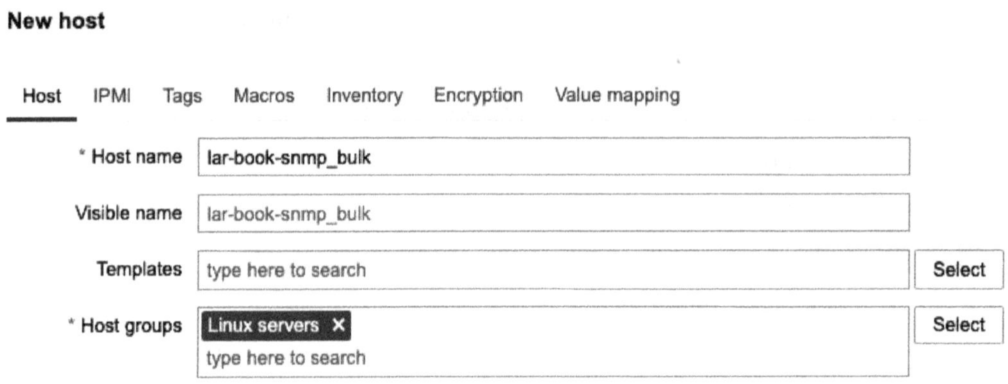

New host

Host	IPMI	Tags	Macros	Inventory	Encryption	Value mapping

* Host name   `lar-book-snmp_bulk`

Visible name   `lar-book-snmp_bulk`

Templates   `type here to search`   Select

* Host groups   Linux servers ✕   Select
              `type here to search`

Figure 3.17 – Zabbix host configuration page for lar-book-snmp_bulk

8.   Before adding the host, make sure you click on the small dotted underlined **Add** button in the **Interfaces** section and select **SNMP**:

Figure 3.18 – Zabbix interface configuration for lar-book-snmp_bulk

Make sure you fill in the right IP address and credentials for the host you are going to monitor.

9. Switch to the **Macros** tab and add the following information:

Figure 3.19 – Zabbix host configuration Macros tab for lar-book-snmp_bulk

10. Let's also go to **Value mapping** and create the following value map. We'll use this later:

Figure 3.20 – Zabbix host configuration Value mapping tab for lar-book-snmp_bulk

11. Now, you can click on the big **Add** button at the bottom of the page and the host will be created.

12. At this point, we have to start building our SNMP checks. But before we do that, we should decide which checks to build. Let's do a quick SNMP walk from the Zabbix server Linux CLI:

```
snmpwalk -On -v3  -l authPriv -u snmpv3user -a SHA -A
"my_authpass"  -x AES -X "my_privpass" 192.168.1.86
.1.3.6.1.2.1.2.2.1.2
```

This SNMP walk will show us an output similar to the following:

```
.1.3.6.1.2.1.2.2.1.2.1 = STRING: lo
.1.3.6.1.2.1.2.2.1.2.2 = STRING: ens192
```

These are the SNMP interfaces that are available within our system and I want to add them for monitoring. Let's say I want to monitor the `ens192` interface. Remember that the index for the `ens192` interface is the number 2; we will need it later.

To add all the interface information in bulk to our Zabbix environment, I will use a lower OID. However, note that `.1.3.6.1.2.1.2.2.1` contains all our interface information.

13. Test the SNMP walk with all interface information:

```
snmpwalk -On -v3  -l authPriv -u snmpv3user -a SHA -A
"my_authpass"  -x AES -X "my_privpass" 192.168.1.86
.1.3.6.1.2.1.2.2.1
```

You should see a lot more output now.

14. Let's go back to the Zabbix frontend by going to **Data collection | Hosts**, choosing the `lar-book-snmp_bulk` host, and going to **Item**.

15. In the top-right corner, click **Create item** and add the following information:

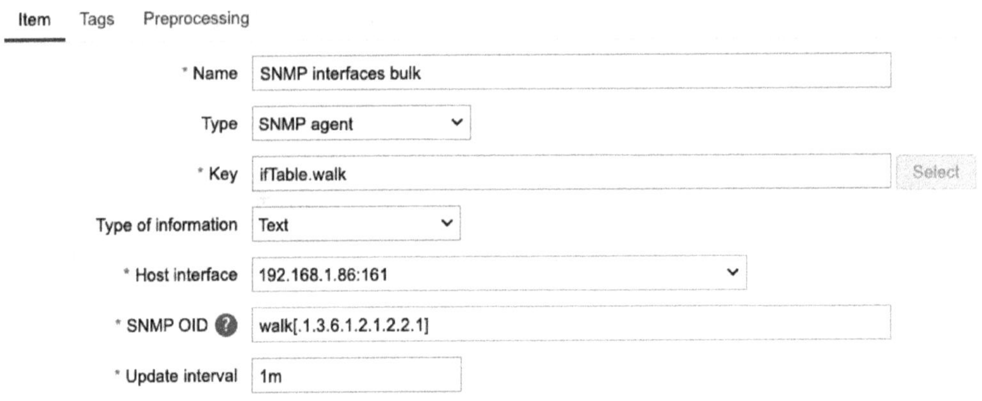

Item Tags Preprocessing	
* Name	SNMP interfaces bulk
Type	SNMP agent
* Key	ifTable.walk Select
Type of information	Text
* Host interface	192.168.1.86:161
* SNMP OID ?	walk[.1.3.6.1.2.1.2.2.1]
* Update interval	1m

Figure 3.21 – Zabbix item configuration for ifTable.walk

16. Don't forget to switch to the **Tags** tab and add the following:

Item   **Tags** 1   Preprocessing

Item tags Inherited and item tags		
Name	Value	
component	raw	Remove
Add		

Figure 3.22 – Zabbix item configuration ifTable.walk Tags tab

17. Click on the big **Add** button at the bottom of the window to add this item to the host.

18. This item will now collect our SNMP data in bulk. At this point, we can create dependent items to get specific values. You should be back on the **Items** page for the host, where we can once again click on **Create item**.

19. My interface was called `ens192`, so let's get the operational status for that interface. Add the following information:

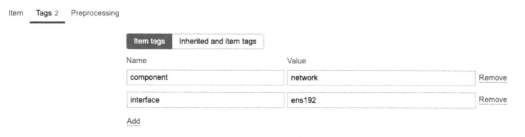

Figure 3.23 – Zabbix item configuration for ifOperStatus[ens192]

20. Don't forget to switch to the **Tags** tab and add the following:

Item	**Tags** 2	Preprocessing

Item tags	Inherited and item tags

Name	Value	
component	network	Remove
interface	ens192	Remove

Add

Figure 3.24 – Zabbix item configuration ifOperStatus[ens192] Tags tab

21. Last, but certainly not least, we will need to go to the **Preprocessing** tab. This is where we will decide which value to extract from the bulk. Remember the index from *Step 12*? Let's use it now by adding the OID for the interface operational status (`1.3.6.1.2.1.2.2.1.8`) with an index of 2:

Figure 3.25 – Zabbix item configuration ifOperStatus[ens192] Preprocessing tab

22. Now, click the big **Add** button at the bottom of the page and let's see if it all worked.

23. Go to **Monitoring** | **Latest data** and find your host – that is, `lar-book-snmp_bulk`:

	Host	Name ▲	Last check	Last value	Change	Tags
☐	lar-book-snmp_bulk	Interface ens192: Operational status	50s	Up (1)		`component: network` `interface: ens192`
☐	lar-book-snmp_bulk	SNMP interfaces bulk	50s	.1.3.6.1.2.1.2.2....		`component: raw`

Figure 3.26 – Zabbix lar-book-snmp_bulk under Monitoring | Latest data

As you can see, we are now collecting the SNMP information in bulk and then collecting a single value from the bulk information.

> **Important note**
>
> It is always recommended to use the **Do not keep history** option on items collecting values in bulk. That way, we aren't storing duplicate values for no reason. Once you finish building all your SNMP items, don't forget to make that change.

How it works...

The new SNMP walk might seem like a bit of a headscratcher at first. Why do we need this new change? The way the internals work in Zabbix before 6.4 is that it will collect each SNMP OID separately. There is a smart mechanism that combines requests to make it a bit more efficient, but it was never officially a bulk request (even though that's what the frontend called it).

Now, with the new `walk[]` item key, we are collecting all the SNMP values in a single SNMP `GetBulk` request. This makes the entire process a lot more efficient and stresses the SNMP devices a lot less.

There's also a new addition in Zabbix 7.0 that is for the following three pollers:

- Agent poller
- HTTP agent poller
- SNMP poller (for `walk[OID]` and `get[OID]` items)

These processes now execute checks asynchronously. What this means for our SNMP checks using `walk[]` or `get[]` is that they can execute multiple (item) checks at the same time. In older versions of Zabbix, these pollers could only execute a single check at a time.

It's still possible to add multiple of these processes with **StartSNMPPollers**, for example, but it now functions differently. They will execute a maximum of 1,000 checks per poller, something that can be configured with the **MaxConcurrentChecksPerPoller** parameter.

So, what did we use? Well, we started with a simple request, which was to get all of the values under the SNMP interface's OID – that is, `.1.3.6.1.2.1.2.2.1`. This contains all the information for our SNMP interfaces, as shown in the following screenshot:

```
Timestamp           Value

2023-03-18 10:31:30  .1.3.6.1.2.1.2.2.1.1.1 = INTEGER: 1
                     .1.3.6.1.2.1.2.2.1.1.2 = INTEGER: 2
                     .1.3.6.1.2.1.2.2.1.2.1 = STRING: lo
                     .1.3.6.1.2.1.2.2.1.2.2 = STRING: ens192
                     .1.3.6.1.2.1.2.2.1.3.1 = INTEGER: 24
                     .1.3.6.1.2.1.2.2.1.3.2 = INTEGER: 6
                     .1.3.6.1.2.1.2.2.1.4.1 = INTEGER: 65536
                     .1.3.6.1.2.1.2.2.1.4.2 = INTEGER: 1500
                     .1.3.6.1.2.1.2.2.1.5.1 = Gauge32: 10000000
                     .1.3.6.1.2.1.2.2.1.5.2 = Gauge32: 4294967295
                     .1.3.6.1.2.1.2.2.1.6.1 = STRING:
                     .1.3.6.1.2.1.2.2.1.6.2 = STRING: 0:50:56:9a:25:79
                     .1.3.6.1.2.1.2.2.1.7.1 = INTEGER: 1
                     .1.3.6.1.2.1.2.2.1.7.2 = INTEGER: 1
                     .1.3.6.1.2.1.2.2.1.8.1 = INTEGER: 1
                     .1.3.6.1.2.1.2.2.1.8.2 = INTEGER: 1
                     .1.3.6.1.2.1.2.2.1.9.1 = 0
                     .1.3.6.1.2.1.2.2.1.9.2 = 0
                     .1.3.6.1.2.1.2.2.1.10.1 = Counter32: 0
                     .1.3.6.1.2.1.2.2.1.10.2 = Counter32: 22457056
                     .1.3.6.1.2.1.2.2.1.11.1 = Counter32: 0
                     .1.3.6.1.2.1.2.2.1.11.2 = Counter32: 30150
                     .1.3.6.1.2.1.2.2.1.12.1 = Counter32: 0
                     .1.3.6.1.2.1.2.2.1.12.2 = Counter32: 455
                     .1.3.6.1.2.1.2.2.1.13.1 = Counter32: 0
                     .1.3.6.1.2.1.2.2.1.13.2 = Counter32: 6316
                     .1.3.6.1.2.1.2.2.1.14.1 = Counter32: 0
                     .1.3.6.1.2.1.2.2.1.14.2 = Counter32: 0
                     .1.3.6.1.2.1.2.2.1.15.1 = Counter32: 0
                     .1.3.6.1.2.1.2.2.1.15.2 = Counter32: 0
                     .1.3.6.1.2.1.2.2.1.16.1 = Counter32: 0
                     .1.3.6.1.2.1.2.2.1.16.2 = Counter32: 2153620
                     .1.3.6.1.2.1.2.2.1.17.1 = Counter32: 0
                     .1.3.6.1.2.1.2.2.1.17.2 = Counter32: 13936
                     .1.3.6.1.2.1.2.2.1.18.1 = Counter32: 0
                     .1.3.6.1.2.1.2.2.1.18.2 = Counter32: 0
                     .1.3.6.1.2.1.2.2.1.19.1 = Counter32: 0
                     .1.3.6.1.2.1.2.2.1.19.2 = Counter32: 0
                     .1.3.6.1.2.1.2.2.1.20.1 = Counter32: 0
                     .1.3.6.1.2.1.2.2.1.20.2 = Counter32: 0
                     .1.3.6.1.2.1.2.2.1.21.1 = Gauge32: 0
                     .1.3.6.1.2.1.2.2.1.21.2 = Gauge32: 0
                     .1.3.6.1.2.1.2.2.1.22.1 = OID: .0.0
                     .1.3.6.1.2.1.2.2.1.22.2 = OID: .0.0
```

Figure 3.27 – Zabbix lar-book-snmp_bulk raw SNMP walk on Monitoring | Latest data

After, we extracted a single value from the bulk we collected with a preprocessing step:

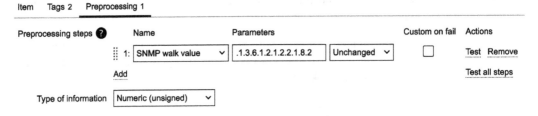

Figure 3.28 – Zabbix lar-book-snmp_bulk SNMP walk value preprocessing

We can do this to extract any OID from the already completed SNMP walk manually. This is super useful if you have a lot of information from an SNMP walk item but you only need a few static values from the walk.

This becomes extra apparent once we collect big pieces of information and start using LLD rules to automate the task later, as well as when we still need to get some specific values, something that might not be an option for LLD. All that information can be collected in a single call to the SNMP device and then split into LLD rules, as well as separate items.

We'll continue working with this kind of new monitoring in *Chapter 7, Using Discovery for Automatic Creation*, in the *Setting up Zabbix SNMP low-level discovery the new way* recipe.

Creating Zabbix simple checks and the Zabbix trapper

In this recipe, we will go over two checks that can help you build some more customized setups. Zabbix simple checks provide you with an easy way to monitor some specific data, while the Zabbix trapper combines with the Zabbix sender to get data from your hosts into the server, providing you with some scripting options. Let's get started.

Getting ready

To create these checks, we will need a Zabbix server and a Linux host to monitor. We can use the host with a Zabbix agent and SNMP monitoring from the previous recipes.

Note that we do not need the Zabbix agent for these checks.

How to do it...

As the name suggests, working with simple checks is quite simple. So, let's get started.

Creating simple checks

We will create a simple check to monitor whether a service is running and accepting TCP connections on a certain port:

1. To get this done, we will need to create a new host on the Zabbix frontend. Go to **Data collection | Hosts** in your Zabbix frontend and click **Create host** in the top-right corner.

2. Create a host with the following settings:

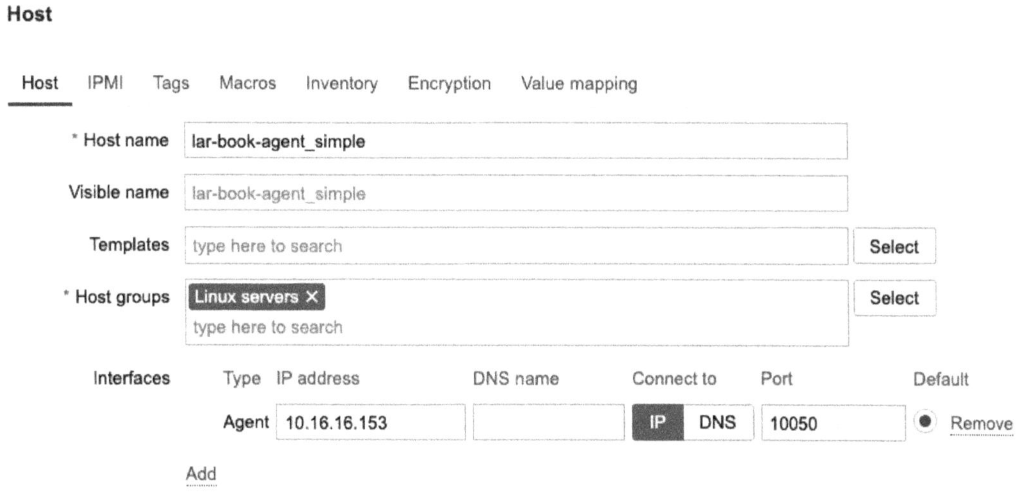

Host

| Host | IPMI | Tags | Macros | Inventory | Encryption | Value mapping |

* Host name	lar-book-agent_simple
Visible name	lar-book-agent_simple
Templates	type here to search	Select
* Host groups	Linux servers ✕ type here to search	Select

| Interfaces | Type | IP address | DNS name | Connect to | Port | Default |
| | Agent | 10.16.16.153 | | IP DNS | 10050 | ● Remove |

Add

Figure 3.29 – Zabbix host configuration page for lar-book-agent_simple

3. Now, go to **Data collection | Hosts** and go to **Items** for the newly created host. We want to create a new item here by clicking the **Create item** button.

We must create a new item with the following values. After doing so, click the **Add** button at the bottom of the page:

Figure 3.30 – The Zabbix item configuration page for the port 22 check on the lar-book-agent_simple host

4.   Make sure that you also add a tag to the item since we need this in several places to filter and find our item when we're working with Zabbix. Set it up like this:

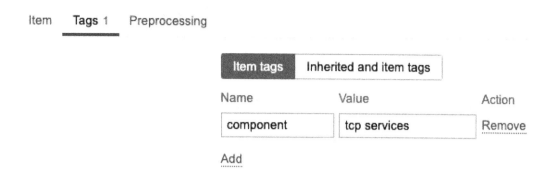

Figure 3.31 – Zabbix SSH port item, Tag tab

> **Important note**
>
> We are adding the `net.tcp.services[ssh,,22]` item key here. The port in this case is optional as we can specify the service SSH with a different port if we want to.

5. Now, we should be able to see whether our server is accepting SSH connections on port 22 on our **Latest data** screen. Navigate to **Monitoring** | **Hosts** and check the **Latest data** screen for our new value:

Figure 3.32 – Zabbix Latest data page for lar-book-agent_simple, item port 22 check

6. There is one more thing wrong here. As you can see, we do not currently have a value mapping setup. Here, **Last value** is just displaying **1** or **0**, making it hard to distinguish what this means. To change this, navigate back to **Data collection | Hosts** and edit the `lar-book-agent_simple` host.

7. Click on the **Value mapping** tab and click the small **Add** button to add a value mapping, like so:

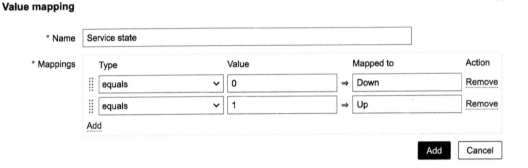

Figure 3.33 – lar-book-agent_simple, Value mapping window

8. Click on the blue **Add** button and click on the blue **Update** button.

9. Then, back at the full **Data collection | Hosts** list, navigate to our `lar-book-agent_simple` host and click on **Items** for this host.

10. Edit the **Check if port 22 is available** item and add the following value mapping:

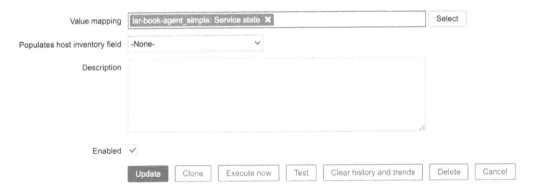

Figure 3.34 – lar-book-agent_simple, edit item window

That's all there is to creating simple checks in Zabbix. The latest data page will now look like this:

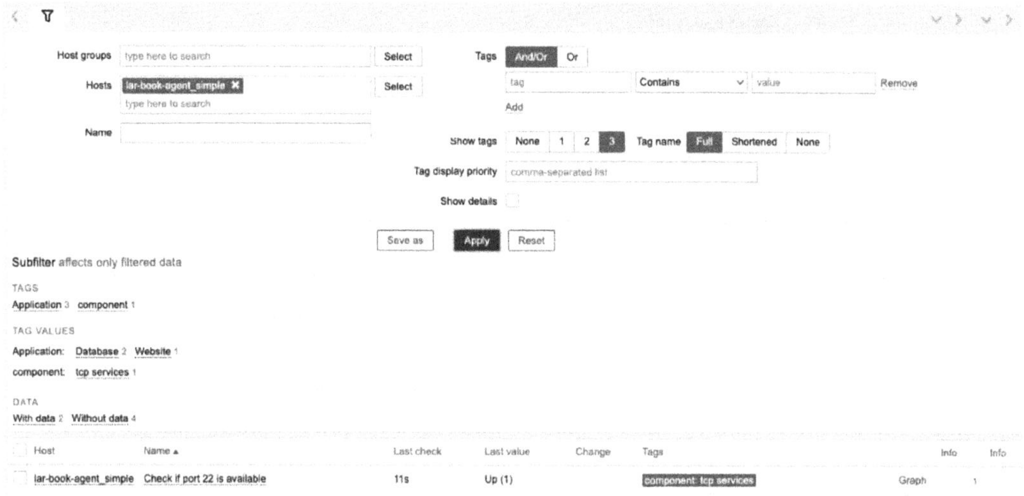

Figure 3.35 – Latest data page for our port 22 check item

As you can see, there is a human-readable value now displaying either **Up** or **Down**, giving us a human-readable entry that's easier to understand. Now, let's look at the Zabbix trapper item.

Creating a trapper

We can do some cool stuff with Zabbix trapper items once we get more advanced setups. But for now, let's create an item on our `lar-book-agent_simple` host:

1. Go to **Data collection** | **Hosts** and click on the host, then go to **Items**. We want to create a new item here by clicking the **Create item** button.

    So, let's create the following item and click the **Add** button:

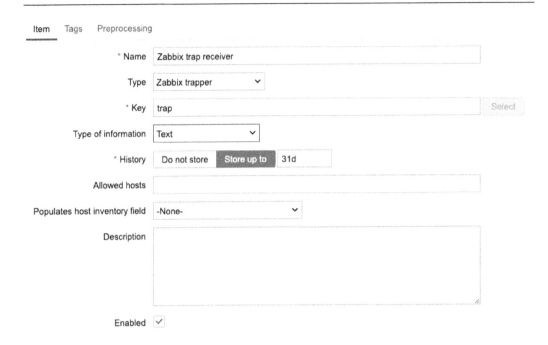

Figure 3.36 – Zabbix item trap receiver configuration screen for lar-book-agent_simple

2. Make sure you also navigate to the **Tags** tab and add a tag. We will use this later for filtering:

Figure 3.37 – Zabbix item trap receiver tag configuration screen for lar-book-agent_simple

3. If we go to the CLI of our monitored server, we can install Zabbix sender.

    Run the following command for RHEL-based systems:

    ```
    dnf -y install zabbix-sender
    ```

    Run the following command for Ubuntu systems:

    ```
    apt install zabbix-sender
    ```

4. After installation, we can use Zabbix sender to send some information to our server (make sure you use your Zabbix server IP when using the -z option):

```
zabbix_sender -z 10.16.16.152 -s "lar-book-agent_simple" -k trap
-o "Let's test this book trapper"
```

Now, we should be able to see whether our monitored host has sent out the Zabbix trap and the Zabbix server has received this trap for processing.

5. Navigate to **Monitoring | Hosts** and check the **Latest data** screen for our new value:

Host	Name ▲	Last check	Last value
lar-book-agent_simple	Zabbix trap receiver	18s	Let's test this boo…

Figure 3.38 – Zabbix Latest data page for lar-book-agent_simple, item trap receiver

There it is – our Zabbix trap is in our Zabbix frontend.

How it works...

Now that we have built our new items, let's see how they work by diving into the theoretical side of Zabbix simple checks and trappers.

Simple checks

Zabbix simple checks are a list of built-in checks that are made for monitoring certain values. There is a list and descriptions available for all the simple checks that are available in the Zabbix documentation: https://www.zabbix.com/documentation/current/manual/config/items/itemtypes/simple_checks.

All of these checks are performed by the Zabbix server to collect data from a monitored host. For example, when we do the Zabbix simple check to check whether a port is open, our Zabbix server requests whether it can reach that port and turns that into a status we can then see in our Zabbix frontend.

This means that if your monitored host's firewall is blocking port 22 from the Zabbix server, we'll get a service *down* value. However, this doesn't necessarily mean that SSH isn't running on the server; it simply means SSH is down as seen from the side of the Zabbix server or proxy:

Figure 3.39 – Zabbix server-to-host communication diagram

> **Tip**
>
> Keep in mind that working with simple checks is dependent on external factors such as the firewall settings on the monitored host. When you build simple checks, make sure to check these factors as well.

There's one more thing to note here. In Zabbix 6.4, the ability to add simple checks without an interface on the host was added. This means you can simply add the item with the connection details as parameters in the item key instead of selecting an interface.

Trappers

When working with Zabbix sender, we are doing exactly the opposite of most checks – we are building an item on our Zabbix server, which allows us to capture trap items. This allows us to build some custom checks so that we can send data to our Zabbix server from a monitored host:

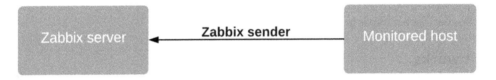

Figure 3.40 – Zabbix server trap receiver diagram

Let's say, for instance, that we want to build a custom Python script that, at the end of running the scripts, sends output to the Zabbix server. We could ask Python to send this data using the Zabbix sender utility, at which point we'd have this data available for processing on the Zabbix server.

This process is used by some companies who write software to completely integrate their software into Zabbix. As you can see, we can greatly extend our options with Zabbix trappers and customize our Zabbix server even further. Amazingly, this also works with low-level discovery, so long as we send the correct data formats (JSON).

Working with calculated and dependent items

Calculated and dependent items are used in Zabbix to produce additional values from existing values. Sometimes, we have already collected a value and we need to do more with the values created by that item. We can do exactly that by using calculated and dependent items.

Getting ready

To work with calculated items and dependent items, we are going to need the Zabbix server and monitored hosts from the previous recipes. We will add the items to the `lar-book-agent_passive` host and our Zabbix server (or any MySQL server) host so that we already have some items available to calculate and make dependent.

How to do it...

Let's see how we can extend our items. We'll start by looking at calculated items.

Working with calculated items

Follow these steps:

1. Let's navigate to our host configuration by going to **Data collection | Hosts** and clicking on our `lar-book-agent_passive` host's **Items** area. In the **Name** filter field, enter memory; you will get the following output:

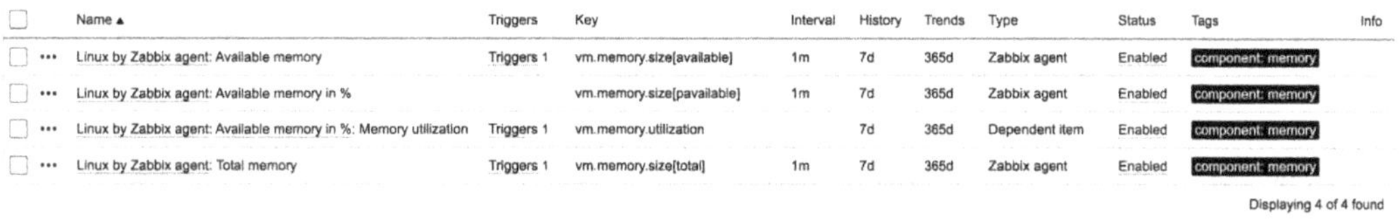

	Name ▲	Triggers	Key	Interval	History	Trends	Type	Status	Tags	Info
⬜ •••	Linux by Zabbix agent: Available memory	Triggers 1	vm.memory.size[available]	1m	7d	365d	Zabbix agent	Enabled	component: memory	
⬜ •••	Linux by Zabbix agent: Available memory in %		vm.memory.size[pavailable]	1m	7d	365d	Zabbix agent	Enabled	component: memory	
⬜ •••	Linux by Zabbix agent: Available memory in %: Memory utilization	Triggers 1	vm.memory.utilization		7d	365d	Dependent item	Enabled	component: memory	
⬜ •••	Linux by Zabbix agent: Total memory	Triggers 1	vm.memory.size[total]	1m	7d	365d	Zabbix agent	Enabled	component: memory	

Displaying 4 of 4 found

Figure 3.41 – Zabbix item page for lar-book-agent_passive

2. Now, we can create a calculated item that is going to show us the average memory utilization over 15 minutes. We can use this value to determine how busy our host was during that period, without having to look at the graphs.

3. Let's click the **Create item** button and start creating our new calculated item. We want our item to have the following values:

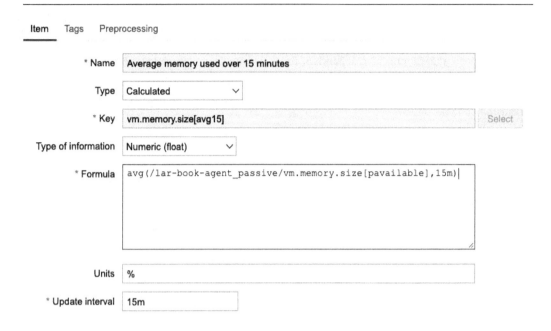

Figure 3.42 – Zabbix item configuration page, average memory used

4. Make sure you also navigate to the **Tags** tab and add a tag that we will use later for filtering:

Figure 3.43 – Calculated item Tags tab

5. Now, if we go to check our **Monitoring | Hosts** page and select **Latest data**, we can check out our value. Make sure you filter the **Name** field for memory so that you see the correct values:

Host	Name ▲	Last check	Last value	Change
lar-book-agent_passive	Average memory used over 15 minutes	12m 35s	47.6076 %	+2.0391 %

Figure 3.44 – Zabbix Latest data page for lar-book-agent_passive, memory items

As we can see, we are calculating the 15-minute average of the memory utilization on our newly created item.

Working with dependent items

It's time to make our first dependent item. I'll use the `lar-book-rocky` host or our (as it's called by default) **Zabbix server** host, but any MySQL database server should work. Let's say we want to request some variables from our MySQL database in one big batch. In this case, we can create dependent items on top of the first item to further process the data:

1. Let's start by creating the main check. Navigate to **Data collection | Hosts**, select our host, and click the **Create item** button to start creating our first new item. We want an item with the following variables:

Figure 3.45 – Zabbix item configuration page, database status

This item is an SSH check that logs in to our Zabbix server host using SSH and then executes the code that was entered in the **Executed script** field. This code will then log in to our MariaDB database and will print its status. Make sure you enter your credentials correctly.

> **Tip**
>
> Instead of using plaintext credentials in the MySQL command, which is not recommended, use macros in the **Executed script** field. This way, you can use the **Secret text** macro type to make sure no one can read your password from the frontend.

2. Before saving this new item, make sure you also add a tag, like this:

Figure 3.46 – Zabbix master item configuration page, Tags tab

3. Now, click the blue **Add** button to save this new item.

4. Go back to the list of items and click on this host's hostname, then **Macros**. Create a new {$USERNAME} and {$PASSWORD} macro with your SSH username and password under **Value**.

5. Next, go to **Monitoring | Latest data** and check out the data for our new check. There should be a long list of MariaDB values. If so, we can continue creating the dependent item.

6. To create the dependent item, navigate to **Data collection | Hosts**, select our host, and click the **Create item** button. We want this item to have the following variables:

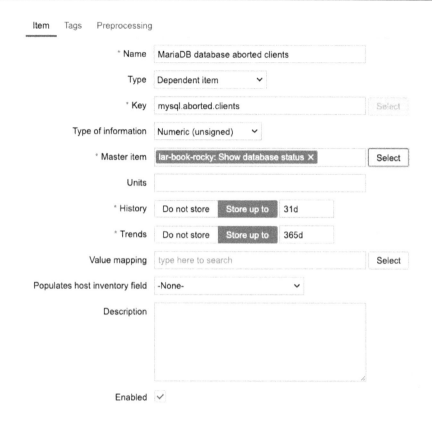

Figure 3.47 – Zabbix item configuration page, MariaDB aborted clients

7. Make sure you add the following tag:

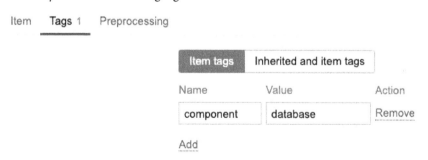

Figure 3.48 – Zabbix dependent item configuration page, Tags tab

8.   It's very important to add preprocessing to this item; otherwise, we will simply get the same data as our master item. So, let's add the following:

Figure 3.49 – Zabbix item Preprocessing page, MariaDB aborted clients

With the preprocessing added, the result will be the number of aborted clients for our MariaDB instance:

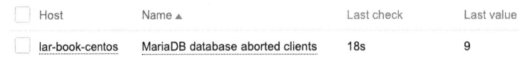

Figure 3.50 – Zabbix Latest data page, MariaDB aborted clients

As you can see, using a dependent item, we can use already available information from other Zabbix items and split them up into dependent items.

How it works...

The calculated and dependent items we worked with in the *How to do it...* section can be quite complicated, so let's go over how they work.

Calculated items

Working with calculated items can be a great way to get even more statistics out of your existing data. Sometimes, you just need to combine multiple items into one specific value.

What we did just now works by taking several values of one item every 15 minutes and calculating the average, as follows:

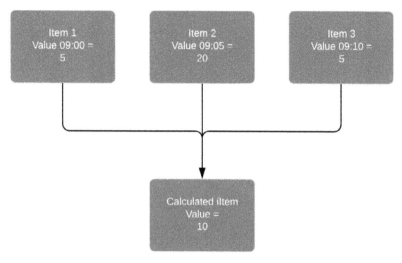

Figure 3.51 – Zabbix dependent item diagram

We're taking those values and calculating the average every 15 minutes. It gives us a nice indication of what we are doing over a set period.

Dependent items

Dependent items work by taking the data from a master item and processing that data into other data. This way, we can structure our data and keep our check interval for all these items the same since the dependent items will receive their data on the update interval as the master item. That means that dependent items don't have (and don't need) an update interval:

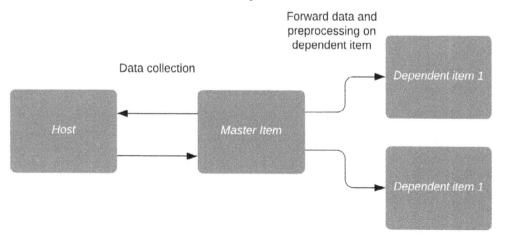

Figure 3.52 – Dependent item diagram

As we can see, dependent items work as duplicators, on which we use preprocessing options to get specific values. Note that preprocessing must be used to extract data from the master items since without preprocessing, our data will be the same as it is for the master item.

> **Tip**
>
> Often, we don't require our master item to be saved in our database since we already have the information in our dependent items. When we don't want the master item to be saved, we can simply select the **Do not keep history** option on that master item. This will save us some storage space.

Creating external checks

To further extend our Zabbix functionality, we can use custom scripts that can be executed as Zabbix external checks. Not everything that we want to monitor will always be standard in Zabbix, although a lot is. There's always something that could be missing, and external checks are just a way to bypass some of these.

Getting ready

In this recipe, we are going to need just our Zabbix server. We can create an item on our `lar-book-rocky` host, which is our Zabbix server-monitored host.

How to do it...

Follow these steps:

1. First, let's make sure our Zabbix server configuration is set up correctly. Execute the following on the Zabbix server CLI:

    ```
    cat /etc/zabbix/zabbix_server.conf | grep ExternalScripts=
    ```

2. This should show us the path where we will place the script that's used by the Zabbix external check. By default, this is `/usr/lib/zabbix/ externalscripts/`. Let's create a new script called `test_external` in this folder with the following command:

    ```
    vim /usr/lib/zabbix/externalscripts/test_external
    ```

    Add the following code to this file and save it:

    ```
    #!/bin/bash
    echo $1
    ```

3. Make sure our Zabbix server can execute the script by adding the right permissions to the file. The `zabbix` user on your Linux server needs to be able to access and execute the file:

    ```
    chmod +x /usr/lib/zabbix/externalscripts/test_external
    chown zabbix:zabbix /usr/lib/zabbix/externalscripts/test_
    external
    ```

4. Now, we are ready to go to our host to create a new item. Navigate to **Data collection | Hosts**, select our host, `lar-book-rocky`, and click the **Create item** button. We want this item to be created as follows:

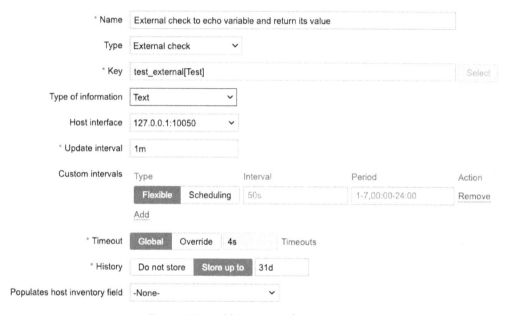

Figure 3.53 – Zabbix item configuration page

5. Now that we've added this new item, let's navigate to **Monitoring | Hosts** and check the **Latest data** page for our host. Our **Test** variable should be returned by our script as **Last value** in Zabbix, as shown in the following screenshot:

Figure 3.54 – Zabbix Latest data page

> **Tip**
>
> Use the macros in the frontend as variables to send data from your frontend to your scripts. You can further automate your checks with this to enhance your external checks.

How it works...

External checks seem like they have a steep learning curve, but they are quite simple from the Zabbix side. All we do is execute an external script, at which point we will receive the standard result output (STDOUT) and error (STDERR):

Figure 3.55 – Zabbix server external script communication diagram

In our example, we sent a value of Test to our script, which the script then echoed back to us as $1.

When you have good knowledge of a programming language such as Python, you can use this function to build a lot more expansions on top of the current existing Zabbix feature set – a simple yet powerful tool to work with.

Setting up JMX monitoring

Zabbix has JMX monitoring built into it so that we can monitor our Java applications. In this recipe, we'll learn how to monitor Apache Tomcat with Zabbix JMX so that we can get a feel for what this monitoring option is all about.

Getting ready

To get ready for this recipe, we are going to need our Zabbix server so that we can monitor our JMX application.

I used a CentOS 7 machine for this recipe, with Tomcat installed. It can be quite tricky to use Tomcat on later CentOS versions due to package dependencies, so I recommend sticking with CentOS 7 for this example. You can add the following to your Tomcat configuration after installing it to get it working for this recipe:

```
JAVA_OPTS="-Djava.rmi.server.hostname=10.16.16.155
-Dcom.sun.management.jmxremote
-Dcom.sun.management.jmxremote.port=12345
-Dcom.sun.management.jmxremote.authenticate=false
-Dcom.sun.management.jmxremote.ssl=false"
```

If you want to set up JMX monitoring in your production environment, you can use the settings you have probably already set up there. Simply change the port and IP address accordingly.

How to do it...

To set up JMX monitoring, we are going to add a host to our Zabbix server that will monitor our Apache Tomcat installation. But first, we will need to add some settings to our `/etc/zabbix/zabbix_server.conf` file:

1. Let's edit the `zabbix_server.conf` file by logging in to our Zabbix server and executing the following command:

   `vim /etc/zabbix/zabbix_server.conf`

2. Now, we need to add the following lines to this file:

   ```
   JavaGateway=127.0.0.1
   StartJavaPollers=5
   ```

> Tip
>
> It's possible to install your Java gateway on a host that's separate from your Zabbix. This way, you can spread the load and scale more. Simply install it on a separate host and add the IP address of that host to the `JavaGateway` parameter. So long as your Zabbix server or proxy can reach the gateway on port `10052` over the network, this should work. We won't be doing this in this example, so keep the Java gateway set up on the Zabbix server host itself.

3. We will also need to install the `zabbix-java-gateway` application on our Zabbix server with the following command.

   RHEL-based systems:

   ```
   dnf install zabbix-java-gateway
   systemctl enable zabbix-java-gateway
   systemctl start zabbix-java-gateway
   systemctl restart zabbix-server
   ```

   Ubuntu systems:

   ```
   apt install zabbix-java-gateway
   systemctl enable zabbix-java-gateway
   systemctl start zabbix-java-gateway
   systemctl restart zabbix-server
   ```

   That is all we need to do on the server side of things to get JMX monitoring to work. Zabbix doesn't include these settings by default, so we need to add the respective text to our file and install the application.

4. To start monitoring our JMX host, go to **Data collection | Hosts** in your Zabbix frontend and click **Create host** in the top-right corner.

Add a host with the following settings:

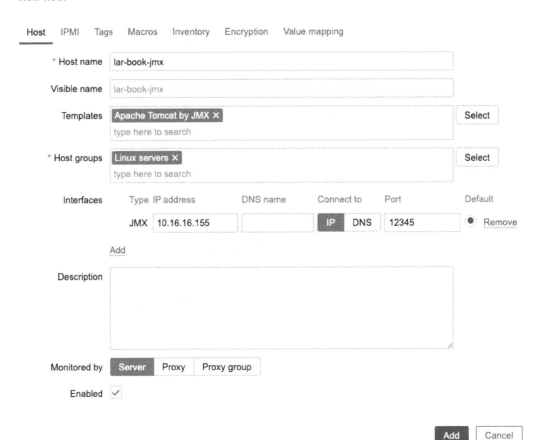

Figure 3.56 – Zabbix item configuration page

5. After this, our JMX icon should turn green; let's check this under **Monitoring | Hosts**. It should look like this:

lar-book-jmx   10.16.16.155:12345   JMX

Figure 3.57 – Monitoring | Hosts

6. If we click on **Latest data** for our new JMX-monitored host, we should also see our incoming data. Check it out; it should return stats like these:

Host	Name ▲	Last check	Last value
lar-book-jmx	Tomcat: Version [?]	4s	Apache Tomcat/7....

Figure 3.58 – Zabbix Latest data page

How it works...

Zabbix utilizes a Java gateway either hosted on the Zabbix server itself or hosted on another server (proxy) to monitor JMX applications:

Figure 3.59 – Communication diagram between the Zabbix server and Java

Zabbix polls the Java gateway and the Java gateway, in turn, communicates with our JMX application, as it does with Tomcat in our example. The data is then returned through the same path, at which point we can see our data in our Zabbix server.

See also

There are loads of applications that can be monitored through Zabbix JMX. Check out the Zabbix monitoring and integrations page for more uses of Zabbix JMX monitoring: `https://www.zabbix.com/integrations/jmx`.

Setting up database monitoring

Databases are a black hole to a lot of engineers; data is being written to them and something is being done with this data. But what if you want to know more about the health of your database? That's where Zabbix database monitoring comes in – we can use it to monitor the health of our database.

Getting ready

For convenience, in this recipe, we'll be monitoring our Zabbix database. This means that all we are going to need is our installed Zabbix server with our database on it. We'll be using MariaDB in this example, so if you have a PostgreSQL setup, make sure you install a MariaDB instance on a Linux host (although the same kind of setup could be created on PostgreSQL if you change some of the ODBC parameters).

How to do it...

Before getting started with the item configuration, we'll have to do some stuff on the CLI side of the server:

1. Let's start by installing the required modules on our server.

    RHEL-based systems:

    ```
    dnf install unixODBC mariadb-connector-odbc
    ```

    Ubuntu systems:

    ```
    apt install odbc-mariadb unixodbc unixodbc-dev odbcinst
    ```

2. Now, let's verify whether our **Open Database Connectivity** (**ODBC**) configuration files exist:

    ```
    odbcinst -j
    ```

    Your output should look similar to this:

    ```
    unixODBC 2.3.7
    DRIVERS............: /etc/odbcinst.ini
    SYSTEM DATA SOURCES: /etc/odbc.ini
    FILE DATA SOURCES..: /etc/ODBCDataSources
    USER DATA SOURCES..: /root/.odbc.ini
    SQLULEN Size.......: 8
    SQLLEN Size........: 8
    SQLSETPOSIROW Size.: 8
    ```

3. If the output is correct, we can go to the Linux CLI and continue by editing odbc.ini so that we can connect to our database:

    ```
    vim /etc/odbc.ini
    ```

    Now, fill in your Zabbix database information. It will look like this:

    ```
    [book]
    Description = MySQL book test database
    Driver      = MariaDB
    Server      = 127.0.0.1
    Port        = 3306
    Database    = zabbix
    ```

4. Let's also check that our driver exists:

    ```
    vim /etc/odbc.ini
    ```

5. You should see the driver:

```
# Driver from the mariadb-connector-odbc package
# Setup from the unixODBC package
[MariaDB]
Description     = ODBC for MariaDB
Driver          = /usr/lib/libmaodbc.so
Driver64        = /usr/lib64/libmaodbc.so
FileUsage       = 1
```

6. Now, let's test whether our connection is working as expected by executing the following command:

```
isql -v book
```

You should get a message saying `Connected`; if you don't, check your configuration files and try again.

7. Now, let's move to the Zabbix frontend to configure our first database check. Navigate to **Data collection** | **Hosts** and click the `lar-book-rocky` host; note that it might still be called **Zabbix server**. Now go to **Items**; we want to create a new item here by clicking the **Create item** button.

> Tip
>
> If you haven't already, a great way to keep Zabbix structured is to keep all hostnames in Zabbix equal to the real server hostname. Rename your default **Zabbix server** host in the frontend to what you've called your server.

We want to add an item with the following parameters:

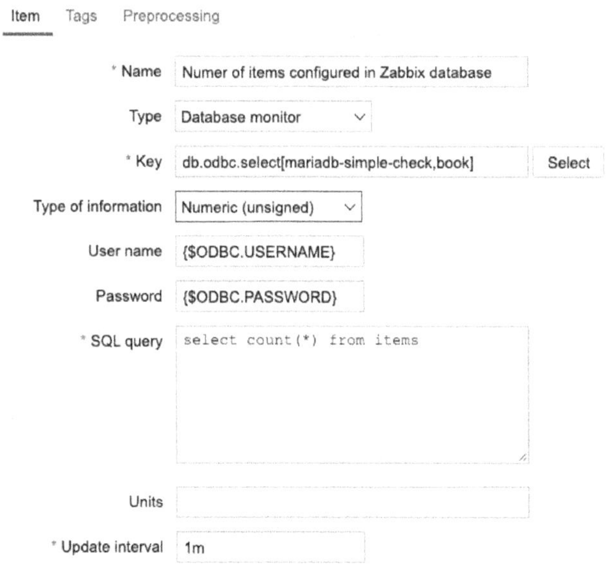

Item	Tags	Preprocessing

* Name	Numer of items configured in Zabbix database
Type	Database monitor ⌄
* Key	db.odbc.select[mariadb-simple-check,book] Select
Type of information	Numeric (unsigned) ⌄
User name	{$ODBC.USERNAME}
Password	{$ODBC.PASSWORD}
* SQL query	select count(*) from items
Units	
* Update interval	1m

Figure 3.60 – Zabbix item configuration page, items in Zabbix database

8. Make sure you also add a tag to the item:

Figure 3.61 – Zabbix item configuration page, items in Zabbix database, Tags tab

9. Now, click the **Add** button and click on the name of the host to add the macros, as follows:

Figure 3.62 – Zabbix host macro configuration page

10. Now, if you go to **Monitoring | Hosts** and click on **Latest data** for our host, you'll see the following:

Figure 3.63 – Zabbix Latest data page for lar-book-rocky, items in Zabbix database

From here, we can see how many items have been written to the database directly.

How it works...

Zabbix database monitoring works by connecting to your database with the ODBC middleware API. Any database supported by ODBC can be queried with Zabbix database monitoring:

Figure 3.64 – A diagram showing communication between the Zabbix server and ODBC

Your Zabbix server sends a command with, for instance, your MySQL query to the ODBC connector. Your ODBC connecter sends this query to the database through the ODBC API, which, in turn, returns a value to ODBC. ODBC then forwards the value to the Zabbix server and hey presto: we have a value under our item.

There's more...

You can do loads of queries to your databases with Zabbix database monitoring, but keep in mind that you are working with actual queries. Querying a database takes time and processing power, so keep your database monitoring structured and define the right execution times.

Alternatively, we can use Zabbix Agent 2 to monitor most databases natively. This can improve security and performance and keep complexity lower.

Setting up HTTP agent monitoring

With the Zabbix HTTP agent, we can monitor a web page or API by retrieving data from it. For instance, if there's a counter on a web page and we want to keep an eye on that counter value, we can do so with the Zabbix HTTP monitor.

Getting ready

For this recipe, we're going to need a web page to monitor, as well as our Zabbix server. For this lab, we will use Zabbix update v1: `https://services.zabbix.com/updates/v1`.

Please note that your Zabbix server will need an active internet connection for this recipe.

How to do it...

Let's poll the web page from Zabbix so that it shows the latest version of Zabbix 7.0 that's currently available:

1. Navigate to your Zabbix frontend and navigate to **Data collection | Hosts**. Then, click the `lar-book-agent_simple` host.

2. Now, go to **Items**; we want to create a new item here by clicking the **Create item** button. We are going to need to create an **HTTP agent** item, as shown in the following screenshot:

Figure 3.65 – Zabbix Item configuration page, visitor count on the oicts.com page

3. Make sure you fill in the query fields as follows:

    *   type: software_update_check

    *   version: 1.0

    *   software_update_check_hash: A randomly generated 64-character string with lowercase letters and numbers

4. We also need to add a tag to this item:

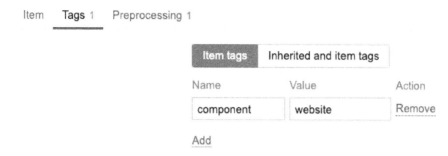

Figure 3.66 – Zabbix Item configuration page, visitor count on the oicts.com page, Tags tab

5. Use the following preprocessing steps:

Figure 3.67 – Zabbix Item configuration page, visitor count on the oicts.com page, Preprocessing tab

6. Now, navigate to **Monitoring | Hosts** and open the **Latest data** page for our `lar-book-agent_simple` host. If everything is working as it should, we should now be requesting the latest Zabbix 7.0 version:

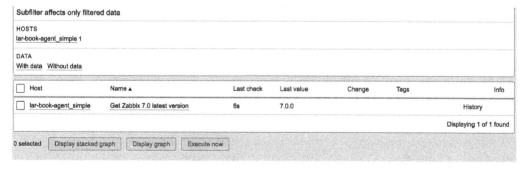

Figure 3.68 – Zabbix Latest data page

How it works...

Here, we request the complete web page from Zabbix by navigating to the page with the HTTP agent and downloading it. Once we have the complete content of the page – in this case, an HTML/PHP page – we can process the data:

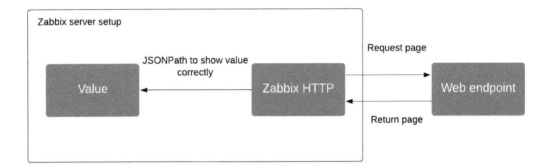

Figure 3.69 – Diagram showing Zabbix HTTP agent communication

We ask our preprocessor to go through the requested code via a JSONPath and only show the version for the `latest_release` node.

All that's left is the number, ready for us to use in graphs and other types of data visualization.

Using Zabbix browser items to simulate a web user

Zabbix now includes the ability to monitor web pages in a brand-new way. It's now possible to use the new Zabbix **Browser items** functionality to simulate the things a browser user would do when navigating your web page. This makes it possible to navigate to pages, simulate clicks, get results, and much more.

Getting ready

For this recipe, we will only need our Zabbix server and Zabbix frontend. Keep in mind that we will be running Selenium in Docker on our Zabbix server to get this new type of monitoring working.

We'll also use some pre-prepared JavaScript that you can find here: `https://github.com/PacktPublishing/Zabbix-7-IT-Infrastructure-Monitoring-Cookbook/blob/main/chapter03/browser_item_script.txt`.

How to do it...

First, we'll log in to the CLI of our server and start preparing the environment:

1. We'll use a lightweight Docker container to run Selenium, which will handle the browser emulation. Issue the following on the Zabbix server CLI.

   For RHEL-based systems:

   ```
   dnf install docker-ce
   ```

   For Ubuntu systems:

   ```
   dnf install docker-ce
   ```

2. Make sure you start Docker as well:

   ```
   systemctl enable docker --now
   ```

3. Now, let's download and run our Docker container:

   ```
   docker run -d -p 4444:4444 -p 7900:7900 --shm-size="2g"
   selenium/standalone-chrome:latest
   ```

4. Next, we must edit the Zabbix server configuration:

   ```
   vim /etc/zabbix/zabbix_server.conf
   ```

5. Two new parameters have been added to the Zabbix server configuration file that we can edit. Let's connect to the container and add some browser pollers:

   ```
   WebDriverURL=http://localhost:4444
   StartBrowserPollers=2
   ```

6. Restart your Zabbix server to make the changes take effect:

   ```
   systemctl restart zabbbix-server
   ```

7. With the Zabbix server side of things done, let's move on to the Zabbix frontend. Navigate to **Data collection | Hosts**.

8. Let's add a new host to monitor our Zabbix frontend website:

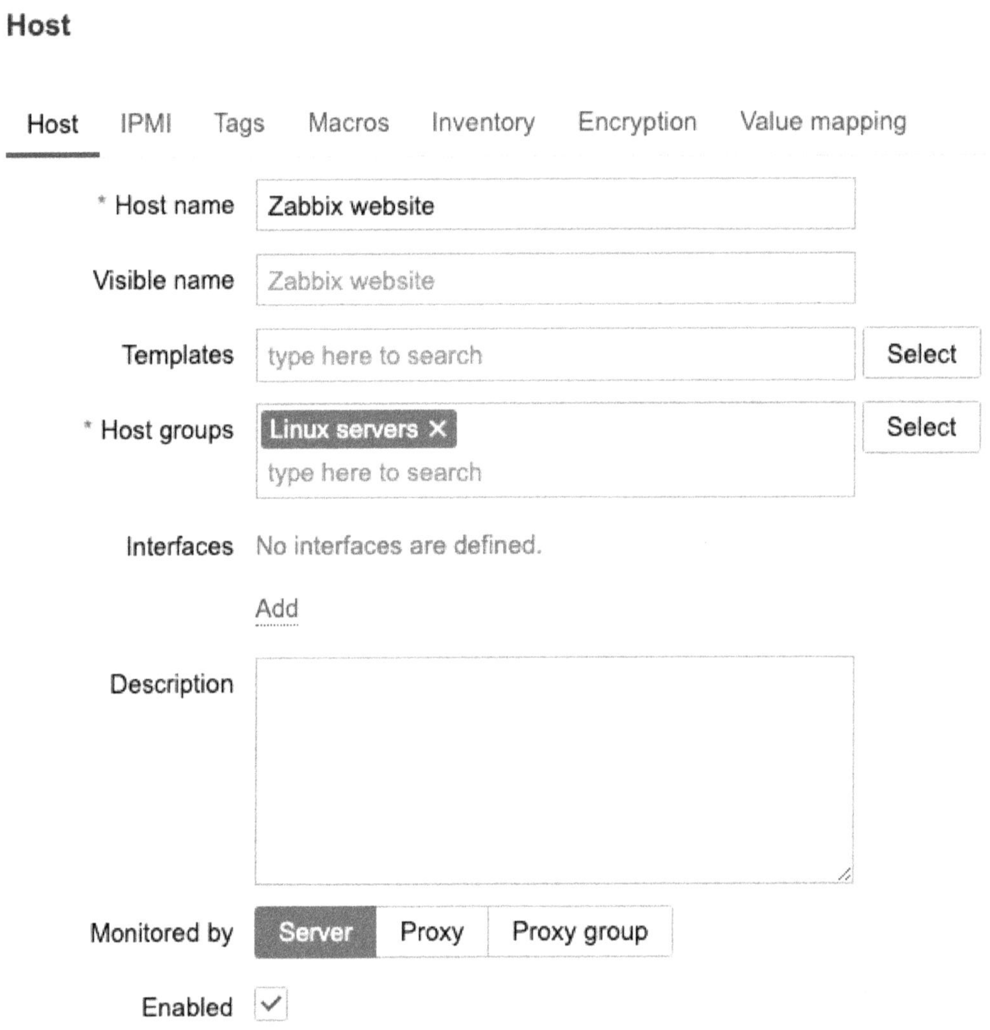

Host

| Host | IPMI | Tags | Macros | Inventory | Encryption | Value mapping |

* Host name	Zabbix website
Visible name	Zabbix website
Templates	type here to search **Select**
* Host groups	Linux servers ✕ type here to search **Select**
Interfaces	No interfaces are defined. Add
Description	
Monitored by	Server Proxy Proxy group
Enabled	✓

Figure 3.70 – Zabbix website host configuration window

9. Click on the **Add** button at the bottom of the page to add this new host.

10. Now, let's add an item to this host. Click **Items** next to the **Zabbix website** host, then click on the **Create item** button in the top-right corner.

11. To create the item, we will need to download a bit of JavaScript from the Packt GitHub repo. You can find it here: `https://github.com/PacktPublishing/Zabbix-7-IT-Infrastructure-Monitoring-Cookbook/blob/main/chapter03/browser_item_script.txt`.

12. Now, let's fill in the new item fields. Make sure you place the aforementioned script in the **Script** field. The item should now look like this:

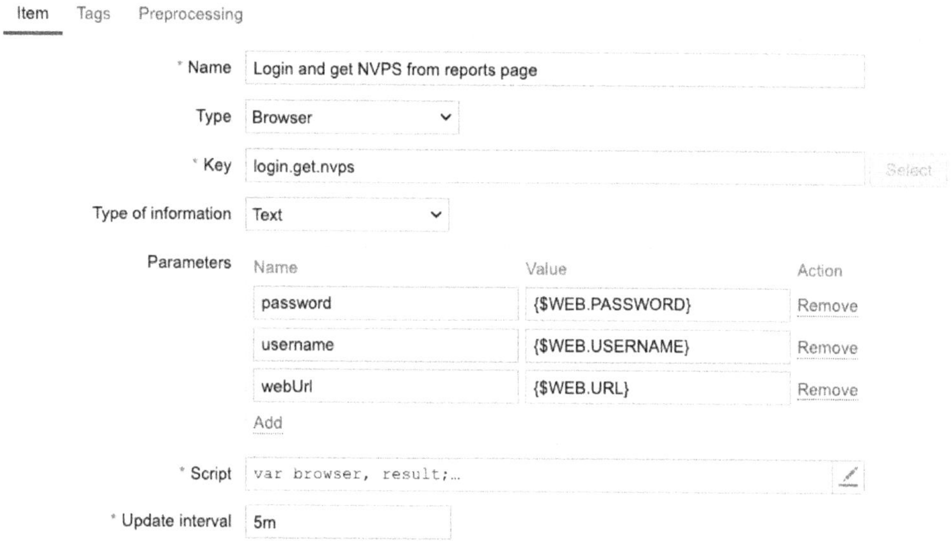

Figure 3.71 – Zabbix website host item configuration window

13. Make sure you add a tag as well:

Figure 3.72 – Zabbix website host item tag configuration window

14. Now, save the item by clicking on the **Add** button at the bottom of the window.

15. Since we're using macros, make sure you add them to the host. Click on the **Zabbix website** hostname. Then, add the following macros:

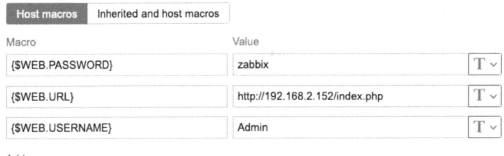

Figure 3.73 – Zabbix website host macros configuration window

Make sure you fill in the correct URL. Also, don't forget to set the password macro to the **Secret text** type.

16. Navigating to **Monitoring | Latest data** should now show us the value for our new host:

Figure 3.74 – Zabbix website item result

17. As we learned earlier in this chapter, we can extract data from this bulk metric item using dependent items. This is what we'll do next.

18. Let's go back to **Data collection | Hosts** and click on **Items** for the **Zabbix website** host.

19. Click **Create item** in the top-right corner and create a new item to get the total duration:

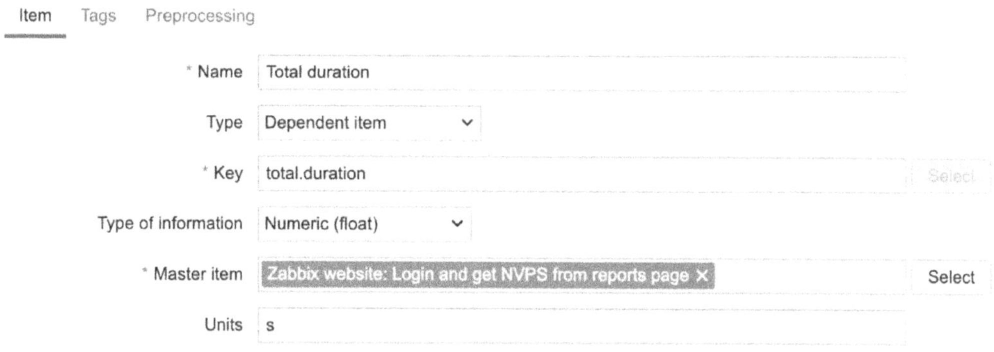

Figure 3.75 – Browser monitoring total duration item

20. Click on **Preprocessing** so that we can add preprocessing details as well:

Figure 3.76 – Browser monitoring total duration item preprocessing

21. Click on **Tags** – we can't forget to add a tag:

Figure 3.77 – Browser monitoring total duration item tag

22. Click **Add** to finish creating the item.

23. Now, click **Create item** in the top-right corner again and create a new item to get the number of enabled hosts:

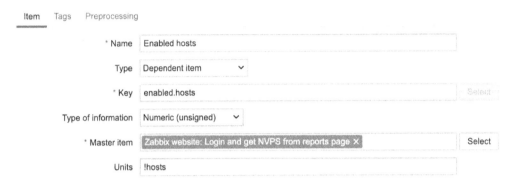

Figure 3.78 – Browser monitoring enabled hosts item

24. Click on **Preprocessing** and add some preprocessing details:

Figure 3.79 – Browser monitoring enabled hosts item preprocessing

25. Click on **Tags** – we can't forget to add a tag:

Figure 3.80 – Browser monitoring enabled hosts item tag

26. Click **Add** to finish creating the item.

> **Tip**
>
> Feel free to add more dependent items yourself to get even more statistics from the raw JSON that was collected by the master item. Don't forget to set **History** to **Do not store** on the master item once you've finished adding dependent items to save some disk space.

27. Let's have a look at how this code works.

How it works...

With the new Zabbix browser monitoring, we can use advanced JavaScript in combination with Selenium, for example. This gives us the option to do almost anything a normal browser user can. This provides us with endless opportunities to monitor what our end users are doing.

Let's have a look at some of the steps in our JavaScript:

```
try {
    var params = JSON.parse(value); // Parse the JSON string passed
from Zabbix
    var webUrl = params.webUrl;
    var username = params.username;
    var password = params.password;

    browser.navigate(webUrl);
    browser.collectPerfEntries("open page");
```

We start by parsing the parameters that we defined in the Zabbix frontend. We don't want to use just hardcoded usernames and passwords – we want to use other values that can be dynamic in this script, such as URLs, which are useful to parse. This way, we create flexibility, which can also be useful later when we template the item:

```
    // Find and fill username
    var el = browser.findElement("xpath", "//input[@id='name']");
    if (el === null) {
        throw Error("cannot find name input field");
    }
    el.sendKeys(username);

    // Find and fill password
    el = browser.findElement("xpath", "//input[@id='password']");
    if (el === null) {
        throw Error("cannot find password input field");
```

```
        }
        el.sendKeys(password);

        // Find and click the login button
        el = browser.findElement("xpath", "//button[@id='enter']");
        if (el === null) {
            throw Error("cannot find login button");
        }
        el.click();
```

Then, we have a `browser.findElement` function. We will be using this to find the correct field to fill in with the username and password before we click on the login button. With this type of monitoring, we are translating JavaScript to what looks like what a user could also be doing:

```
        // Collect performance data after login
        browser.collectPerfEntries("login");
```

We also collect some performance statistics so that we can find how quickly the login was performed:

```
        // Navigate to Reports -> System information
        el = browser.findElement("xpath",
    "//a[contains(text(),'Reports')]");
        if (el === null) {
            throw Error("cannot find Reports menu");
        }
        el.click();

        el = browser.findElement("xpath", "//a[contains(text(),'System
    information')]");
        if (el === null) {
            throw Error("cannot find System information submenu");
        }
        el.click();

        // Find the required server performance row and get the value
        nvps = browser.findElement("xpath", "//tr[th[contains(text(),
    'Required server performance, new values per second')]]/td[1]");
        totalHosts = browser.findElement("xpath", "//
    tr[th[contains(text(), 'Number of hosts (enabled/disabled)')]]/
    td[1]");
        enabledHosts = browser.findElement("xpath", "//
    tr[th[contains(text(), 'Number of hosts (enabled/disabled)')]]/td[2]/
    span[1]");
        disabledHosts = browser.findElement("xpath", "//
    tr[th[contains(text(), 'Number of hosts (enabled/disabled)')]]/td[2]/
    span[2]");
```

```
    numberOfTemplates = browser.findElement("xpath", "//
tr[th[contains(text(), 'Number of templates')]]/td[1]");
    totalItems = browser.findElement("xpath", "//tr[th/
span[contains(text(), 'Number of items (enabled/disabled/not
supported)')]]/td[1]");

    enabledItems = browser.findElement("xpath", "//tr[th/
span[contains(text(), 'Number of items (enabled/disabled/not
supported)')]]/td[2]/span[1]");
    disabledItems = browser.findElement("xpath", "//tr[th/
span[contains(text(), 'Number of items (enabled/disabled/not
supported)')]]/td[2]/span[2]");
    nsItems = browser.findElement("xpath", "//tr[th/
span[contains(text(), 'Number of items (enabled/disabled/not
supported)')]]/td[2]/span[3]");
    if (el === null) {
        throw Error("cannot find required server performance row");
    }
    var performanceValue = nvps.getText();
    var totalHosts = totalHosts.getText();
    var enabledHosts = enabledHosts.getText();
    var disabledHosts = disabledHosts.getText();
    var numberOfTemplates = numberOfTemplates.getText();
    var totalItems = totalItems.getText();
    var enabledItems = enabledItems.getText();
    var disabledItems = disabledItems.getText();
    var nsItems = nsItems.getText();
```

Then, we have a few `browser.findElement` functions to navigate to the **Reports | System information** menu. On this page, we want to find specific rows from the table. Something important to note here is that apart from using browser monitoring to gather performance data or see if a user's functionality is still working, we can also use it to extract metrics. We'll come back to this shortly:

```
    // Find and click the logout button
    el = browser.findElement("xpath", "//a[contains(text(),'Sign
out')]");
    if (el === null) {
        throw Error("cannot find logout button");
    }
    el.click();

    // Collect performance data after logout
    browser.collectPerfEntries("logout");

    // Set result with the performance value
    result = browser.getResult();
```

```
        result.performanceValue = performanceValue;
        result.totalHosts = totalHosts;
        result.enabledHosts = enabledHosts;
        result.disabledHosts = disabledHosts;
        result.numberOfTemplates = numberOfTemplates;
        result.totalItems = totalItems;
        result.enabledItems = enabledItems;
        result.disabledItems = disabledItems;
        result.nsItems = nsItems;

    } catch (err) {
        if (!(err instanceof BrowserError)) {
            browser.setError(err.message);
        }
        result = browser.getResult();
        result.error.screenshot = browser.getScreenshot();
    } finally {
        return JSON.stringify(result);
    }
```

We must ensure we log out (it's best practice to end any session correctly), at which point we can do some more error and result-catching. At this point, I would love to show you the JSON result, but let's not cut down any more trees than necessary. Open your **Latest data** page and have a look at some of the JSON entries:

```
    "duration":1.8088712692260742,
```

We can see the total duration of the whole item execution at the top of the JSON result:

```
        {
            "mark":"open page",
            "navigation":{
    ….
                "transfer_size":4220,
                "duration":0.08759999996423722,
    ….
```

For each step we did, we can see a mark value, which will give us more information about the action we executed through the browser:

```
"performanceValue":"9.46","totalHosts":"23",
"enabledHosts":"18","disabledHosts":"5",
"numberOfTemplates":"287","totalItems":
"1108","enabledItems":"650",
"disabledItems":"241","nsItems":"217"}
```

Last, but not least, at the bottom, we have our extracted Zabbix system report values. We gathered all this information through this single browser item monitoring type:

	Host	Name ▲	Last check	Last value	Change
	Zabbix website	Enabled hosts	1s	18 hosts	
	Zabbix website	Login and get NVPS from reports page	1s	{"duration":1.48424...	
	Zabbix website	Total duration	1s	1s 484.24ms	-32.7ms

Figure 3.81 – Zabbix

Using the dependent items we learned about earlier, we can now extract useful data from our bulk metric collection, which was done by the browser item type. This gives us nice and clean metrics in a single item, which can also easily be used in triggers. For example, we can use these triggers to specify whether the duration is too long or whether there are fewer hosts enabled now than before.

As you can see, this single new item type opens up a world of possibilities for us. I can only imagine the Zabbix community finding more and more use cases for this and sharing their amazing new templates.

Using Zabbix preprocessing to alter item values

Preprocessing item values is an important functionality in Zabbix; we can use it to create all kinds of checks. We've already done some preprocessing in this chapter, but let's take a deeper dive into it and what it does.

Getting started

We are going to need a Zabbix server to create a check for. We will also need a passive Zabbix agent on a Linux host to get our values from and preprocess them. We can use the agent that's running on our Zabbix server for this. In my case, this is `lar-book-rocky`.

How to do it...

Follow these steps:

1. Let's start by logging in to our Zabbix frontend and going to **Data collection | Hosts**.

2. Click on your Zabbix server host; in my case, it's called `lar-book-rocky`.

3. Now, go to **Items** and click on the blue **Create item** button in the top-right corner. Let's create a new item with the following information:

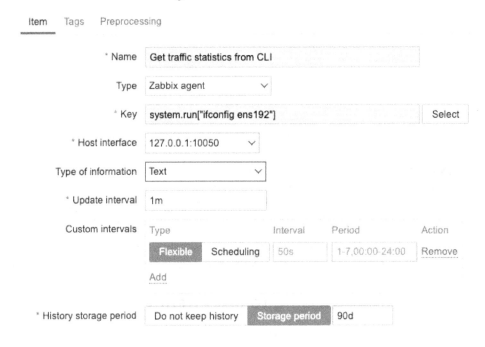

Figure 3.82 – New item creation screen, Get traffic statistics from CLI

4. Don't forget to add your tag:

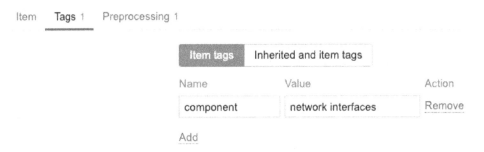

Figure 3.83 – New item creation screen, Get traffic statistics from CLI, Tags tab

5. Make sure you change ens192 to your primary network interface. You can find your primary network interface by logging in to the Linux CLI and executing the following command:

```
Ifconfig
```

6. Back on the item creation screen, click on the blue **Add** button. This item will use the Zabbix agent to execute a remote command on the Linux CLI.

7. When we navigate to this new item, we'll see that the item becomes unsupported. This is because when we use the `system.run` key, we need to allow it in the Zabbix agent configuration:

Figure 3.84 – Unsupported item information, Unknown metric system.run

8. Log in to the Linux CLI of the monitored host and edit the Zabbix agent configuration by running the following command:

```
vim /etc/zabbix/zabbix_agent2.conf
```

9. Go to the `Option: AllowKey` line and add `AllowKey=system.run[*]`, as shown here:

```
### Option: AllowKey
#       Allow execution of item keys matching pattern.
#       Multiple keys matching rules may be defined in combination with DenyKey.
#       Key pattern is wildcard expression, which support "*" character to match any
number of any characters in certain position. It might be used in both key name and
key arguments.
#       Parameters are processed one by one according their appearance order.
#       If no AllowKey or DenyKey rules defined, all keys are allowed.
#
# Mandatory: no
AllowKey=system.run[*]
```

Figure 3.85 – Zabbix agent configuration file, AllowKey=system.run[*]

10. Save the file and restart the Zabbix agent, like so:

```
systemctl restart zabbix-agent2
```

11. Back at the Zabbix frontend, the error we noticed in *Step 7* should be gone after a few minutes.

12. Navigate to **Monitoring** | **Latest data** and filter your Zabbix server host, `lar-book-rocky`, and the name of the new **Get traffic statistics from CLI** item.

13. The value should now be pulled from the host. If we click on **History**, we can see the full value; it should look as follows:

Timestamp
Value
2020-12-08 11:47:24
ens192: flags=4163<UP,BROADCAST,RUNNING,MULTICAST> mtu 1500
inet 10.16.16.152 netmask 255.255.255.0 broadcast 10.16.16.255
inet6 fe80::c462:d30e:b24a:b31d prefixlen 64 scopeid 0x20<link>
ether 00:0c:29:5e:c8:2c txqueuelen 1000 (Ethernet)
RX packets 128297172 bytes 24030338556 (22.3 GiB)
RX errors 0 dropped 783 overruns 0 frame 0
TX packets 134639844 bytes 42556882891 (39.6 GiB)
TX errors 0 dropped 0 overruns 0 carrier 0 collisions 0

Figure 3.86 – Zabbix agent system.run command executing ifconfig ens192 results

14. The information we can see here is way too much for just one item, so we need to split it up. We'll use preprocessing to get the number of RX bytes from the information.
15. Go back to **Data collection** | **Hosts** and click on your Zabbix server host. Go to **Items** on this host.
16. Click on the **Get traffic statistics from CLI** item to edit it. Change its name to `Total RX traffic in bytes for ens192` and add `B` to **Units**, where **B** stands for **bytes**. It will look like this:

Item
Tags
Preprocessing
* Name
Total RX traffic in bytes for ens192
Type
Zabbix agent
* Key
system.run["ifconfig ens192"]
Select
* Host interface
127.0.0.1:10050
Type of information
Numeric (unsigned)
Units
B
* Update interval
1m
Custom intervals
Type
Interval
Period
Action
Flexible
Scheduling
50s
1-7,00:00-24:00
Remove
Add

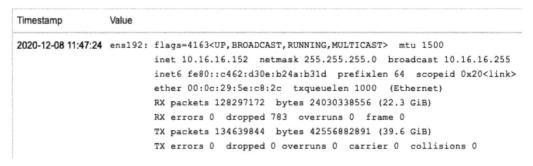

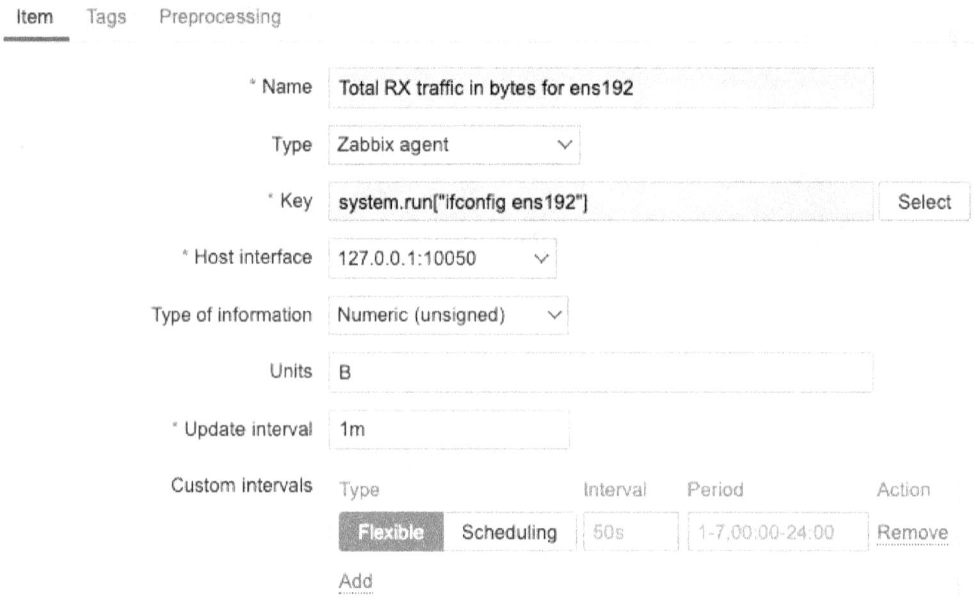

Figure 3.87 – Zabbix agent system.run item

17. Add your tag:

Figure 3.88 – New item creation screen, Get traffic statistics from CLI, Tags tab

18. Now, click on **Preprocessing** and click on the underlined **Add** button.

19. A **Regular expression** (regex) field will be added. We are going to fill this so that it matches the total number of bytes for your interface:

Figure 3.89 – Zabbix agent system.run item preprocessing

20. Make sure you also select the **Discard value** box under **Custom on fail**.

21. Now, click on the underlined **Add** button again and use the drop-down menu for this new step to select **Discard unchanged**. The result will look like this:

Figure 3.90 – Zabbix agent system.run item preprocessing

22. We can now click the blue **Update** button to finish editing this item.

23. Navigate back to **Monitoring | Latest data** and filter on your host and the new item name, **Total RX traffic in bytes for ens192**. Make sure you use your own interface name.

24. We can now see our value coming in. Here, we have an item displaying our total RX traffic for our main interface:

	Host	Name ▲	Last check	Last value	Change	Tags
	lar-book-centos	Total RX traffic in bytes for ens192	12s	65.3 GB		Application: Network i...

Figure 3.91 – Zabbix Total RX traffic item latest data

How it works...

We did some preprocessing in the *Working with calculated and dependent items* recipe to get data from a master item. We also used preprocessing in the *Setting up HTTP agent monitoring* recipe to get a specific value from a web page. We didn't go over the preprocessing concepts used in those recipes, though, so let's go over them here.

When working with preprocessing, it's important to know the basic setup. Let's take a look at the incoming data before we use preprocessing:

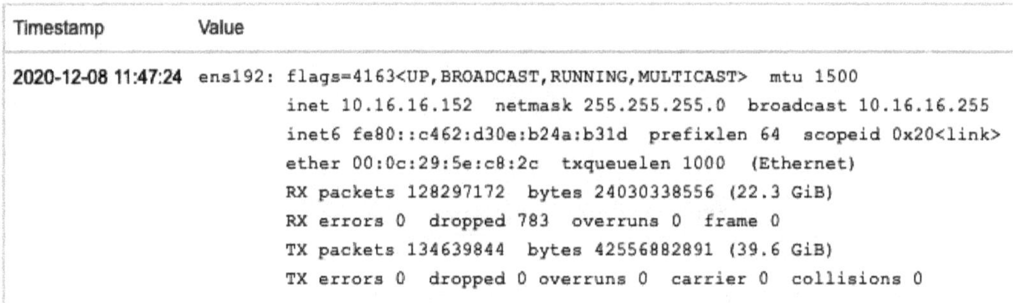

Timestamp	Value
2020-12-08 11:47:24	ens192: flags=4163<UP,BROADCAST,RUNNING,MULTICAST> mtu 1500
	inet 10.16.16.152 netmask 255.255.255.0 broadcast 10.16.16.255
	inet6 fe80::c462:d30e:b24a:b31d prefixlen 64 scopeid 0x20<link>
	ether 00:0c:29:5e:c8:2c txqueuelen 1000 (Ethernet)
	RX packets 128297172 bytes 24030338556 (22.3 GiB)
	RX errors 0 dropped 783 overruns 0 frame 0
	TX packets 134639844 bytes 42556882891 (39.6 GiB)
	TX errors 0 dropped 0 overruns 0 carrier 0 collisions 0

Figure 3.92 – Zabbix agent system.run command executing ifconfig ens192 results

This is a lot of information. When we look at how Zabbix items are used, we try to put graspable information in a single item. Luckily, we can preprocess this item before we store the value in Zabbix. In the following figure, we can see the preprocessing steps we added to our item:

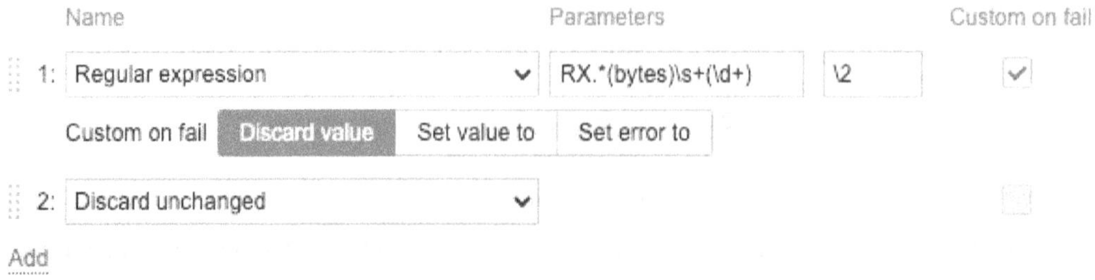

	Name	Parameters		Custom on fail
1:	Regular expression	RX.*(bytes)\s+(\d+)	\2	✓
	Custom on fail **Discard value** Set value to Set error to			
2:	Discard unchanged			

Add

Figure 3.93 – Zabbix agent system.run item preprocessing with two steps

Our first step is a regex. This step will make sure we only use the numbers we need. We match on the word RX, then the word `bytes`, and a sequence of numbers after them. This way, we end up with the total number of RX bytes in capture group 2. This is why we fill in \2 in the output field. We also specify **Custom on fail**, which will discard any value if the regex doesn't match.

Our second step is to discard any values that are the same as the value received before. Instead of storing duplicate values, we simply discard them and save some space in our Zabbix database.

> **Tip**
> It's a lot easier to build a regex when using an online tool such as `https://regex101.com/`. You can see what number your capture groups will get, and there's a lot of valuable information in the tools as well.

It's important to note that steps are executed in the sequence they are defined in the frontend. If the first step fails, the item becomes unsupported unless **Custom on fail** is set to do something else.

By adding preprocessing to Zabbix, we open up a whole range of options for our items, and we can alter our data in almost any way required. These two steps are just the beginning of the options that are available when diving into the world of Zabbix preprocessing.

See also

Preprocessing in Zabbix is an important subject, and it's impossible to cover every aspect of it in a single recipe. The two preprocessing steps in this recipe's example are just two of the many options we can use. Check out the official Zabbix documentation to see the other options we can use: `https://www.zabbix.com/documentation/current/en/manual/config/items/preprocessing`.

4

Working with Triggers and Alerts

Now, what use would all of that collected data in Zabbix be without actually doing some alerting with it? Of course, we can use Zabbix to collect our data and just go over it manually, but Zabbix gets a lot more useful when we actually start sending out notifications to users. This way, we don't have to always keep an eye on our Zabbix frontend, but we can just let our triggers and alerts do the work for us, redirecting us to the frontend only when we need it.

In Zabbix 7, you will find a new trigger expressions syntax compared to Zabbix 5. This syntax has been available since Zabbix 5.4, so if you skipped some versions, this might be the first time you're working with it. If you've been working with a Zabbix version before version 5.4, keep in mind that you might need to get used to this new syntax. If you have Zabbix 5.4 or higher running already, the syntax will be the same in Zabbix 7.

We will learn all about setting up effective triggers with the new expression format and about alerts in the following recipes:

- Setting up triggers
- Setting up advanced triggers
- Setting up alerts
- Keeping alerts effective
- Customizing alerts

Technical requirements

For this chapter, we will need a Zabbix server, for instance, the one used in the previous chapter.

- The Zabbix server installed on a Linux distribution of your choice. We will use the server set up in *Chapter 1*.

- MariaDB set up to work with your Zabbix server.

- NGINX or Apache set up to serve the Zabbix frontend.

- We will also need a Linux host to monitor so that we can actually build some cool triggers to use.

Setting up triggers

Triggers are important in Zabbix because they notify you as to what's going on with your data. We want to get a trigger when our data reaches a certain threshold or when we receive a certain value.

So, let's get started with setting up some cool triggers. There are loads of different options for defining triggers, but after reading this recipe, you should be able to set up some of the most prominent triggers. Let's take your trigger experience to the next level.

Getting ready

For this recipe, we will need our Zabbix server ready and we will need a Linux host. I will use the `lar-book-agent_simple` host from the previous chapter because we already have some items on that.

We'll also need one more host that is monitored by the Zabbix agent with the Zabbix agent template. We'll use one of the items on this host to create a trigger. This will be the `lar-book-agent_passive` host from the previous chapter.

On this host, we will already have some triggers available, but we will extend these triggers further to inform us even better.

How to do it...

In this section, we are going to create three triggers to monitor state changes. Let's get started by creating our first trigger.

Trigger 1 – SSH service monitoring

Let's create a simple trigger on the `lar-book-agent_simple` host. We made a simple check on this host called `Check if port 22 is available`, but we haven't created anything to notify us about this yet:

1. First, let's get started by going to **Data collection | Hosts**, then clicking the host and going to **Triggers**. This is where we will find our triggers and where we can create them. We want to create a new trigger here by clicking the blue **Create trigger** button in the top-right corner.

2. Let's create a new trigger with the following information:

* Name	Service unreachable: Port 22 (SSH)
Event name	Service unreachable: Port 22 (SSH)
Operational data	
Severity	Not classified Information **Warning** Average High Disaster
* Expression	`last(/lar-book-agent_simple/net.tcp.service[ssh,,22])=0` Add
	Expression constructor
OK event generation	**Expression** Recovery expression None
PROBLEM event generation mode	**Single** Multiple
OK event closes	**All problems** All problems if tag values match
Allow manual close	
URL	
Description	
Enabled	✓

Add   Cancel

Figure 4.1 – The Zabbix trigger creation page – Service unreachable

3. Click on **Add** and finish creating the trigger. This will create a trigger for us that will fire when our **Secure Shell** (**SSH**) port goes down.

4. Let's test this by navigating to our host **command-line interface** (**CLI**) and executing some commands to shut our Zabbix server off from port 22. We will add an `iptables` rule to block off all incoming traffic on port 22 (SSH):

```
iptables -A INPUT -p tcp -i ens192 -s 10.16.16.152
--destination-port 22 -j DROP
```

5. Make sure to change the `ens192` network card and the IP address `10.16.16.152` to your own values. You can use the following command to get that information:

```
ip addr
```

6. Now, if we click on **Dashboards** in the navigation bar, after a while we should see the following:

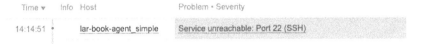

Figure 4.2 – Zabbix problems on a dashboard – port 22 down

Trigger 2 – triggering when there is a new Zabbix version

Now, to create our second trigger, let's ramp it up a bit. If you followed *Chapter 3, Setting Up Zabbix Monitoring*, in the recipe titled *Setting up HTTP agent monitoring*, we created an item that polls the Zabbix website for the latest release of Zabbix, Zabbix 7.0. Now, what we probably want to do ourselves is keep an eye out for any new version of Zabbix being released:

1. Let's navigate to **Data collection | Hosts** and click on the **lar-book-agent_simple** host.

2. Now, go to **Triggers** and click the **Create trigger** button. We will build our trigger with the following settings:

Figure 4.3 – The Zabbix trigger creation page – new Zabbix 7.0 release trigger

3. Click on **Add** and finish creating the trigger.

Now, this might not actually trigger for you in the frontend, but I'll explain to you just how this trigger works in the *How it works...* section of this recipe.

Trigger 3 – using multiple items in a trigger

We have seen triggers that use one item, but we can also use multiple items in a single trigger. Let's build a new trigger by using multiple items in the same expression:

1. Let's navigate to **Data collection | Hosts** and click on the **lar-book-agent_ passive** host. Now, go to **Triggers** and click the **Create trigger** button.

2. We are going to create a trigger with the following settings:

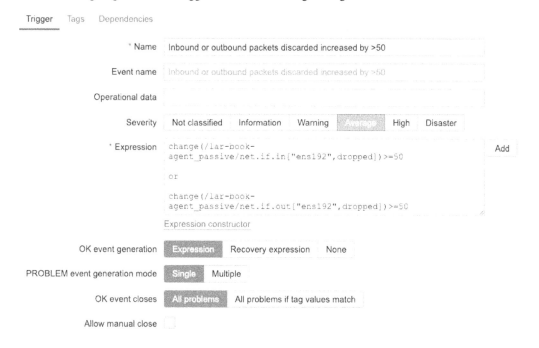

Figure 4.4 – The Zabbix trigger creation page – inbound or outbound packets trigger

3. Please note that your item keys might need different interface names. In my case, the interface is called ens192, so use the correct name for your interface in its place. Use the following Linux command to get the interface on your host:

```
ip addr
```

4. Click on **Add** and finish creating the trigger.

Here's a tip - On the trigger creation page, use the **Add** button next to the **Expression** field to add a condition and build your expression easily. For example, we can use the **Select** button to pick an item from a list. Something that's also very useful, when using the **Function** drop-down menu, is the short explanation for every trigger function that's included:

Figure 4.5 – Trigger creation page

That's all we need to do to build a trigger that will function on two items.

How it works...

We need a good understanding of how to build triggers and how they work so we can create a well-set-up monitoring platform. Especially important here is that we make sure that our triggers are set up correctly and we test them well. Triggers are a very important part of Zabbix as they will be vital to inform you about things going on with your monitoring targets. Configure your triggers too loosely and you will be missing things. Configure them too strictly and you will be overloaded with information.

In all of these triggers, we have also included a trigger severity, as we can see in the following screenshot:

Figure 4.6 – A Zabbix trigger severity selector

These severities are important to make sure your alerts will be correctly defined by importance. We can also filter on these severities in several places in the Zabbix frontend and even in things such as actions.

Now, let's discover why we built our triggers as we did.

Trigger 1 – SSH service monitoring

This is a very simple but effective trigger to set up in Zabbix. When our value returns either 1 for UP or 0 for DOWN, we can easily create triggers such as these—not just for monitoring logical ports that are up or down, but for everything that returns a simple value change from, for example, 1 to 0 and vice versa.

Now, if we break down our expression, we have the following:

Figure 4.7 – A Zabbix trigger expression – port 22 (SSH)

When building an expression, we have four parts:

- **Trigger function**: The trigger function is the part of the expression that determines what we expect of the value, such as whether we want just the last value or, for example, an average value over a period of time.

- **Host**: The host part of the expression is where we define which host we are using to trigger on. Most of the time, it's simply just the host (or template) we are working on.

- **Item key**: The item key is the part of the expression where we define which item key we'll be using to retrieve the value(s) on a host and feed it into the trigger function.

- **Operator**: The operator determines how our function will be calculated based on the trigger expression—against a constant or another expression, for example. The operator can be anything, such as the following:

=	Equal to.
<>	Not equal to.
>	Bigger than.
<	Smaller than.
>=	Bigger than or equal to.
<=	Smaller than or equal to.
+	Add to.
-	Subtract from.

/	Divide by.
*	Multiply by.
and	Logical AND. Used to, for example, equal both one and another expression.
or	Logical OR. Used to, for example, equal either one or another epxression.
not	Logical NOT. Used to, for example, specifically not equal an expression.

- **Constant**: The constant is the actual constant (often a value) that our trigger function uses to determine whether the trigger should be in an OK or PROBLEM state. We can also use macros here.

Now, for our first trigger, we defined our host and the item that gives us the SSH status. What we are saying in the trigger function is that we want the last value to be 0 before triggering it.

For this item, that would mean it would trigger within a minute because in our item, we specified the following:

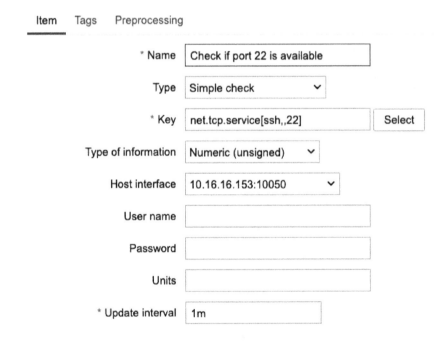

Figure 4.8 – The Zabbix item configuration page – port 22 availability item

Looking at the **Update interval** field on the **Item** configuration page, we can determine that when building this trigger, we are expecting our value to be 0 and that it will take a maximum of one minute of SSH port 22 downtime due to the 1m interval.

Trigger 2 – triggering when there is a new Zabbix version

Now, for our second trigger, we did something different. We not only made an expression for triggering this problem but also one for recovering from the trigger. What we do in the **Problem expression** option is define a trigger function, telling our host to compare the last value with the value before it. We did this by using the **change** trigger function.

Figure 4.9 – A Zabbix trigger expression – HTTPS check

So, our trigger will only be activated when the latest Zabbix version has been changed. We could just let the trigger resolve the first time the current value and the value before that are the same again, but I want to keep this trigger in the PROBLEM state just a little longer.

Therefore, I defined a recovery expression as well. I'm telling it that this problem can only be recovered if the last received value and the fifth last are received. Check out the recovery expression up close:

Figure 4.10 – Another Zabbix trigger expression – HTTPS check with different value

Recovery expressions are powerful when you want to extend your trigger functionality with just a bit more control over when it comes back into the OK state.

> **Tip**
>
> You can use the recovery expression to extend the trigger's PROBLEM state beyond what you defined in the **Problem expression** option. This way, we know we are still close to the PROBLEM state. We define that we only want the trigger to go back to the OK state after we've reached another threshold as defined in the recovery expression. This will work by evaluating both the problem and recovery expressions, where the problem expression has to be FALSE and the recovery expression TRUE.

Trigger 3 – using multiple items in a trigger

Now, trigger 3 might seem complicated because we've used more than one item, but it's basically the same setup:

Figure 4.11 – A Zabbix trigger expression using several items

We have the same setup for the expression, with the function, host, item key, and value. Yet when we are working with multiple items, we can add an or statement between the items. This way, we can say we need to match one of the items before triggering the PROBLEM state. In this case, we trigger when either item exceeds the threshold.

> **Important note**
>
> In this trigger expression, we have some empty lines between the different item expressions. Empty lines between item expressions are totally fine and actually make for good readability. Use this wisely when building triggers.

Old versus new trigger expression syntax

Now, if you've worked with Zabbix before version 5.4, the next part might be interesting to you. As mentioned in our introduction, there has been a big update to expressions within Zabbix. Trigger expressions now work in a new way, which is the same way as you will see in calculated items and other places for a unified experience.

Let's take a look at the old expression syntax as seen in Zabbix 5.2 and older versions:

Figure 4.12 – A Zabbix trigger expression using the old syntax

In the old syntax, we always started with a *curly bracket* and then the hostname or template name. Between the hostname or template name and the item key, we had a colon. Marking the end of the item key, we had a *dot*, but item keys can also include dots themselves. Then, after the *dot*, we have the trigger function followed by the ending *curly bracket*. Then, all we have left is the operator and constant we want to compare the expression against.

As you can imagine, this could become confusing at times, especially when using dots in item keys. Now let's check out the new trigger syntax:

Figure 4.13 – A Zabbix trigger expression using the new syntax

Our new trigger syntax starts off right away with our trigger function; no hassle, just immediately showing you what we're doing with this line. This is followed by a *bracket* and a *forward slash* before entering the host or template name. We then use another *forward slash* to divide the hostname or template name and the item key. We end with a *bracket*, and then all we have left is the operator and value we want to hold the expression against.

Starting with the trigger function makes for a clear indicator of what your line is doing. Putting the hostname or template name into brackets and then dividing it with forward slashes from the item key makes for a more cohesive experience when writing expressions. We also don't have confusing extra dots any longer. Altogether a very nice change to the trigger syntax, which in all honesty might take a bit of time to get used to.

It's the small stuff that makes the entire software feel more professional and well thought out. Zabbix including changes such as these really helps with that.

There's more...

Not only can we match one of the items in a trigger expression, but we can also use an `and` statement. This way, we can make sure our trigger only goes into a `PROBLEM` state when multiple items are reaching a certain value. Triggers are very powerful like this, allowing us to define our own criteria in great detail. There's no predefinition—we can add as many `and`, `not`, or `or` statements and different functions as we like in the trigger expressions. Customize your triggers to exactly what you need, and suddenly you are going to have a lot more peace of mind because you know your triggers will notify you when something is up.

See also

To know more about trigger expressions, check out the Zabbix documentation. There's a lot of information on which functions you can use to build the perfect trigger. For more details, go to `https://www.zabbix.com/documentation/current/en/manual/config/triggers/expression`.

Setting up advanced triggers

Triggers in Zabbix keep getting more advanced and it might be hard to keep up. For people working with Zabbix 5.2 or older and upgrading to Zabbix 7, not only is there a new Zabbix trigger syntax but there's also a whole new array of functions.

Let's dive into setting up some more advanced triggers in Zabbix 7.

Getting ready

For this recipe, we will need our Zabbix server ready and we'll need one host that is monitored by a Zabbix agent with the Zabbix agent template. We'll use the items on this host to create triggers. Let's use the `lar-book-agent_passive` host from the previous chapter.

If you don't have this host from the previous chapter, simply hook up a new host with the default passive Linux monitoring template called `Linux by Zabbix agent`.

We'll also be touching on some more advanced topics that are discussed later in the book. If you don't know how to use **Low-Level Discovery** (**LLD**), for example, it might be a good idea to dive into *Chapter 7, Using Discovery for Automatic Creation*, first.

How to do it...

Let's take a look at three *more advanced* triggers compared to the three we've seen in the previous recipe: `trendavg` for going through trend data, `timeleft` to predict values in the future, and **time shifting** to compare to the past.

Advanced trigger 1 – trendavg function

First, we'll take a look at one of the newer trigger functionalities, the trend average function:

1. Let's start by creating a new trigger in our frontend. Navigate to **Data collection | Hosts** and select `lar-book-agent_passive`.

2. Navigate to **Triggers** and click on the blue **Create trigger** button in the top-right corner.

3. Next to the **Expression** field, click on the white **Add** button. Fill out the trigger using the expression builder:

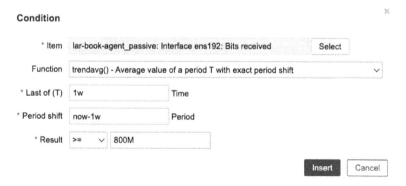

Figure 4.14 – trendavg trigger expression builder

4. Click on **Insert** and add a name. It will look like this if done correctly.

| Trigger | Tags | Dependencies |

* Name	Average incoming interface usage last week >800Mbps
Event name	Average incoming interface usage last week >800Mbps
Operational data	
Severity	Not classified Information Warning Average High Disaster
* Expression	`trendavg(/lar-book-agent_passive/net.if.in["ens192"],1w:now-1w)>=800M` Add

Expression constructor

Figure 4.15 – trendavg trigger form filled out

5. Now let's click the blue **Add** button at the bottom of the page to finish creating this trigger.

That's all for creating this trigger. Check out the *How it works…* section of this recipe to get more information about the trigger.

Advanced trigger 2 – timeleft function

Next up is our `timeleft` function, which is very useful for things such as space utilization. Let's take a look:

1. We'll create a new trigger in our Zabbix frontend. Navigate to **Data collection | Hosts** and select `lar-book-agent_passive`.

2. Navigate to **Discovery rules** and click on **Trigger prototype** next to **Mounted filesystem discovery**.

> **Important note**
> In this case, we are creating the trigger prototype directly on the host, using an existing template discovery rule. If you want to apply a trigger like this to every host using a template, make sure to create the trigger on a template level. Furthermore, discovery rules are explained further in *Chapter 7, Using Discovery for Automatic Creation*, of this book.

3. Click on **Create trigger prototype**.

4. Next to the **Expression** field, click on the white **Add** button. Fill out the trigger using the expression builder:

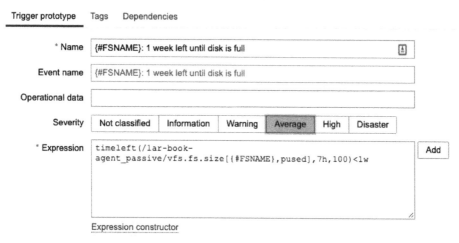

Condition   ×

* Item	lar-book-agent_passive: (#FSNAME): Space utilization Select Select prototype
Function	timeleft() - Time to reach threshold estimated based on period T ⌄
* Last of (T)	7h ⚠ Time ⌄
Time shift	now-h Time
* Threshold	100
Fit	
* Result	< ⌄ 1w

Insert   Cancel

Figure 4.16 – timeleft trigger expression builder

> **Important note**
> Using short intervals in predictive triggers to predict long time periods is not recommended. Make sure to use the right dataset for the time period we want to use in relation to the time we want to predict.

5. Click the blue **Insert** button and the finished trigger will look like this.

Trigger prototype   Tags   Dependencies

* Name	(#FSNAME): 1 week left until disk is full ⚠
Event name	(#FSNAME): 1 week left until disk is full
Operational data	
Severity	Not classified Information Warning **Average** High Disaster
* Expression	`timeleft(/lar-book-agent_passive/vfs.fs.size[{#FSNAME},pused],7h,100)<1w` Add

Expression constructor

Figure 4.17 – timeleft trigger form filled out

6. Click the blue **Add** button at the bottom of the page to finish setting up the trigger.

We now have a new trigger using the `timeleft` function to tell us when hard disks are filling up within a week. Check out the *How it works…* section of this recipe to get more information about the trigger.

Advanced trigger 3 – time shifting using mathematical functions

Lastly, we are going to work with time shifting, and in this case, we'll do so in combination with a mathematical function. Time shifting is a little bit of a difficult example, so bear with me:

1. Let's navigate to **Data collection** | **Hosts** and select our host, **lar-book-agent_passive**.

2. Go to **Triggers** and click the blue **Create trigger** button.

3. Add the following trigger, as seen in the screenshot:

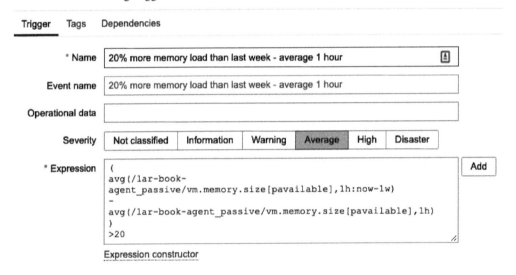

Figure 4.18 – Time shifting average trigger form filled out

This is a very complex trigger to set up, so let's dive right into how it's set up in the *How it works…* section.

How it works...

Advanced triggers can get very complex. The triggers we have just set up are just the tip of the iceberg. Do not worry if these triggers seem intimidating, as there is plentiful documentation out there to help you set them up, which you can find here: `https://www.zabbix.com/documentation/current/en/manual/config/triggers`.

It's near impossible to cover every single use case in this book, so the triggers we set up will show you what's possible. Use what you have learned in the examples in your own scenarios, but make sure to apply your own thinking to it.

Advanced trigger 1 – trendavg function

Let's start off the *How it works…* section with the trend average. Trend average is one of the few trigger functions that use trend data instead of history data. Let's do a short crash course on history and trend data in Zabbix. History data is the exact value every time an item receives data from a monitored host. Trend data is the average, minimum, and maximum values over one hour created from the history data and a count of the number of values.

Now, let's look at the available functions for creating triggers using trend data:

- `trendavg`: To get the average value from trend values within a time period
- `trendmax`: To get the maximum value from trend values within a time period
- `trendmin`: To get the minimum value from trend values within a time period
- `trendcount`: To get the number of retrieved trend values within a time period
- `trendsum`: To get a sum of trend values within a time period

As I said, all of these will use our trend values. The values used are stored in a special Zabbix trend cache in memory, for use in our trigger. We've used the `trendavg` function. Let's check out how we used it in our trigger expression again:

Figure 4.19 – trendavg trigger expression

We start off our trigger with the `trendavg` function and then the *host/template* and *item key* as we saw earlier in the last recipe. What's new here is the part where we state `1w:now-1w`. This is the time period; here we've stated to use a value from one week ago.

What this means is that if the average value from our trends one week ago is above 800 Mbps, then this trigger will go into a problem state.

Advanced trigger 2 – timeleft function

`timeleft` is another very interesting trigger function. We can use `timeleft` to create triggers that only fire when it expects something to reach a certain threshold in the future. This is called a predictive trigger, as it makes a prediction based on older data.

Let's check out our trigger expression again.

Expression `timeleft(/lar-book-agent_passive/vfs.fs.size[{#FSNAME},pused],7h,100)<1w`   Add

Figure 4.20 – timeleft trigger expression

As we can see, we start our expression as usual: the *trigger function, host/template,* and our *item key.* In this case, we combined that with a time period we want to use for our predictive trigger to define its prediction. We used 7h, to tell this expression to use seven hours of historic data. Combine that with a threshold of 100, to make sure this will trigger if we expect to reach 100% disk space usage. Now we only need one more element to complete this, the expected result, which in this case is <1w.

To sum it all up, this trigger expression looks at *seven hours* of historic data and if it expects to reach *100%* disk space in *less than one week*, it will go into a problem state, alerting you that you will need to make sure your disks don't run out of space.

A tip would be to combine the `timeleft` trigger function with other functions to limit how many times you get alerted. For example, with disk space, we might expect a disk to fill up in a week, but you might not want to see that unless the used space is at least less than 50 Gigabytes. Add another expression and you are golden:

```
last(/Linux filesystems by Zabbix agent/vfs.fs.size[{#FSNAME},pused])>90%
and
timeleft(/Linux filesystems by Zabbix agent/vfs.fs.size[{#FSNAME},pused],1h,100)<1d)
```

Figure 4.21 – timeleft trigger function expression

Advanced trigger 3 – time shifting using mathematical functions

As a Zabbix trainer, time shifting trigger expressions are where my students and I always need to spend some additional time on what they all do exactly. This makes sense, as it is one of the more complex expressions, and in this example, we even combined it with some mathematical functions.

So, let's take another look at our expression and break it down.

```
1  (
2  avg(/lar-book-agent_passive/vm.memory.size[pavailable],1h:now-1w)
3  -
4  avg(/lar-book-agent_passive/vm.memory.size[pavailable],1h)
5  )
6  >20
```

Figure 4.22 – Time shifting trigger expression

I've added line numbers for our convenience. Now we can go over each line and explain what they mean:

1. This is the opening bracket for our mathematical statement, using the operator between two items.

2. Our first item, using the time shift function. This item will get our memory availability as a percentage from one week ago starting from this moment exactly. If the current date and time are Monday 24th November at 14:00, it will get the one-hour average value for Monday 17th November between 13:00 and 14:00.

3. Our mathematical operator, stating a minus. This means we'll subtract the result of the first expression from the result of the second expression.

4. This is our second item, not using a time shift. This item will be filled with a one-hour average value of the last hour.

5. The closing bracket ends our mathematical statement.

6. Finally, an operator and constant. This states that this trigger will only trigger if the mathematical result is higher than 20.

Now that we know what each of the lines does, let's take a look at how it works in a real-life scenario. We're going to fill out the values manually and see whether the expression is TRUE or FALSE. TRUE means that there is a problem and FALSE means everything is fine. So, the math is as follows:

```
(Last week - This week) = Result
If the Result is higher than 20 then the expression is True
This expression is: TRUE/FALSE
```

Filling it out with 80% memory available last week and only 50% available this week, we can see the following happening:

```
(80 - 50) = 30
If 30 is higher than 20 the expression is TRUE
This expression is TRUE
```

Let's do it one more time but with 80% memory available last week and 70% this week:

```
(80 - 70) = 10
If 10 is higher than 20 the expression is TRUE
This expression is FALSE
```

This is how you should go about setting up your time shifting expressions. Simply use a notebook or whatever you like, write down your expression in simple text for yourself, and do the calculations.

There's more...

Trigger expressions can also be tested within Zabbix itself. If we go to **Data collection | Hosts | Triggers** and select any of our three advanced triggers, we can do a little test. For example, using the time shifting trigger, we can click **Expression constructor**.

Figure 4.23 – Time shifting trigger expression constructor

Here, we can select **Test** and then fill out our values. Let's use the same 80% and 50% we did in the earlier example:

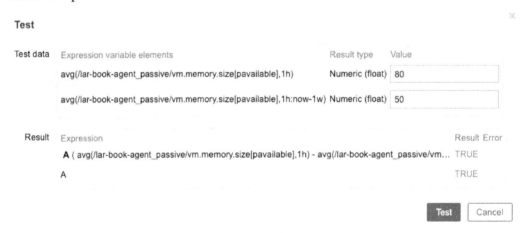

Figure 4.24 – Time shifting trigger expression constructor – Test

As you can see, this will tell us whether our expression ends up being **TRUE** or **FALSE**, using any values we want to fill. In short, if you want to be sure your math on paper is doing the same thing directly in Zabbix, use **Expression constructor** to test it.

Setting up alerts

Alerting can be a very important part of your Zabbix setup. When we set up alerts, we want the person on the other end to be informed of just what is going on. It's also important to not spam someone with alerts; we want it to be effective.

So, in this recipe, we will go over the basics of setting up alerts, so we know just how to get it right from the start.

Getting ready

For this recipe, we will only need two things. We will have to use our Zabbix server to create our alerts and we will need some triggers, such as the triggers from the previous recipe. The triggers will be used to initiate the alerting process to see just how the Zabbix server will convey this information.

How to do it...

1. Let's start by setting up our action on the Zabbix frontend. To do this, we will navigate to **Alerts | Actions | Trigger actions** and we will be served with this screen:

Figure 4.25 – The Zabbix Trigger actions page with one trigger action

There is already one action set up to notify **Zabbix administrators** of problem events. In Zabbix 7, a lot of features, such as **Actions** and **Media**, are predefined. Most of the time, all we need to do is enable them and fill out some information.

2. We will set up our own action, so let's create a new action to notify a user in the **Zabbix administrators** group of our new triggers. Click the blue **Create action** button to be taken to the next page:

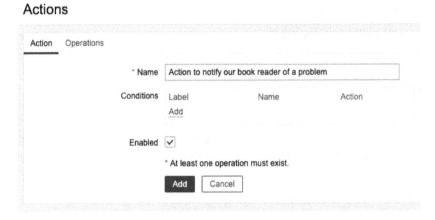

Figure 4.26 – The Zabbix action creation page – notify book reader

3. On this page, check the **Enabled** checkbox to make sure that this action will actually do something. Make sure to name your action clearly so that you won't have any issues with differentiating between actions.

4. Now, move on to the **Operations** tab:

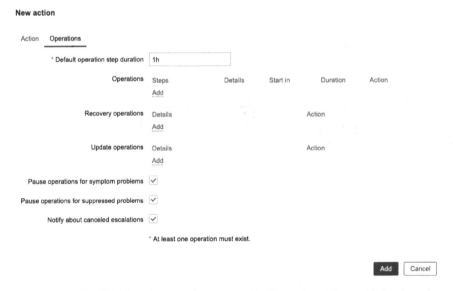

Figure 4.27 – The Zabbix action creation page at the Operations tab – notify book reader

5. The **Operations** tab is empty by default, so we are going to want to create some operations here. There are three forms of operation that we are going to create here—let's start with the **Operations** operation by clicking **Add**:

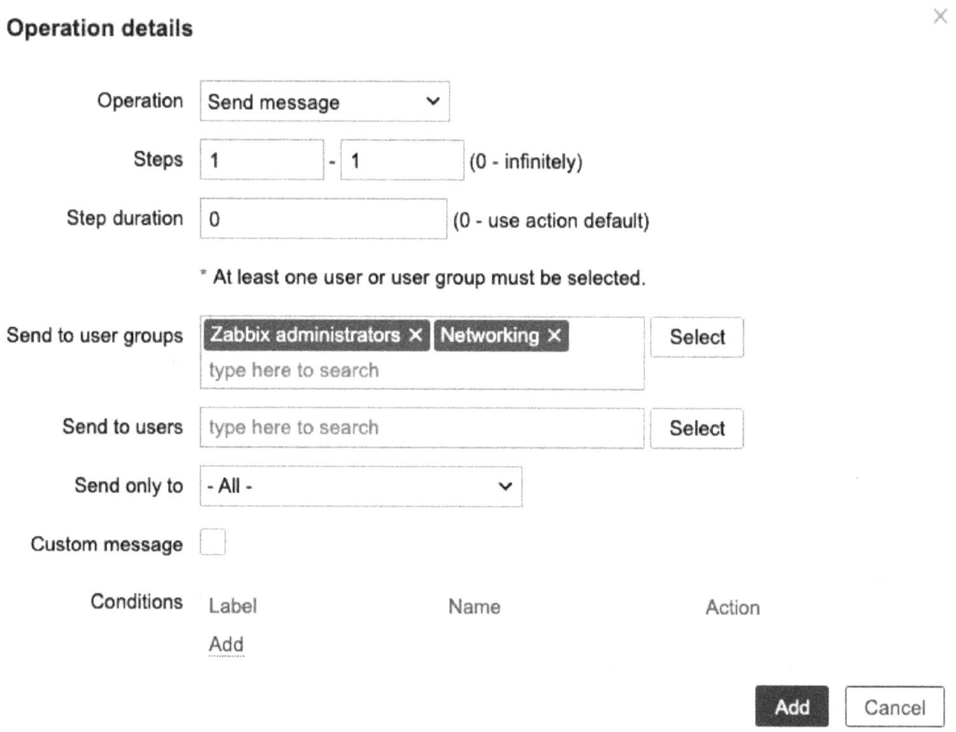

Operation details   ✕

Operation	Send message ▾
Steps	1 - 1 (0 - infinitely)
Step duration	0 (0 - use action default)

* At least one user or user group must be selected.

Send to user groups	Zabbix administrators ✕ Networking ✕ [Select] type here to search
Send to users	type here to search [Select]
Send only to	- All - ▾
Custom message	☐
Conditions	Label Name Action Add

[Add] [Cancel]

Figure 4.28 – Zabbix Operation details page – notify book reader

6. We have the option to add users and/or user groups here that we want to alert. If you've followed along with *Chapter 1, Installing Zabbix and Getting Started Using the Frontend*, you can just select the **Networking** user group here. If not, selecting just the **Zabbix administrators** group is fine.

7. After clicking the blue **Add** button at the bottom of the form, we will be taken back to the **Actions** screen.

8. Now, we will create the next operation, named **Recovery operations**. What we do here is create an operation, as follows:

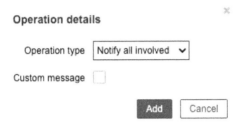

Figure 4.29 – Zabbix Recovery operations details – notify book reader

9. This option will notify all users involved in the initial operation defined earlier. All users that got a PROBLEM generation notification will also receive the recovery this way. Click **Add**, and let's continue.

    Now, if you're like me and you want to stay on top of things, you can create an update notification. This way, we know that—for instance—someone acknowledged a problem and is working on it. Normally, I would select different channels for stuff such as this—for instance, using **SMS** for high-priority alerts and a Slack or Teams channel for everything else.

10. Let's click **Add** under **Update operations** to add the following to our setup:

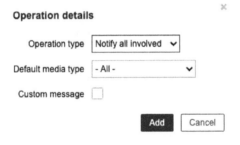

Figure 4.30 – Zabbix update operation details – notify book reader

11. We will do the same thing here as we did for the **Recovery operations** option and notify all users involved of any update to this problem. After clicking **Add** here, click the blue **Add** button again on the **Actions** screen to finish creating the action.

12. Now, the next thing we want to do is create a media type for actually notifying our users of the issue. Go to **Alerts | Media types**, and you will be presented with the following screen:

	Name ▲	Type	Status
☐	Discord	**Webhook**	Enabled
☐	Email	**Email**	Enabled
☐	Email (HTML)	**Email**	Enabled
☐	Jira	**Webhook**	Enabled
☐	Jira ServiceDesk	**Webhook**	Enabled
☐	Jira with CustomFields	**Webhook**	Enabled
☐	Mattermost	**Webhook**	Enabled
☐	MS Teams	**Webhook**	Enabled
☐	Opsgenie	**Webhook**	Enabled
☐	PagerDuty	**Webhook**	Enabled
☐	Pushover	**Webhook**	Enabled
☐	Redmine	**Webhook**	Enabled
☐	ServiceNow	**Webhook**	Enabled
☐	SIGNL4	**Webhook**	Enabled
☐	Slack	**Webhook**	Enabled
☐	SMS	**SMS**	Enabled
☐	Telegram	**Webhook**	Enabled
☐	Zammad	**Webhook**	Enabled
☐	Zendesk	**Webhook**	Enabled

Figure 4.31 – The Zabbix Media types page with predefined media options

As you can see, there are quite a lot of predefined media types in Zabbix 7. We have them for Slack, Opsgenie, and even Telegram. Let's start with something almost everyone has, though: email.

13. Click the **Email** media type and we'll edit it to suit our needs:

New media type

| Media type | Message templates | Options |

* Name	Email
Type	Email ∨
Email provider	Generic SMTP ∨
* SMTP server	smtp.office365.com
SMTP server port	587
* Email	nathan@oicts.nl
SMTP helo	oicts.nl
Connection security	None **STARTTLS** SSL/TLS
SSL verify peer	☐
SSL verify host	☐
Authentication	None **Username and password**
Username	nathan@oicts.nl
Password	••••••••••••••••••••••••
Message format	HTML **Plain text**
Description	
Enabled	☑

Figure 4.32 – The Zabbix Media type creation page for email

14. I set it up to reflect my Office 365 settings, but any **Simple Mail Transfer Protocol (SMTP)** server should work. Fill in your SMTP settings, and we should be able to receive notifications.

15. Be sure to also check the next tab, **Message templates**. For example, the message template for a **Problem** generation event looks like this:

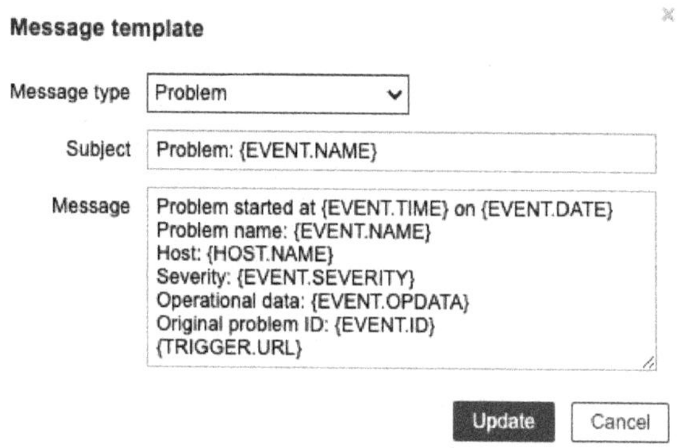

Message template
Message type
Problem
Subject
Problem: {EVENT.NAME}
Message
Problem started at {EVENT.TIME} on {EVENT.DATE}
Problem name: {EVENT.NAME}
Host: {HOST.NAME}
Severity: {EVENT.SEVERITY}
Operational data: {EVENT.OPDATA}
Original problem ID: {EVENT.ID}
{TRIGGER.URL}
Update
Cancel

Figure 4.33 – The default Zabbix Message template page for problems

We set this up like this so that we get a message telling us just what's going on. This is fully customizable as well, to reflect just what we want to know.

16. Let's keep the default settings for now. Last but not least, go to **Users** | **Users** and edit a user in the **Zabbix administrators** or **Networking** group. I will use the **Admin** user as an example.

Users
User
Media
Permissions
Media
Type
Send to
When active
Use if severity
Status
Action
Add
Update
Delete
Cancel

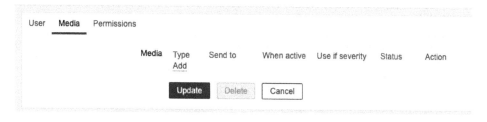

Figure 4.34 – The Zabbix user Media page for the Admin user

17. On this window, go to **Media** and click the **Add** button. We want to add the following to notify us of all trigger severities at our email address:

Figure 4.35 – The Zabbix user media creation page for the Admin user

18. Now, click on the blue **Add** button and finish creating this user's media.

How it works...

Now, that's how we set up alerts in Zabbix. You will now receive alerts at your email address, as shown in the following flowchart:

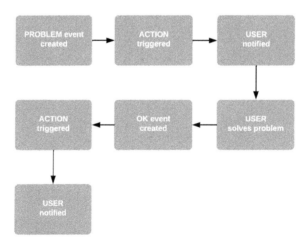

Figure 4.36 – Diagram showing Zabbix problem flow

When something breaks, a **problem** in Zabbix is triggered by our trigger configuration. Our **action** will then be triggered by our problem event and it will use the media type and user media configuration to notify our user. Our user then fixes the issue (for instance, rebooting a stuck server), and then an **OK** event will be generated. We will then trigger the action again and get an OK message.

> Tip
>
> Before building alerts such as this, make a workflow (as shown in *Figure 4.36*) for yourself, specifying just which user groups and users should be notified. This way, it's clear just how you will use Zabbix for alerting.

There's more...

There are loads of media types and integrations, and we've just touched the tip of the iceberg by seeing a list of predefined ones. Make sure to check out the Zabbix integration list (`https://www.zabbix.com/integrations`) for more options or build your own using the Zabbix webhooks and other extensions available.

Keeping alerts effective

It's important to keep our alerts effective to make sure we are neither overwhelmed nor underwhelmed by notifications. To do this, we will change our trigger and the **Email** media type to reflect just what we want to see.

Getting ready

We will be using trigger 1 from the first recipe and the default email media type in Zabbix.

Furthermore, of course, we'll also be using our Zabbix server.

How to do it...

To create effective alerts, let's follow these steps:

1. Let's get started on trigger 1, which we created in this chapter's *Setting up triggers* recipe. Navigate to the `lar-agent-simple` host by going to **Data collection** | **Hosts** and clicking **Trigger** for the host.

2. Here, sometimes people use a different trigger name like the one we see here:

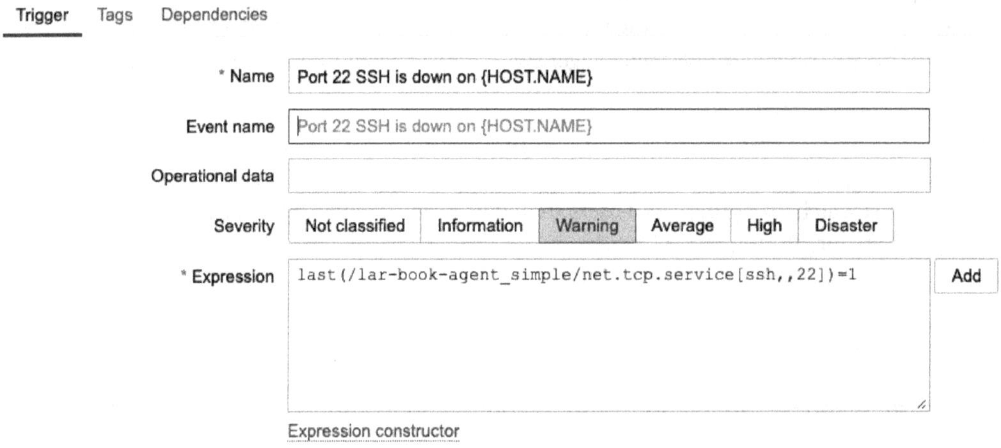
Trigger
Tags
Dependencies
* Name
Port 22 SSH is down on {HOST.NAME}
Event name
Port 22 SSH is down on {HOST.NAME}
Operational data
Severity
Not classified
Information
Warning
Average
High
Disaster
* Expression
last(/lar-book-agent_simple/net.tcp.service[ssh,,22])=1
Add
Expression constructor

Figure 4.37 – Trigger 1 from the previous recipe

Even when you've used the {HOST.NAME} macro in the trigger, it's quite simple, so fortunately there isn't a lot to change here. If you've used the hostname in the trigger name, we can change the name to reflect a message that is clearer.

3. Make sure to use a short and descriptive trigger name, such as the following:

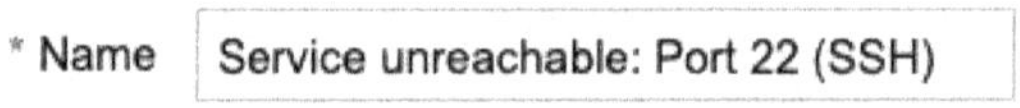
* Name
Service unreachable: Port 22 (SSH)

Figure 4.38 – New Trigger 1 name

4. Next, navigate to the **Tags** page to add a tag to keep the triggers organized. Let's add the following:

Trigger
Tags 1
Dependencies
Trigger tags
Inherited and trigger tags
Name
Value
Action
scope
availability
Remove
Add

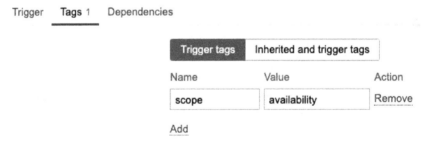

Figure 4.39 – Trigger 1 tags

5. Another great way of keeping everything organized is changing the message on media types. Let's change the media type to reflect our needs. Navigate to **Alerts** | **Media types** and select our media type named **Email**.

6. Select **Message templates** and click **Edit** next to our first problem. This will bring us to the following window:

Message template

Message type Problem

Subject Problem: {EVENT.NAME}

Message Problem started at {EVENT.TIME} on {EVENT.DATE}
Problem name: {EVENT.NAME}
Host: {HOST.NAME}
Severity: {EVENT.SEVERITY}
Operational data: {EVENT.OPDATA}
Original problem ID: {EVENT.ID}
{TRIGGER.URL}

Update   Cancel

Figure 4.40 – Standard email media type message

Currently, Zabbix uses the default configured message under the media type when we do not use a custom message. But if we want to change that message, we can do that here by creating a custom message. The default message under the **Email** media type looks as in the previous screenshot.

7. We can change the message on the media type. For instance, if you don't want to see the original problem ID or want a more customized message, simply remove that line, as shown in the following screenshot:

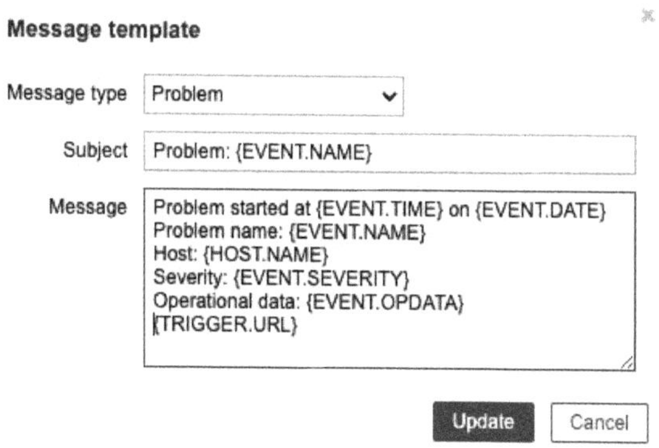

Message template

Message type Problem

Subject Problem: {EVENT.NAME}

Message Problem started at {EVENT.TIME} on {EVENT.DATE}
Problem name: {EVENT.NAME}
Host: {HOST.NAME}
Severity: {EVENT.SEVERITY}
Operational data: {EVENT.OPDATA}
{TRIGGER.URL}

Update   Cancel

Figure 4.41 – Custom Email media type message

How it works...

We've done two things in this recipe: changed our trigger name and added a tag to our trigger.

Keeping trigger names clear and defined in a structured way is important to keep our Zabbix environment structured. Instead of just naming our trigger `Port 22 SSH down on {HOST.NAME}`, we've added standardization to our setup and can now create cool structures such as this with our future triggers:

Service unreachable: Port 22 (SSH)
Service unreachable: Port 3306 (MYSQL)
Service unreachable: Port 443 (HTTPS)

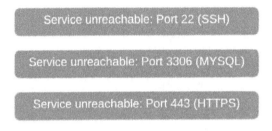

Figure 4.42 – Trigger structure diagram

Our triggers are all clear and we can immediately see which host, port, and service are down.

On top of that, we've added a tag for the service that is down, which will now immediately display our service in a clear way, alerting us to exactly what is going on:

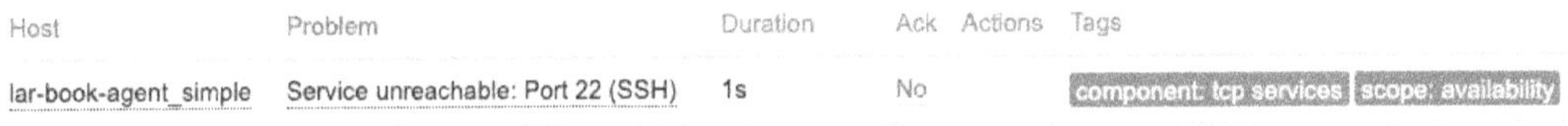

Host
Problem
Duration
Ack
Actions
Tags
lar-book-agent_simple
Service unreachable: Port 22 (SSH)
1s
No
component: tcp services
scope: availability

Figure 4.43 – Trigger down – structured

In Zabbix 6, a new tag policy was introduced. As we created the item used in the trigger with a **component** tag and we just added a trigger tag for **scope**, we followed the new standard. In the problem view in the preceding screenshot, it becomes immediately apparent that we have a problem affecting the availability of the TCP service SSH. The **scope** tag generally contains either one of five options: `availability`, `performance`, `notification`, `security`, and `capacity`.

For more information about the new Zabbix tag policy that we still use in Zabbix 7, check out this link:

`https://blog.zabbix.com/tags-in-zabbix-6-0-lts-usage-subfilters-and-guidelines/19565/`

Another thing we've done is removed the {HOST.NAME} macro if we've used it before. As we can already see which host this trigger is on by checking the **Host** field, we do not need to add the {HOST. NAME} macro. We need to keep trigger names short and effective and use the hostname macros in **Media** or simply use the field already available in the frontend.

We've also changed our action in this recipe. Changing a message on media types is a powerful way to keep our problem channels structured. Sometimes, we want to see less or more information on certain channels, and changing media type messages is one way to do this.

We can also create custom messages on an **Action** level, changing all the messages sent to the selected channels.

There's more...

What I'm trying to show you in this recipe is that although it might be simple to set up Zabbix, it is not simple to set up a good monitoring solution with Zabbix—or any monitoring tool, for that matter—if you don't plan. Carefully plan out how you want your triggers to be structured before you build everything in your Zabbix installation.

An engineer who works in a structured way and takes the time to build a good monitoring solution will save a lot of hours in the future because they will understand the problem before anyone else.

Customizing alerts

Alerting is very useful, especially in combination with some of the tricks we've learned in this book so far to keep everything structured. But sometimes, we need a little more from our alerts than what we are already getting from Zabbix out of the box.

In this recipe, we'll do a small bit of customization to make the alerts more our own.

Getting ready

For this recipe, all we are going to need is our current Zabbix server installation.

How to do it...

To customize alerts, follow these steps:

1. Let's create some custom severities in our Zabbix server to reflect our organization's needs. Navigate to **Administration | General** and select **Trigger displaying options** from the side menu:

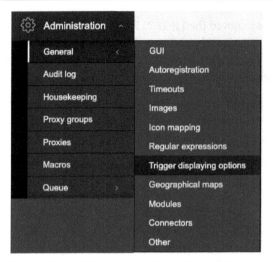

Figure 4.44 – Administration | General from side menu

After selecting this, we'll be taken to the next page. This window contains the default Zabbix trigger severities, as shown in the following screenshot:

Figure 4.45 – Default trigger severities page

2. Next up, we can customize the default trigger severities, as follows:

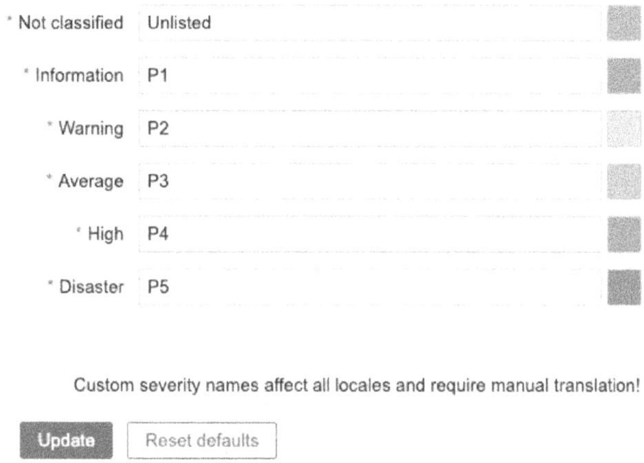

* Not classified	Unlisted	
* Information	P1	
* Warning	P2	
* Average	P3	
* High	P4	
* Disaster	P5	

Custom severity names affect all locales and require manual translation!

Update   Reset defaults

Figure 4.46 – Custom trigger severities page

3. Do not forget to click the blue **Update** button at the bottom of the page to save the changes.

How it works...

Not all companies like using terms such as **High** and **Disaster**, but prefer using different severities such as **P1** and **P2**. Using custom severities, we can customize Zabbix to make it more our own and reflect the terms we've already been using in different tools, for example.

Changing custom severities is not a necessity by any means, but it can be a good way to adopt Zabbix more easily if you are used to something different.

5

Building Your Own Structured Templates

It's time to start one of the most important tasks in Zabbix: building structured templates. A good Zabbix setup relies heavily on templating, and there is a huge difference between a good and a bad template. So, if you're new to Zabbix or you haven't started building your own templates yet, then pay close attention to this chapter.

In this chapter, we will go over how to set up your templates, and how to fill them with the right items and triggers. Also, it is important to make use of macros and **Low-Level Discovery** (**LLD**) in the right way. After following these recipes, you will be more than ready to build solid Zabbix templates with the right format and even LLD.

In this chapter, we'll cover the following recipes:

- Creating your Zabbix template
- Setting up template-level tags
- Creating template items
- Creating template triggers
- Setting up different kinds of macros
- Using LLD on templates
- Nesting Zabbix templates

Technical requirements

We will need our Zabbix server from *Chapter 4, Working with Triggers and Alerts*, to monitor our **Simple Network Management Protocol** (**SNMP**) host. For the SNMP host, we can use the host we set up in the *Working with SNMP monitoring* recipe in *Chapter 3, Setting Up Zabbix Monitoring*.

Creating your Zabbix template

In this recipe, we will start with the basics of creating a Zabbix template. We will go over the structure of Zabbix templating and why we need to pay attention to certain aspects of templating.

Getting ready

All you will need in this recipe is your Zabbix server.

How to do it...

Now, let's get started with building our structured Zabbix template:

1. Open your Zabbix frontend and navigate to **Data collection** | **Templates**.

2. On this page, click the **Create template** button in the top-right corner. This will take you to the following page:

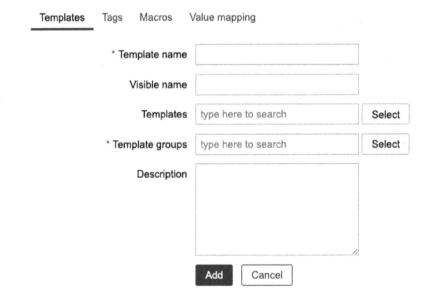

Figure 5.1 – The Create template page, empty

At this point, we are going to need to name our template and assign a template group to it. We will be creating an SNMP template to monitor a Linux host. I'll be using SNMP in the example to show how the templates are structured.

Important Note

Use SNMP to monitor network equipment, custom equipment supporting SNMP, and more. SNMP is very versatile and easy to understand, and it is implemented by a lot of hardware manufacturers. For Linux hosts, I'd still recommend the very powerful Zabbix agent, which we covered in the *Setting up Zabbix agent monitoring* recipe in *Chapter 3*, *Setting Up Zabbix Monitoring*.

3. Create your template with the following information:

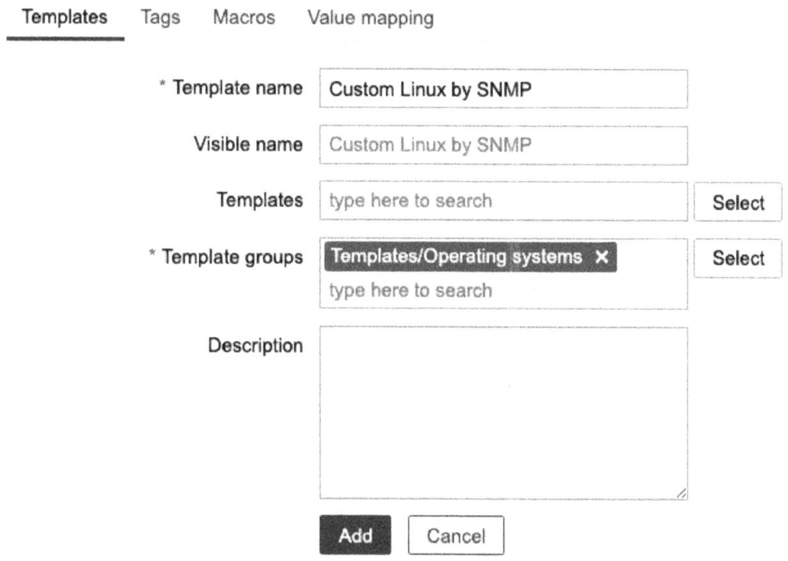

Figure 5.2 – The Create template page filled with information for the SNMP template

We will not link any **Templates**, **Tags**, and **Macros** yet, but we'll address some of these functionalities later. That's all there is to creating our template, but there's nothing in it besides a name, group, and description so far.

How it works...

There's not a lot of work involved in creating our first template—it's quite straightforward. What we need to keep in mind is the right naming convention here.

Now, you might think to yourself: *why is naming a template so important?* Well, we are going to create a lot of templates when working with Zabbix. For example, this is a small part of the list of out-of-the-box templates:

Figure 5.3 – Some out-of-the-box templates

As you can see, this is already a large list, and all of these templates follow a singular straightforward naming convention. If you look at the name of the template we have just built ourselves and, for example, the built-in Apache template, they follow the same convention. Breaking down the convention, it looks like this:

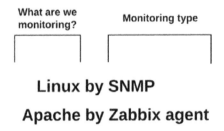

Figure 5.4 – Template naming convention explanation

If we look at the list and compare it to the naming convention we went over in *Figure 5.4*, we can see the following pattern:

- **What are we monitoring?**: (Linux) We name the template—in this case, we'll call it Linux because the OS we monitor will be Linux.

- **Monitoring type**: (by SNMP) We will add our data collection method at the end of the template as we might monitor the Linux OS in other ways besides SNMP, such as the Zabbix agent.

Adhering to the guidelines in this naming convention and thus using the correct template names is our first step in creating the correct structure for our template. This makes it easy to find out which templates we want to use on which hosts.

In our case, we've also added a short custom prefix to make sure we can distinguish our template from others already created in the Zabbix setup. Normally, we can omit this prefix, but for this book, it's useful. As a best practice, it is recommended to clone the default templates you'd like to use and prefix the template name with your company name or shorthand. We do this to not overwrite the default templates and to ensure we can import the official templates later without overwriting possible changes we made. For example, I would clone *Linux by SNMP* and call it *OICTS Linux by SNMP* for use in our company, Opensource ICT Solutions. This would also work in a **Managed Service Provider** (**MSP**) environment where we'd like to have a unique template for each customer.

There's more...

When building templates, adhere to the Zabbix guidelines. That's what we will do in this book as well, combined with our experience in creating templates. If you want to learn more about Zabbix templating guidelines, check the following URL: `https://www.zabbix.com/documentation/guidelines/en/thosts`.

Setting up template-level tags

Our next step in setting up our Zabbix template is setting up template-level tags. Tags on the template level are used to give every single event (problem) created on a host by this template a tag. The tag is then used to filter events in things such as dashboards, actions, and the **Monitoring** | **Problems** view.

Getting ready

To get started with this recipe, you will need a Zabbix server and a template on that server, preferably the template we created in the previous recipe.

How to do it...

Creating template-level tags is a way to make sure that only events created by a certain template will get a configured tag. To get started, the first thing you will need to do is navigate to the template and follow these steps:

1. Go to **Data collection | Templates** and click on our template, which is called **Custom Linux by SNMP**.

2. Then, click the **Tags** tab at the top of the form, and you'll be taken to this tab:

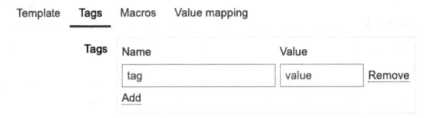

Figure 5.5 – Zabbix Tags tab for the SNMP template

Now, the first thing we can do is create some tags to make sure we know all the events from this template will be Linux-related.

3. The first tag is already ready to be created. Fill out **Name** as class. Then, in the **Value** field, type in os.

4. For the second tag, click the small dotted-underlined **Add** button and set the **Name** for the second tag to target. Then, in the **Value** field, type in linux. It will look like this:

Figure 5.6 – Zabbix Tags tab filled out for the SNMP template

5. Do not forget to click the blue **Update** button to save your tag to this template.

How it works...

Now, there's a lot more to creating tags than it might seem at first through following this recipe. Tags play a key part in keeping your Zabbix environment structured. You will use the template-level tags to filter in a lot of places, such as the **Monitoring | Problems** window, and with a lot of events created by one host, they will improve readability by making problems easy to filter.

For example, once we have configured some triggers later in this recipe, when checking the **Monitoring | Problems** page for our host, we could see something like the following:

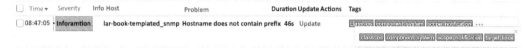

Figure 5.7 – Example Monitoring | Problems page for host lar-book-agent_snmp

> **Note regarding screenshots**
>
> Some screenshots used in the book have been adjusted to fit the margins of the book pages which has resulted in lower readability of the text in the print version. Please refer to the PDF version available here: `https://packt.link/free-ebook/978-1-80107-832-0`; to enlarge the image and view the text with more clarity.

As you can see, the problem we are looking at here is displaying the **target:linux** and **class:os** tags at the end of the page. The event was tagged with the template-level tag, and we can now see that it will always carry that tag, allowing us to filter.

This gives us loads of opportunities because we aren't limited to template-level tags. We also have host-level tags, item-level tags, and trigger-level tags. We could tag everything from a template with **target:linux** and **class:os** or even tag a specific trigger with something like department:architecture.

We could then, for example, create an action that sends out everything Linux-related to a certain Linux engineering email address or Teams/Slack channel based on the **class:linux** tag, but only send specific problems with a trigger-level tag such as department:architecture to a more specific email address or Teams/Slack channel.

For more information regarding the tag policy starting from Zabbix 6, check out the following link:

`https://blog.zabbix.com/tags-in-zabbix-6-0-lts-usage-subfilters-and-guidelines/19565/`

See also

In this chapter, the recipe titled *Using LLD on templates* will also explain **tag prototypes**, where we will create tags automatically based on the LLD settings. Tag prototypes are the recommended way of working with tags when creating discovery and are amazing for keeping templates structured. More about that later.

Creating template items

Let's get started with finally creating some real template items because, in the end, items are what it is all about in Zabbix. Without items, we don't have data, and without data, we do not have anything to work with in our monitoring system.

Getting ready

Now, moving along, we are going to need our Zabbix server and a host that we can monitor with SNMP. In *Chapter 3, Setting Up Zabbix Monitoring*, we monitored a host with SNMP, so we will use this host again. We'll also use the Zabbix template from the previous recipes.

How to do it...

1. First of all, let's log in to our Zabbix server **command-line interface** (CLI) and enter snmpwalk, with the following command:

    ```
    snmpwalk -v3 -l authPriv -u snmpv3user -a SHA -A "my_authpass"
    -x AES -X "my_privpass" 10.16.16.153 .1
    ```

    Make sure to change the IP address 10.16.16.153 to your own value. We will receive an answer such as this:

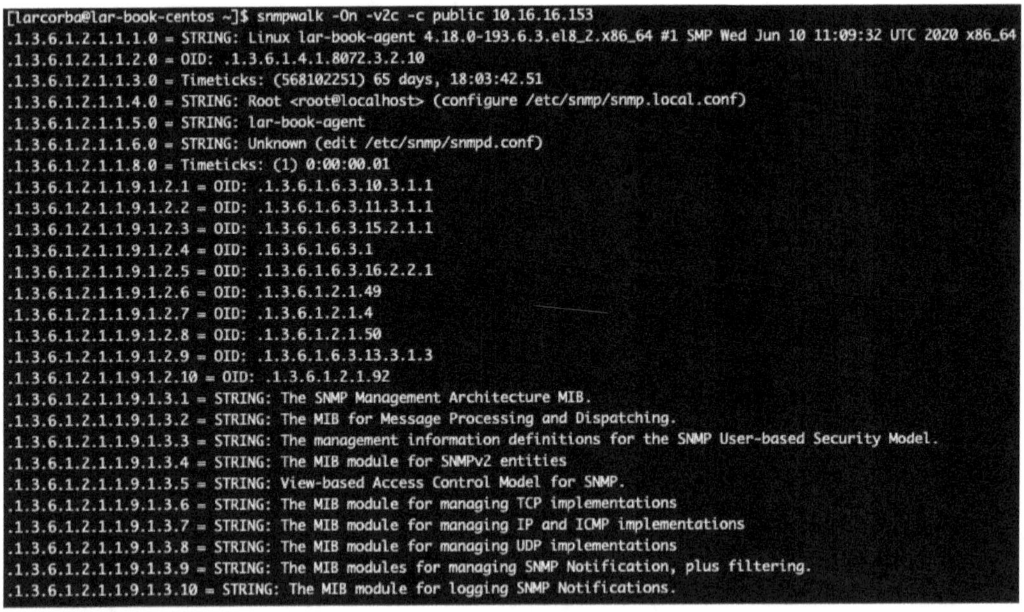

    ```
    [larcorba@lar-book-centos ~]$ snmpwalk -On -v2c -c public 10.16.16.153
    .1.3.6.1.2.1.1.1.0 = STRING: Linux lar-book-agent 4.18.0-193.6.3.el8_2.x86_64 #1 SMP Wed Jun 10 11:09:32 UTC 2020 x86_64
    .1.3.6.1.2.1.1.2.0 = OID: .1.3.6.1.4.1.8072.3.2.10
    .1.3.6.1.2.1.1.3.0 = Timeticks: (568102251) 65 days, 18:03:42.51
    .1.3.6.1.2.1.1.4.0 = STRING: Root <root@localhost> (configure /etc/snmp/snmp.local.conf)
    .1.3.6.1.2.1.1.5.0 = STRING: lar-book-agent
    .1.3.6.1.2.1.1.6.0 = STRING: Unknown (edit /etc/snmp/snmpd.conf)
    .1.3.6.1.2.1.1.8.0 = Timeticks: (1) 0:00:00.01
    .1.3.6.1.2.1.1.9.1.2.1 = OID: .1.3.6.1.6.3.10.3.1.1
    .1.3.6.1.2.1.1.9.1.2.2 = OID: .1.3.6.1.6.3.11.3.1.1
    .1.3.6.1.2.1.1.9.1.2.3 = OID: .1.3.6.1.6.3.15.2.1.1
    .1.3.6.1.2.1.1.9.1.2.4 = OID: .1.3.6.1.6.3.1
    .1.3.6.1.2.1.1.9.1.2.5 = OID: .1.3.6.1.6.3.16.2.2.1
    .1.3.6.1.2.1.1.9.1.2.6 = OID: .1.3.6.1.2.1.49
    .1.3.6.1.2.1.1.9.1.2.7 = OID: .1.3.6.1.2.1.4
    .1.3.6.1.2.1.1.9.1.2.8 = OID: .1.3.6.1.2.1.50
    .1.3.6.1.2.1.1.9.1.2.9 = OID: .1.3.6.1.6.3.13.3.1.3
    .1.3.6.1.2.1.1.9.1.2.10 = OID: .1.3.6.1.2.1.92
    .1.3.6.1.2.1.1.9.1.3.1 = STRING: The SNMP Management Architecture MIB.
    .1.3.6.1.2.1.1.9.1.3.2 = STRING: The MIB for Message Processing and Dispatching.
    .1.3.6.1.2.1.1.9.1.3.3 = STRING: The management information definitions for the SNMP User-based Security Model.
    .1.3.6.1.2.1.1.9.1.3.4 = STRING: The MIB module for SNMPv2 entities
    .1.3.6.1.2.1.1.9.1.3.5 = STRING: View-based Access Control Model for SNMP.
    .1.3.6.1.2.1.1.9.1.3.6 = STRING: The MIB module for managing TCP implementations
    .1.3.6.1.2.1.1.9.1.3.7 = STRING: The MIB module for managing IP and ICMP implementations
    .1.3.6.1.2.1.1.9.1.3.8 = STRING: The MIB module for managing UDP implementations
    .1.3.6.1.2.1.1.9.1.3.9 = STRING: The MIB modules for managing SNMP Notification, plus filtering.
    .1.3.6.1.2.1.1.9.1.3.10 = STRING: The MIB module for logging SNMP Notifications.
    ```

Figure 5.8 – snmpwalk reply

Now, let's capture our hostname in our template first, as it is an important item to have. When working with SNMP, I always like to work with untranslated SNMP **Object Identifiers** (**OIDs**). For our hostname, this is `.1.3.6.1.2.1.1.5.0`.

2. If we have an **Management Information Base** (**MIB**), we can translate this OID to make sure it is actually the system name. Enter the following command at the Zabbix server CLI:

    ```
    snmptranslate .1.3.6.1.2.1.1.5.0
    ```

    This will return the following reply:

    ```
    [root@lar-book-rocky ~]# snmptranslate .1.3.6.1.2.1.1.5.0
    SNMPv2-MIB::sysName.0
    ```

    Figure 5.9 – snmptranslate reply

> **Tip**
>
> Using `-On` in your SNMP command makes sure that we are receiving the OIDs instead of the MIB translation. If we want to work the other way around, we can omit the `-On` in our command and `snmptranslate` the translated OID.

3. Now that we know how to get our hostname, add this to our template. Navigate to **Data collection | Templates** and select our **Custom Linux by SNMP** template.

4. Here, we will go to **Items**. In the upper-right corner, select **Create item** to create the following item:

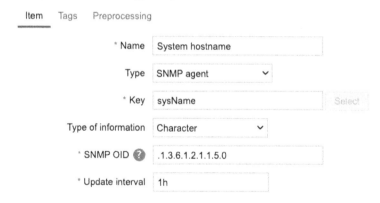

Figure 5.10 – Item for sysName SNMP OID

5. Make sure to also add an item-level tag. These are important for grouping and filtering items. Click the **Tags** tab and add the following:

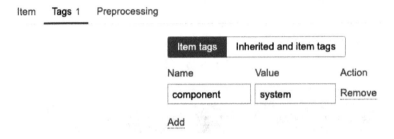

Figure 5.11 – Item for sysName SNMP OID on the Tags tab

Now that we have our first item, let's create a host as well and assign this template to that host.

6. Navigate to **Data collection | Hosts** and click **Create host** in the top-right corner. Create a host with the following settings:

Figure 5.12 – New host with our self-created template

7. Don't forget to add the macros to our new host before clicking the **Add** button. Click on **Macros** and fill in the following information:

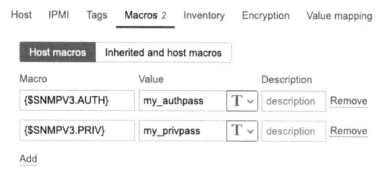

Figure 5.13 – Add macros tab on a host

Do not forget to set your macros to the type secret text to hide the passwords in the frontend.

8. Now, you can click the **Add** button, and our new host will be monitored.

How it works...

When we create items such as this on our template when assigning the template to our hosts, the item will also be created on the host. The great thing about this is that we can assign a template to multiple hosts, meaning we only have to configure the item on the template level once, instead of creating the item on every single host. For instance, our newly created host will show the following latest data:

Figure 5.14 – Monitoring | Latest data for our new host

The value for this item will then be different for all your monitored hosts, depending on the value received by that host.

> **Important Note**
>
> When creating an SNMP item, keep the following in mind. The **Item** field SNMP OID always contains the non-translated OID. This is to ensure that we do not actually need MIB files for our templates to work.

Furthermore, the item key will be based on the translated OID. In our case, the translated OID was sysName, which we then turned into the sysName item key. These are general rules that we should all abide by when creating our templates, to make sure they are structured in the same way for everyone.

See also

To learn more about Zabbix and SNMP OIDs/MIBs, check out this blog post:

`https://blog.zabbix.com/zabbix-snmp-what-you-need-to-know-and-how-to-configure-it/10345/#snmp-oid`

Creating template triggers

Creating templated triggers works in roughly the same way as creating templated items or normal triggers. Let's go over the process to see how we do it and how to keep it structured.

Getting ready

We will need the Zabbix server and the host from the previous recipe for this recipe.

How to do it...

We have configured one item on our template so far, so let's create a trigger for this item:

1. Navigate to **Data collection | Templates** in our Zabbix frontend and select our **Custom Linux by SNMP** template.

2. Now, click **Triggers** and then **Create Trigger** in the top-right corner. This will take us to the next page, where we will enter the following information:

Figure 5.15 – Create trigger window for the SNMP template

3. As discussed in the previous chapter, for triggers there's also the `scope` tag that we need to add:

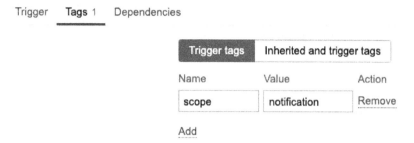

Figure 5.16 – Create trigger window for the SNMP template – tag

4. Last, but not least, let's edit the hostname on our host to see if the trigger is working correctly. Change the hostname entry by executing the following command on the Linux host CLI:

```
hostnamectl set-hostname lar-book-agent-t
```

5. Then, make sure the changes take effect by executing the following command:

```
exec bash
systemctl restart snmpd
```

How it works...

When editing the template, the created trigger will immediately be added to our host named `lar-book-templated_snmp`. This is because when we edited the template, the host was already configured with this template. When we have changed the hostname, the trigger can immediately be triggered after the item is polled again:

Figure 5.17 – Hostname has changed trigger for host lar-book-templated_snmp

Because we used the `change` function in our trigger, the second time we poll this item the problem will automatically go away again. In our case, this will happen after 30 minutes.

> **Important Note**
>
> Like a lot of other Zabbix users, I always like to use the {HOST.NAME} macro in trigger names, but according to Zabbix guidelines, this isn't recommended. If you prefer this you can still use it, but it's a lot more useful to use the Host fields throughout the Zabbix frontend and the built-in macros for notifications. This will keep trigger names short and won't show us redundant information.

Setting up different kinds of macros

When we are working with templates, a very efficient way to make your templates more useful is through the use of macros. In this recipe, we'll discover how to use macros to do this.

Getting ready

We are going to need our Zabbix server and our SNMP-monitored host from the previous recipes. We'll also need our Zabbix template, as created in the previous recipe.

How to do it...

Now, let's start with creating some macros on a template level. We'll be making two different types of macros.

Defining a user macro

1. First, we'll define a user macro on our template. Navigate to **Data collection | Templates** and click on our **Custom Linux by SNMP** template.

2. Here, we will go to **Macros** and fill in the following fields:

Figure 5.18 – Template-level macros

3. Click on **Update**, and let's move to **Trigger** to define a new trigger:

Figure 5.19 – Trigger creation window for the SNMP template

4. Let's also add the trigger tag:

Figure 5.20 – Trigger creation window for the SNMP template tag tab

5. Now, change the hostname entry by executing the following command on the host CLI:

```
hostnamectl set-hostname dev-book-agent
```

6. Then, make sure the changes take effect by executing the following command:

```
exec bash
systemctl restart snmpd
```

7. Our trigger should fire, as shown in the following screenshot:

Host	Problem • Severity
lar-book-templated_snmp	Hostname does not contain prefix

Figure 5.21 – Trigger created problem for a hostname prefix on the lar-book-templated_snmp host

Using built-in macros

1. Now, let's work on defining a built-in macro on our template. Navigate to **Data collection |
   Templates** and click on our **Custom Linux by SNMP** template.

2. Now, click **Triggers** and, in the top-right corner, click on **Create trigger**. Create a trigger with
   the following settings:

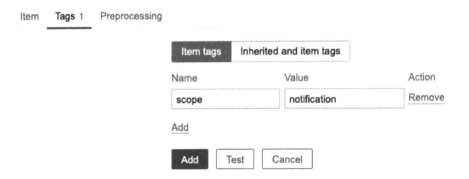

Figure 5.22 – Trigger creation window for hostname match

3. Let's also add the trigger tag:

Figure 5.23 – Trigger creation window for the SNMP template – tag

4. This will then trigger a problem, as expected.

Host	Problem • Severity
lar-book-templated_snmp	Hostname does not match Zabbix hostname

Figure 5.24 – Trigger created problem Hostname does not match

How it works...

There are four types of macros: built-in macros, user macros, expressions macros, and LLD macros. All of these macros can be used on templates, but also directly on hosts and various other locations. Macros are useful for creating unique values in places that would otherwise contain static information.

Let's discover how they work.

How a user macro works

Because we want all of our hosts on this template to contain `lar` as a prefix, we create a user macro at the template level. This way, the user macro that will be used on every host with this template will be the same.

We then define our user macro in our trigger to use the value, which is `lar-` in this case. We can reuse this user macro in other triggers, items, and more. The great thing is that defining a user macro on a template level isn't all we can do. We can override template-level user macros by defining a host-level user macro. So, if we want a single host to contain a different prefix, we simply use a host-level macro to override the template-level macro, like this:

Figure 5.25 – Host-level macros page

If we then look at the inherited and host-level macros screen on our host, we will see the following:

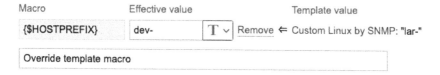

Figure 5.26 – Inherited and host-level macros page

We see the effective value is now `dev-`, not `lar-`, which is exactly what we would be expecting to happen here.

Keep in mind the syntax always starts with a curly bracket and a dollar sign and ends with a curly bracket. You are allowed to break the text in between with either dots or underscores. Here are some examples:

- `{$MACRONAME}`
- `{$MACRO.NAME}`
- `{$MACRO_NAME}`

How a built-in macro works

Now, a built-in macro comes from a predefined list of macros, hardcoded within Zabbix. They are used to get data from your Zabbix system and put them in items, triggers, and more. This means that the built-in macro used in this case already contains a value.

In this case, we used {HOST.HOST}, which is the hostname we defined on our Zabbix host, like this:

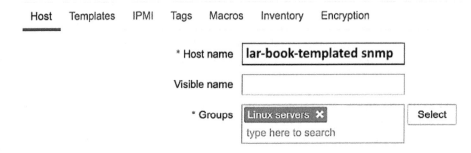

Figure 5.27 – Zabbix host configuration page for host lar-book-templated_snmp

For every single host, this built-in macro would be different as our **Host name** value will be unique. This means that our trigger, although defined on a template level, will always be unique as well. This method is a very powerful way to use built-in macros in triggers, as we'll pull information from Zabbix directly into Zabbix again.

Keep in mind the syntax always starts with a curly bracket and ends with a curly bracket. You are allowed to break the text in between with either dots or underscores. Here are some examples:

- {HOST.NAME}
- {INVENTORY.LOCATION.LAT}

There's more...

A complete list of supported (built-in) macros can be found here:

https://www.zabbix.com/documentation/current/en/manual/appendix/macros/supported_by_location

This list will be updated by Zabbix, just as with every good Zabbix documentation page. This way, you can always use this page as a reference for up-to-date (built-in) macros for building your Zabbix elements.

Using LLD on templates

Now, let's get started on my favorite part of template creation: LLD. I think this is one of the most powerful and most widely used parts of Zabbix.

Getting ready

To get ready for this recipe, you will need your Zabbix server, the SNMP-monitored host from the previous recipes, and our template from the previous recipe.

Working knowledge of the SNMP tree structure is also recommended. So, make sure to read the *Working with SNMP monitoring* recipe in *Chapter 3, Setting Up Zabbix Monitoring*, thoroughly.

How to do it...

1. Let's get started by navigating to **Data collection | Templates** and selecting our **Custom Linux by SNMP** template.

> **Important Note**
>
> First, we will add a value mapping, which we'll use for multiple item prototypes. Keep in mind that value mappings since Zabbix 6 are no longer global, but template- or host-specific. This is to make sure that the templates and hosts (once exported) are even more independent from the global Zabbix settings.

2. Click on the **Value mapping** tab and the dotted **Add** button. Add the following:

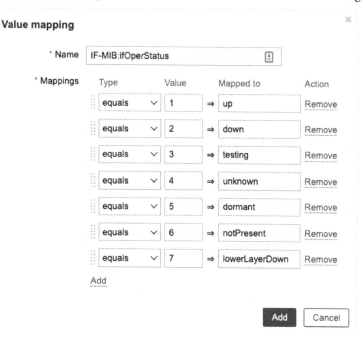

Figure 5.28 – Zabbix add Value mapping page

3.   Make sure to save this change by clicking the blue **Add** button and then the blue **Update** button.

4.   Now, go back to the template and go to **Discovery rules**, and in the top-right corner, click **Create discovery rule**. This will take you to the LLD creation page:

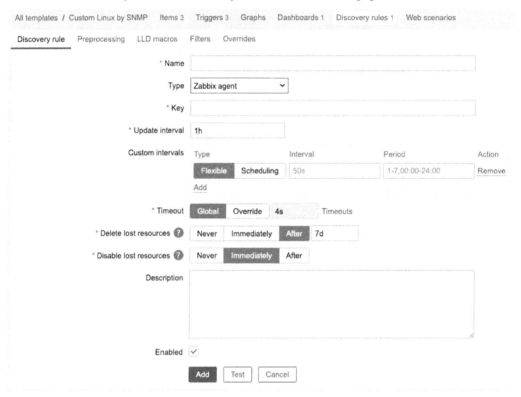

Figure 5.29 – Zabbix LLD creation page, empty

Now, we will be making a discovery rule to discover our interfaces on the Linux host. The Linux SNMP tree for interfaces is at OID .1.3.6.1.2.1.2.

Important Note

Make sure that Linux net-snmp is configured correctly in the /etc/ snmp/snmpd.conf file. It's important to change the view in this file to show everything from .1 and up, like this:
```
view systemview included .1
```

5. Now, let's continue with creating our LLD rule by adding the following to our LLD creation page:

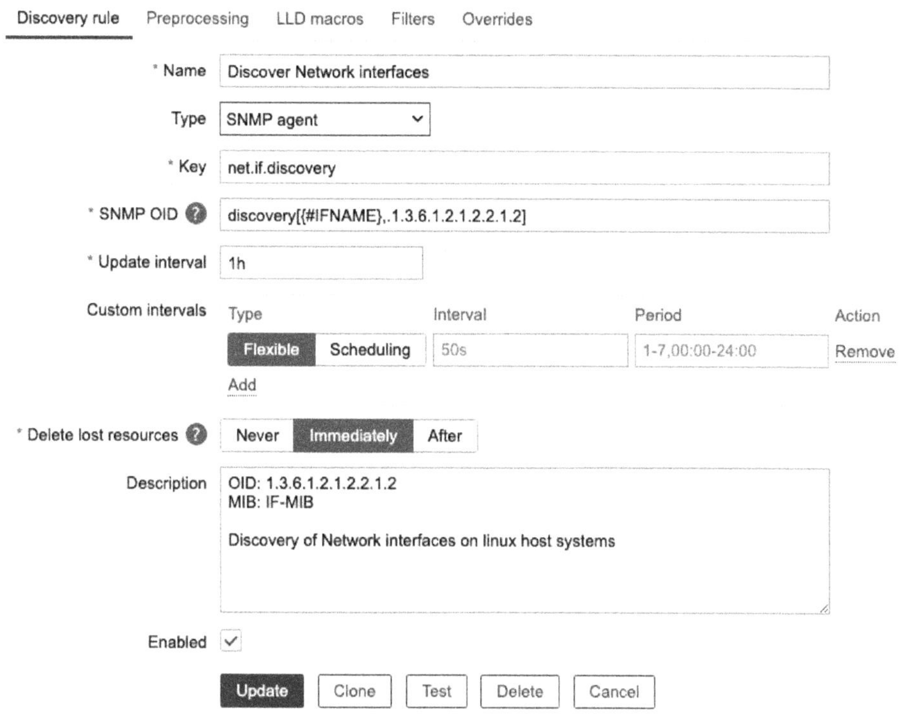

| Discovery rule | Preprocessing | LLD macros | Filters | Overrides |

* Name	Discover Network interfaces
Type	SNMP agent ⌄
* Key	net.if.discovery
* SNMP OID ❓	discovery[{#IFNAME},.1.3.6.1.2.1.2.2.1.2]
* Update interval	1h
Custom intervals	Type Interval Period Action
	[Flexible] Scheduling 50s 1-7,00:00-24:00 Remove
	Add
* Delete lost resources ❓	Never [Immediately] After
Description	OID: 1.3.6.1.2.1.2.2.1.2 MIB: IF-MIB Discovery of Network interfaces on linux host systems
Enabled	✓
	[Update] Clone Test Delete Cancel

Figure 5.30 – Zabbix LLD creation page filled with our information for network interface discovery

6. After clicking the **Add** button, we can navigate back to our template at **Data collection | Templates** and click **Custom Linux by SNMP**.

> **Important Note**
> We define **Delete lost resources** as **immediately**; we do this because this is a test template. This option is used by LLD to remove created resources (such as items and triggers) if they are no longer present on our monitored host. Using **immediately** can lead to lost data because we might get a resource back within a set amount of time, so make sure to adjust this value to your production environment's standard.

7. Go to **Discovery rules** and click our newly created rule, **Discover Network interfaces**.

8.   Now, go to **Item prototypes** and click **Create item prototype** in the top-right corner. This will open the **Item prototype** creation popup, as shown in the following screenshot:

Figure 5.31 – Zabbix LLD Item prototype creation page, empty

Here, we will create our first prototype for creating items from LLD. This means we have to fill it with the information we want our items to contain.

9. Let's start by filling in an item prototype for the interface operational status, like this:

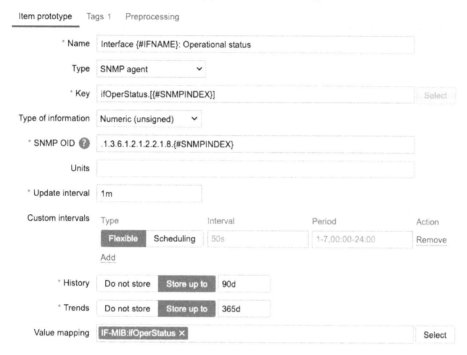

Figure 5.32 – Zabbix LLD item prototype creation page filled with
our information for the interface's operational status

10. On the **Tags** tab, make sure to also add a tag prototype as follows:

Figure 5.33 – Zabbix LLD item prototype tag creation tab

Tip

In the next step, we'll create an item that is very similar to the item we just created. It's super useful to use the **Clone** button instead of filling in the entire form from scratch again.

11. After clicking the **Add** button, let's repeat the process and also add the following item prototype:

Item prototype	Tags 1	Preprocessing		

* Name	Interface {#IFNAME}: Admin status
Type	SNMP agent ⌄
* Key	ifAdminStatus.[{#SNMPINDEX}] Select
Type of information	Numeric (unsigned) ⌄
* SNMP OID ❓	.1.3.6.1.2.1.2.2.1.7.{#SNMPINDEX}
Units	
* Update interval	1m

Custom intervals

Type	Interval	Period	Action
Flexible Scheduling	50s	1-7,00:00-24:00	Remove

Add

* History	Do not store **Store up to**	90d
* Trends	Do not store **Store up to**	365d
Value mapping	IF-MIB:ifOperStatus ×	Select

Figure 5.34 – Zabbix LLD item prototype creation page filled with
our information for the interface admin status

12. Do not forget the **Tags** tab:

Figure 5.35 – Zabbix LLD item 2 prototype tag creation tab

13. Now, move over to the **Trigger prototype** page, click the **Create trigger** prototype button in the top-right corner, and create the following trigger:

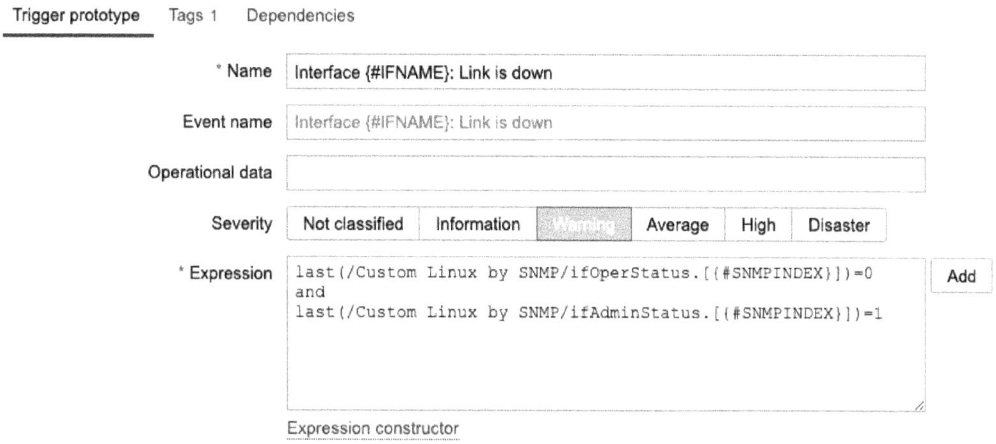

Figure 5.36 – Zabbix LLD trigger prototype creation page filled
with our information for interface link status

14. Last but not least, add the trigger tag:

Figure 5.37 – Zabbix LLD trigger prototype creation page Tags tab

How it works...

LLD is quite an extensive topic in Zabbix, but by following the steps in this recipe you should be able to apply what you learn here to almost every form of LLD there is to configure in Zabbix. First of all, let's look at how discovery works.

In the discovery rule, we just configured the following:

* Key   net.if.discovery

* SNMP OID   discovery[{#IFNAME},1.3.6.1.2.1.2.2.1.2]

Figure 5.38 – Zabbix LLD discovery key and OID for key net.if.discovery

> **Tip**
> Zabbix LLD works by using a specific JSON format. When creating discovery rules, we can
> always go to the discovery rules at the host level and use the **Test** button. This should then
> show us what the JSON Zabbix uses looks like.

What we are basically saying here is that for every interface after OID .1.3.6.1.2.1.2.2.1.2,
we fill in the {#IFNAME} LLD macro. In our case, we will end up with the following OIDs:

```
.1.3.6.1.2.1.2.2.1.2.1 = STRING: lo
.1.3.6.1.2.1.2.2.1.2.2 = STRING: ens192
```

So, we are saving these for use in our prototypes. Now, when we look at what we did to our **Operational
status** prototype, this all comes together:

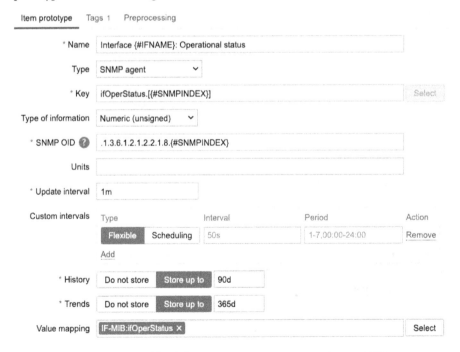

Figure 5.39 – Zabbix LLD item prototype name, type, key, and OID

We are telling our item prototype to create an item for every single {#IFNAME} value using the key defined plus the {#SNMPINDEX} LLD macro. SNMPINDEX is the last number of our SNMP poll. In this case, we would see the following:

```
.1.3.6.1.2.1.2.2.1.8.1 = INTEGER: up(1)
.1.3.6.1.2.1.2.2.1.8.2 = INTEGER: up(1)
```

For all the vendors in the world, there's a set of predefined SNMP rules they should adhere to. Our first interface entry when polling .1.3.6.1.2.1.2.2.1.2 was the .1 SNMPINDEX with the value lo. This means that when polling .1.3.6.1.2.1.2.2.1.8, the .1 SNMPINDEX here should still contain a value for lo.

Zabbix LLD will now create an item with the name Interface lo: Operational status, which will poll the SNMP OID:

```
.1.3.6.1.2.1.2.2.1.8.1 = INTEGER: up(1)
```

It will also create an item with the name Interface ens192: Operational status, which will poll the SNMP OID:

```
.1.3.6.1.2.1.2.2.1.8.2 = INTEGER: up(1)
```

The created items will then look like this:

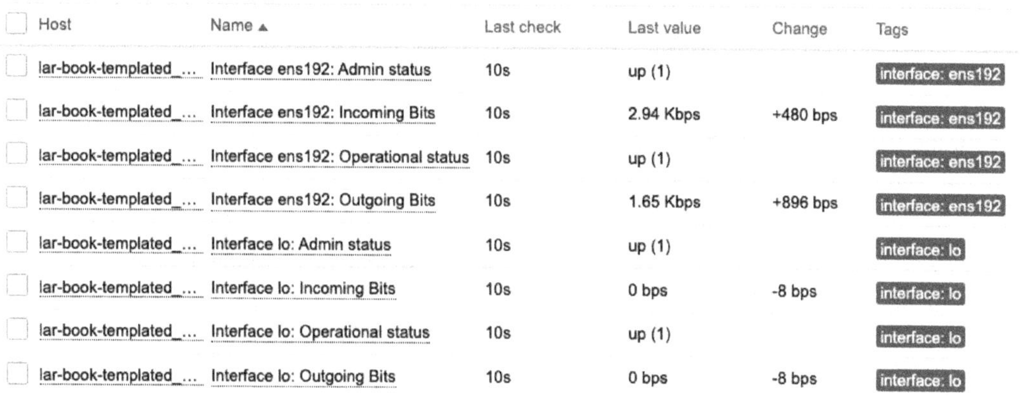

Host	Name ▲	Last check	Last value	Change	Tags
lar-book-templated_...	Interface ens192: Admin status	10s	up (1)		interface: ens192
lar-book-templated_...	Interface ens192: Incoming Bits	10s	2.94 Kbps	+480 bps	interface: ens192
lar-book-templated_...	Interface ens192: Operational status	10s	up (1)		interface: ens192
lar-book-templated_...	Interface ens192: Outgoing Bits	10s	1.65 Kbps	+896 bps	interface: ens192
lar-book-templated_...	Interface lo: Admin status	10s	up (1)		interface: lo
lar-book-templated_...	Interface lo: Incoming Bits	10s	0 bps	-8 bps	interface: lo
lar-book-templated_...	Interface lo: Operational status	10s	up (1)		interface: lo
lar-book-templated_...	Interface lo: Outgoing Bits	10s	0 bps	-8 bps	interface: lo

Figure 5.40 – Zabbix latest data screen for our SNMP-monitored host

Besides creating these LLD items, we also created an LLD trigger prototype. This works in the same manner as item prototypes. If we check our host triggers, we can see two created triggers:

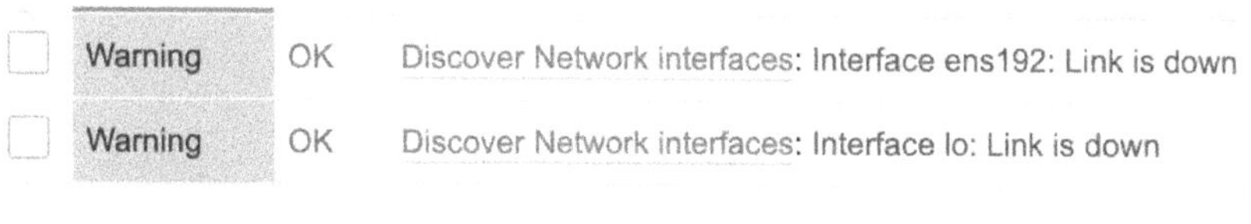

	Warning	OK	Discover Network interfaces: Interface ens192: Link is down
	Warning	OK	Discover Network interfaces: Interface lo: Link is down

Figure 5.41 – Our SNMP-monitored host triggers

These triggers have been created in the same manner as the items and are then filled with the correct items for triggering on:

Figure 5.42 – Our SNMP-monitored host trigger for ens192

We can see that for the interface operation status, we have an SNMPINDEX of 2, and we have the same for the Interface ens192: Admin status item as well. Our trigger will now trigger when the operation status is 0 (*down*) and our admin status is 1 (*up*).

A neat trigger, to make sure we only have a problem when the admin status is *up*; after all, we only want our interface down alert when we configure the interface to be admin *up*.

> **Tip**
>
> It's possible to use discovery filters to only add the interfaces that have admin status *up* to our monitoring. This way, we keep our required Zabbix server performance lower and our data cleaner. Consider using discovery filters for use cases such as this.

See also

Discovery is an extensive subject and takes a while to master. It's something that can be used like we did in this chapter with SNMP, but also with the Zabbix agent, and for a lot of other use cases. Once you start working with Zabbix discovery and you keep it structured, that's when you'll start building the best templates you've seen yet.

Check out the following link for the Zabbix LLD documentation:

```
https://www.zabbix.com/documentation/current/en/manual/discovery/
low_level_discovery
```

Nesting Zabbix templates

Using a simple template per device or group of devices is in most cases the best practice way to create Zabbix templates, but it isn't the only way. We can also use nested templates to break pieces of them apart and put them back together in the highest template in the hierarchy.

In this recipe, we'll go over how to configure this and why.

Getting ready

We are going to need our Zabbix server, our SNMP-monitored host, and the template we created in the previous recipe.

How to do it...

1. Let's start by navigating to our **Data collection** | **Templates** page and clicking the **Create template** button in the top-right corner.

2. We are going to create a new template for monitoring the uptime of our SNMP host. Input the following information:

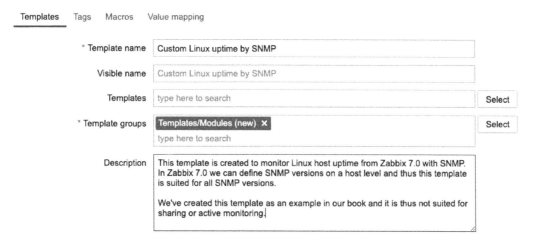

Figure 5.43 – New template creation page for uptime with SNMP

3. Next, we are going to click the **Add** button and click our **Custom Linux uptime by SNMP** template name. This will take us to the template editing screen.

4. Click on **Items** and **Create item** in the top-right corner. We will create an example item here, like this:

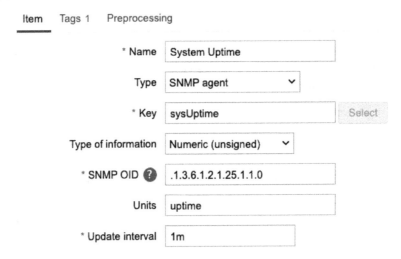

Figure 5.44 – A new item on the template creation page called System Uptime

5. Do not forget to add a tag, as shown in the screenshot, by going to the **Tags** tab:

Figure 5.45 – New item on the template creation page, System Uptime, Tags tab

6. Make sure to click the blue **Add** button to finish adding this item.

7. Now, let's navigate to our original template by going back to the **Data collection | Templates** page and clicking **Custom Linux by SNMP**.

8. On this page, link a template to the current template by adding it in the **Templates** entry field, like this:

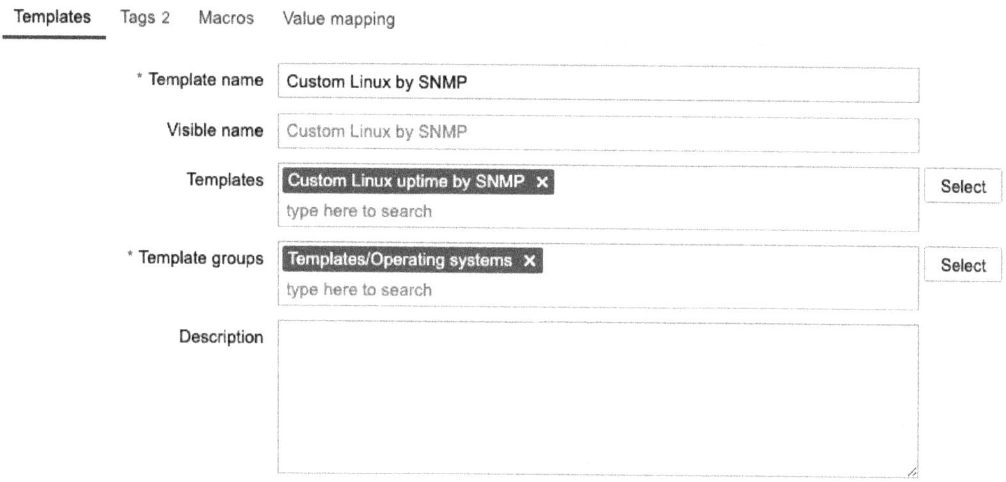

Templates   Tags 2   Macros   Value mapping

* Template name	Custom Linux by SNMP
Visible name	Custom Linux by SNMP
Templates	Custom Linux uptime by SNMP ✕ / type here to search	Select
* Template groups	Templates/Operating systems ✕ / type here to search	Select
Description	

Figure 5.46 – Template link page for master SNMP template

9. Click on the blue **Update** button to finish linking the template.

10. Last, but not least, navigate to **Data collection | Hosts**, click our `lar-book- templated_ snmp` SNMP-monitored host, and check out the **Items** page if the item is present:

	Name ▼	Triggers	Key	Interval	History	Trends	Type	Status	Tags
☐ ⋯	Custom Linux uptime by SNMP: System Uptime		sysUptime	1m	90d	365d	SNMP agent	Enabled	component: system
☐ ⋯	Custom Linux by SNMP: System hostname	Triggers 3	sysName	30m	90d		SNMP agent	Enabled	component: system

Figure 5.47 – Our Hosts | Items page for host lar-book-templated_snmp

The item is present, and it shows it's actually from another template. That's all there is to do to link a template—using these nested templates is easy to work with but harder to keep it structured. Let's see how this works.

How it works...

Nesting templates have a simple tree structure, just like this:

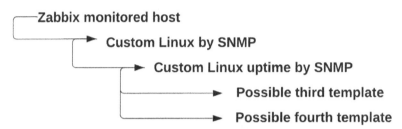

Figure 5.48 – Template nesting tree structure

So, we have our Zabbix-monitored host, which in turn has Custom Linux by SNMP linked as the only template. Now, because we have a nested template on Custom Linux by SNMP (which is, of course, Custom Linux uptime by SNMP), the items on that template will also be linked to our Zabbix-monitored host.

We can use this for a great deal of cases—one of my favorites is for networking equipment. If we have a Juniper EX (or Cisco Catalyst) and a Juniper QFX (or Cisco Nexus) series switch, both series switches use the same SNMP discovery for interfaces. So, we can create a template for interfaces and nest it in the main template of the EX or QFX series, which use different SNMP OIDs for other values.

This way, we don't have to write the same discovery rules, items, graphs, and everything else on a template a hundred times. We can simply do it once and nest the template neatly.

6
Visualizing Data, Inventory, and Reporting

When working with Zabbix, collected data must be put to good use. After all, the data is of no use if we don't have a place to easily access it. Zabbix already puts our data to good use with the **Latest data** page and with problems created from triggers, but we can also put our data to good use by building some stuff ourselves, such as graphs, maps, an inventory, and completely custom dashboards. We can even create reports from the dashboards and use built-in reports in the frontend.

After working through these recipes, you'll be able to set up the most important parts of Zabbix data visualization. You'll also be able to make good use of your inventory and reporting systems to get the most out of their useful features.

In this chapter, we'll cover the following recipes to show you how to achieve good results:

- Creating graphs to access visual data
- Creating maps to keep an eye on infrastructure
- Creating dashboards to get the right overview
- Templating dashboards to work at the host level
- Setting up Zabbix inventory
- Using the Zabbix Geomap widget
- Working through Zabbix reporting
- Setting up scheduled PDF reports
- Setting up improved business service monitoring

Technical requirements

For this chapter, we will need our Zabbix server, our **Simple Network Management Protocol** (SNMP)-monitored host from *Chapter 5*, *Building Your Own Structured Templates*. We'll be doing most of our work in the frontend of Zabbix, so have your mouse at the ready.

The code files for this chapter can be found in this book's GitHub repository: `https://github.com/PacktPublishing/Zabbix-7-IT-Infrastructure-Monitoring-Cookbook/tree/main/chapter06`.

Creating graphs to access visual data

Graphs in Zabbix are a powerful tool to show what's going on with your collected data. You might have already created some ad hoc graphs by using the **Latest data** page, but we can also create o predefined graphs. In this recipe, we will go over doing just that.

Getting ready

Make sure to get your Zabbix server ready, along with a Linux host that we can monitor (with SNMP). If you followed the recipes in *Chapter 5*, *Building Your Own Structured Templates*, you should already have a template.

Alternatively, you can download the templates available at `https://github.com/PacktPublishing/Zabbix-7-IT-Infrastructure-Monitoring-Cookbook/tree/main/chapter06`.

If you're using the downloaded templates, download and import **Custom Linux uptime by SNMP** first, then **Custom Linux by SNMP**. You can import a template by going to **Data collection | Templates** and clicking the blue **Import** button in the top-right corner.

Make sure you put the template on a host and monitor it.

How to do it...

Follow these steps:

1. Let's start by navigating to our templates by going to **Data collection | Templates** and selecting the template. For me, it is still called **Custom Linux by SNMP**.

2. Go to **Items** and create the following item on the template:

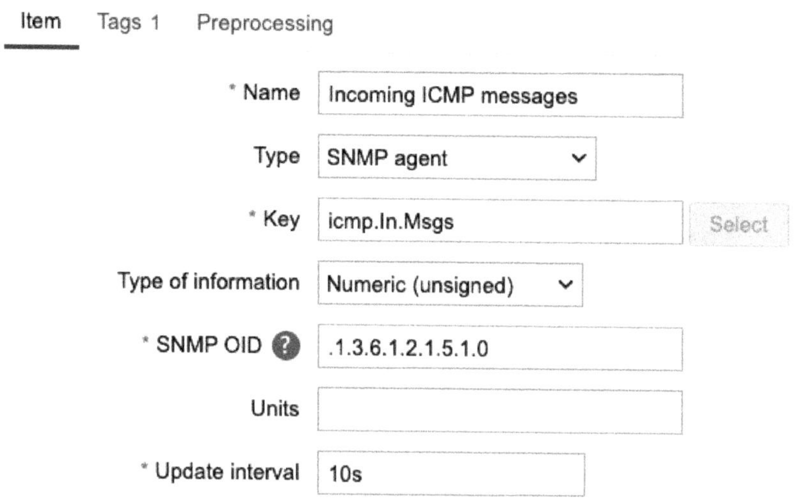
Item
Tags 1
Preprocessing
* Name
Incoming ICMP messages
Type
SNMP agent
* Key
icmp.In.Msgs
Select
Type of information
Numeric (unsigned)
* SNMP OID
.1.3.6.1.2.1.5.1.0
Units
* Update interval
10s

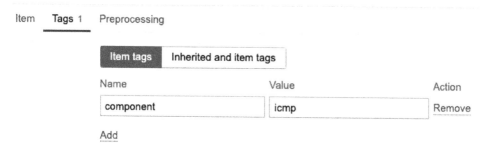

Figure 6.1 – ICMP item creation page

3. Make sure you go to the **Tags** tab and add a tag, as follows:

Item
Tags 1
Preprocessing
Item tags
Inherited and item tags
Name
Value
Action
component
icmp
Remove
Add

Figure 6.2 – ICMP item creation page – the Tags tab

4. Click the blue **Add** button to save this item.
5. Now, back at the template configuration page, go to **Graphs**. This is where we can see all of our configured graphs for this template; at the moment, there are none.

6. Click **Create graph** in the top-right corner. This will take you to the graph creation page:

Figure 6.3 – Graph creation page

This is where we can create graphs for standalone items. Let's create a graph to see our uptime.

7. Fill in the graph creation page with the following information:

Figure 6.4 – Graph creation page filled with our information

> **Tip**
>
> When working with graphs, it's a good idea to keep colorblind people in mind. Worldwide, about 8% of all males and 0.5% of all females are affected by this condition. There are great sources online that explain which colors to use for your production environment. You can find one such source here: `https://www.tableau.com/about/blog/2016/4/examining-data-viz-rules-dont-use-red-green-together-53463`.

8. Now, ping your SNMP-monitored host for a while. Do this from your Zabbix server **command-line interface (CLI)**:

    ```
    ping 10.16.16.153
    ```

9. Afterward, navigate to **Monitoring | Hosts** and click the **Graphs** button next to your host. In my case, the host is still called `lar-book-templated_snmp`.

10. This will immediately take us to an overview of graphs for this host, where we can see our new **Incoming ICMP messages** graph:

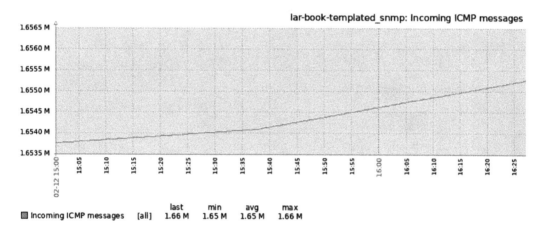

Figure 6.5 –Monitoring | Hosts graph page with our graph

We can also make graphs for discovery items; this is called a graph prototype. They work in about the same way as our item prototypes. Let's create one of these as well:

1. Navigate to **Data collection | Templates** and select our **Custom Linux by SNMP** template.

2. Go to **Discovery rules**. Then, for the **Discover Network Interfaces** discovery rule, click on **Item prototypes**. In the top-right corner, click **Create item prototype** and create the following item prototype:

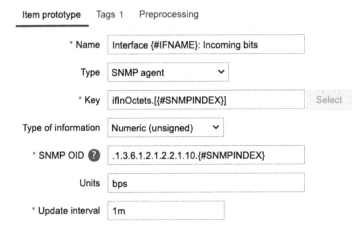

Figure 6.6 – Item prototype – the Incoming bits page filled with our information

3. Now, let's add a tag on the **Tags** tab:

Figure 6.7 – Item prototype Incoming bits – the Tags tab

4. Lastly, make sure you add the following on the **Preprocessing** page:

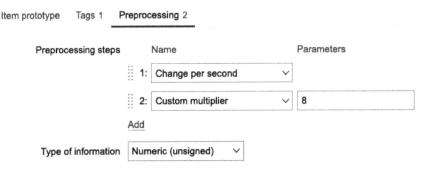

Figure 6.8 – Item prototype – the Incoming bits Preprocessing tab filled with our information

Important note

Preprocessing is quite an extensive topic. In short, the preprocessing in this step will ensure that our data is calculated at a change per second, with the mathematical formula *(value - prev_value)/ (time - prev_time)*, and that our data is multiplied by 8 so that it's changed from bytes to bits.

5. Click the blue **Add** button to finish creating this item prototype.

6. Now, back at our **Discover Network interfaces** discovery rule, click the **Graph prototypes** button.

7. In the top-right corner, click **Create graph prototype** and fill in the next page with the following information:

Figure 6.9 – Graph prototype – the Incoming bits page filled with our information

8. Now, if we go back to **Monitoring | Hosts** and click the **Graphs** button, we'll see two new graphs:

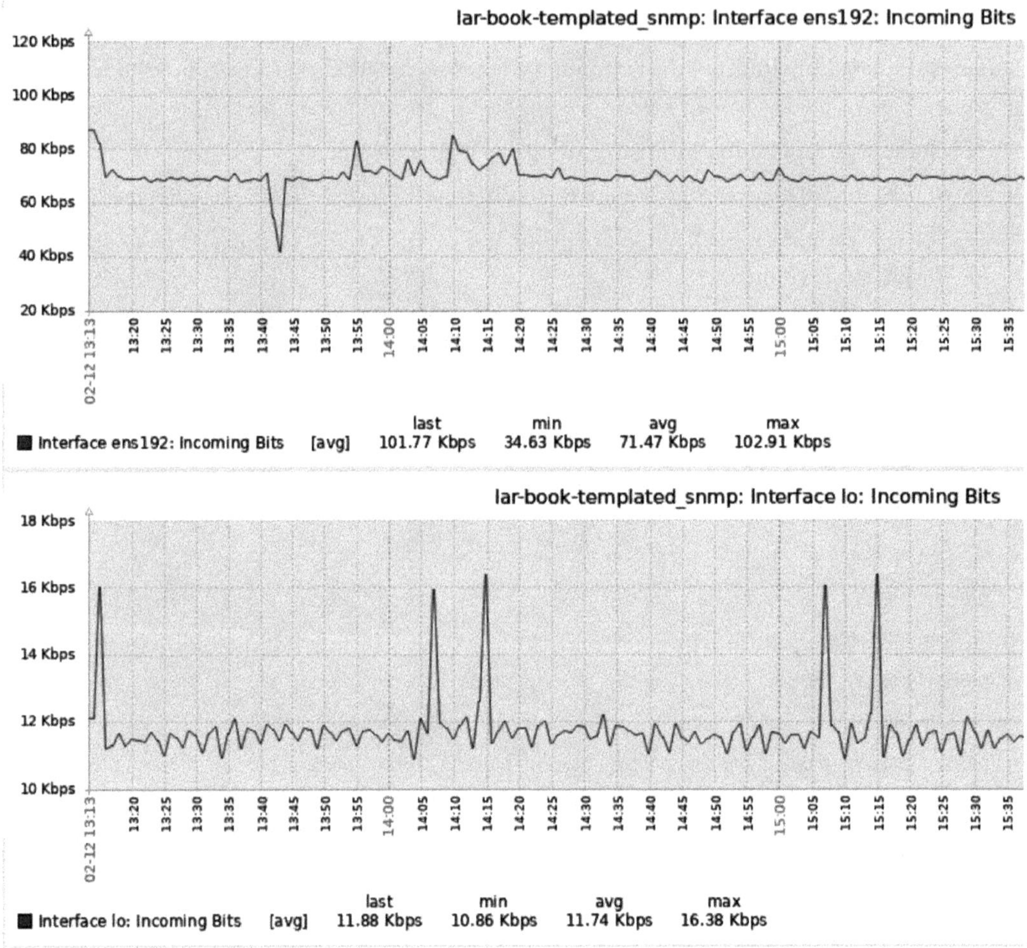

Figure 6.10 – Graphs page for our host

It might take some time for the graph to fill with data since we've only just added the item. Give it some time and you will start to see them fill up.

How it works...

Graphs work by putting your collected values in a visual form. We collect our data from our host – through SNMP, for example – and we put that data in our database. Our graphs, in turn, collect this data from the database and put it in this visual form. For humans, this is a lot better to read, and we can interpret the data easily.

The graph prototype works in almost the same way as our item prototype. For every discovered interface, we create a graph using a name containing the {#IFNAME} **low-level discovery (LLD)** macro. This way, we get a versatile structured environment because when a new interface is created (or deleted), a new graph is also created (or deleted).

Creating maps to keep an eye on infrastructure

Maps in Zabbix are a great way to get an overview of infrastructure. For instance, they're amazing for following traffic flows or seeing where something is going off in your environment. They're not only super-useful for network overviews but also for server management overviews, and even for a lot of cool customization.

Maps are super useful, and we use them a lot in environments we build. Since we love maps so much, we've also taken the liberty of opening a feature request suggesting some collected map improvements to make them even better: `https://support.zabbix.com/browse/ZBXNEXT-7680`.

Getting ready

We will need our Zabbix server, our SNMP-monitored host, and the templates from the previous recipe.

How to do it...

Follow these steps:

1. Let's start this recipe off by navigating to **Data collection | Templates** and selecting our **Custom Linux by SNMP** template.

2. Go to **Discovery rules** and then **Item prototypes**. Create the following item prototype by filling in the fields on the **Item prototype** creation page:

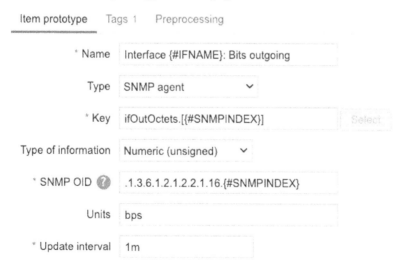

Figure 6.11 – Item prototype creation page

3. We'll also need to go to **Tags** to add a new tag, as shown in the following screenshot:

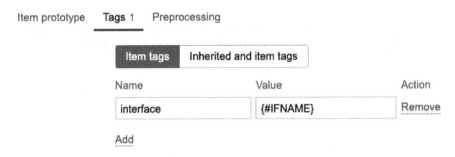

Figure 6.12 – Item prototype creation page – the Tags tab

4. Lastly, don't forget to add preprocessing by going to the **Preprocessing** tab:

Figure 6.13 – Item prototype – the Preprocessing tab

5. Click on the blue **Add** button to finish.

6. Next, navigate to **Monitoring | Maps**. There's already a default map here that's included in all Zabbix server installs called **Local network**. Feel free to check it out:

Figure 6.14 – The default local network map

7. There's not much to see here besides your local Zabbix server host and whether it is in a problem state or not. So, let's click on **All maps**.

8. We're going to create our own map, so click the **Create map** button in the top-right corner. Create the map by filling in the following fields:

Figure 6.15 – Map creation page

9. After clicking the blue **Add** button, the frontend will take you back to the **Map** overview page. Click the newly created `Templated SNMP host map` here.

10. Click **Edit map** in the top-right corner to start editing the map.

11. Now, what we want to do here is select the **Add** button next to **Map element**, which is in the horizontal menu at the top of the map. This will add the following element:

Y
50
5
New element

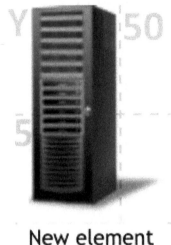

Figure 6.16 – The added element

12. Click the newly added element. This will open the following screen:

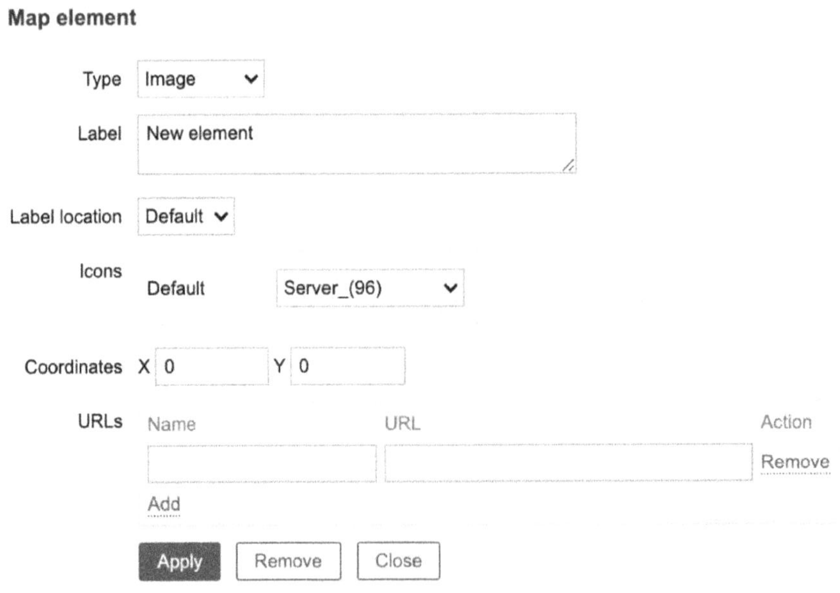
Map element
Type
Image
Label
New element
Label location
Default
Icons
Default
Server_(96)
Coordinates
X
0
Y
0
URLs
Name
URL
Action
Remove
Add
Apply
Remove
Close

Figure 6.17 – The Map element edit window

13. Here, we can fill out our host information. Let's add the following information to the fields:

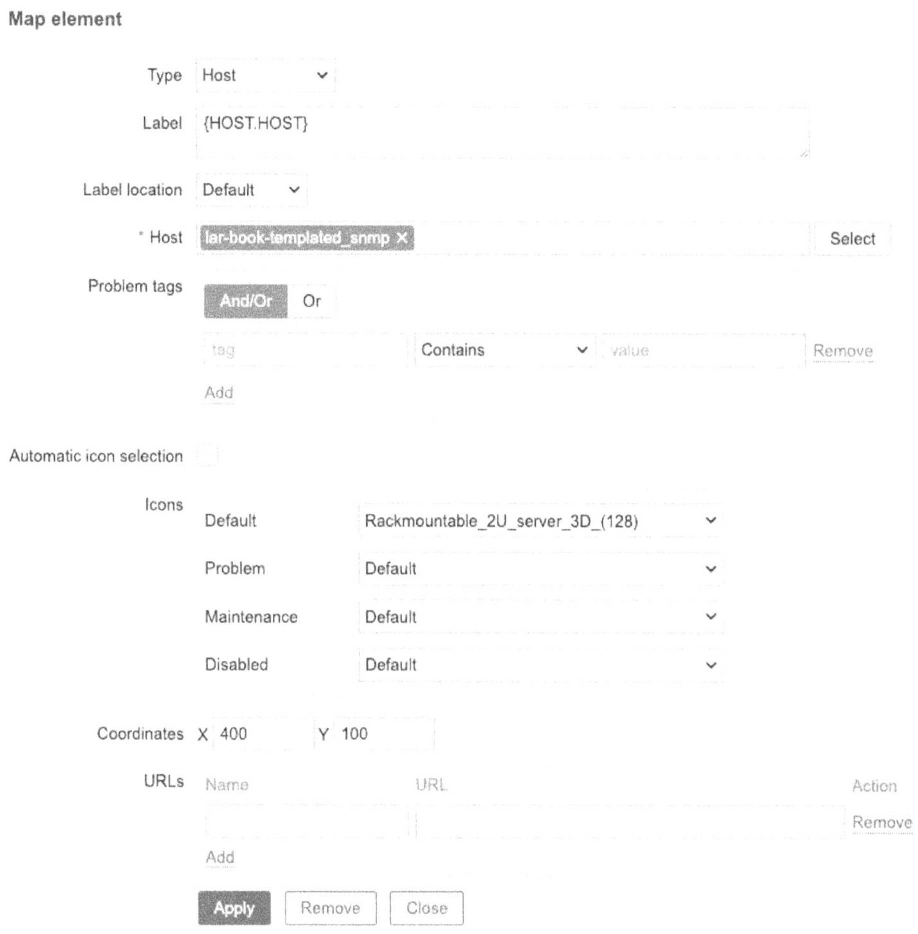

Map element

Type	Host ⌄
Label	{HOST.HOST}
Label location	Default ⌄
* Host	lar-book-templated_snmp ✕ Select
Problem tags	**And/Or** Or
	tag Contains ⌄ value Remove
	Add

Automatic icon selection ☐

Icons		
	Default	Rackmountable_2U_server_3D_(128) ⌄
	Problem	Default ⌄
	Maintenance	Default ⌄
	Disabled	Default ⌄

Coordinates	X 400 Y 100
URLs	Name URL Action
	Remove
	Add

[Apply] [Remove] [Close]

Figure 6.18 – Map element – lar-book-templated_snmp

Click **Apply** and move the element by dragging it to **X**:400 and **Y**:100 (see *Figure 6.20*).

14. Now, add another element by clicking the **Add** button next to **Map element**. Edit the new element and add the following information:

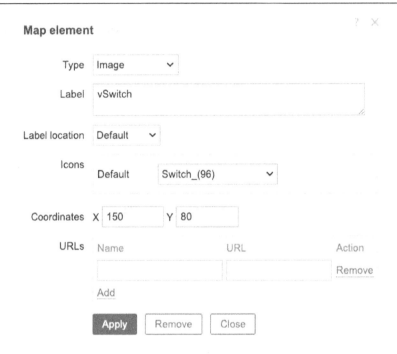

Figure 6.19 – Map element vSwitch edit window filled with information

After creating both elements, move the new switch element to **X:**150 and **Y:**80, as seen in *Figure 6.20*.

15. Now, select both elements by holding the *Ctrl* key (*command* on Mac) on your keyboard.

16. Then, click **Add** next to **Link** to add a link between the two elements. It should now look like this:

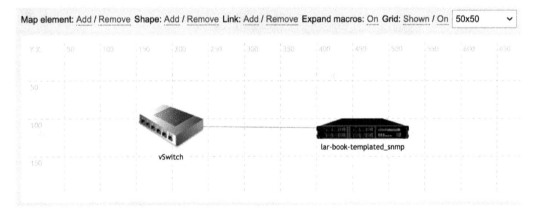

Figure 6.20 – Our newly created map

17. Edit the information for our server again after creating the link by clicking on our icon. Click on **Edit** next to the newly created link, as shown in the following screenshot:

Links	Element name	Link indicators	Action
	Switch_(96)		Edit

Figure 6.21 – The Edit link in the Map element edit window

18. Add the following information to the window:

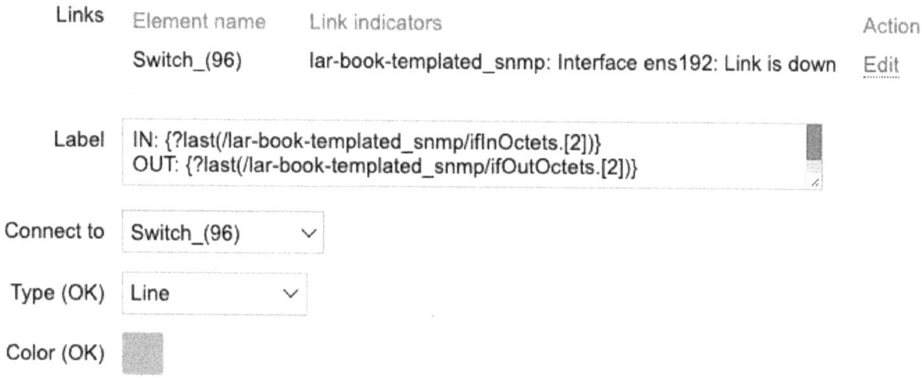

Links	Element name	Link indicators	Action
	Switch_(96)	lar-book-templated_snmp: Interface ens192: Link is down	Edit

Label	IN: {?last(/lar-book-templated_snmp/ifInOctets.[2])} OUT: {?last(/lar-book-templated_snmp/ifOutOctets.[2])}
Connect to	Switch_(96)
Type (OK)	Line
Color (OK)	

Figure 6.22 – Editing the link in the Map element edit window with our information

> **Important information**
>
> Make sure the hostname (in this example, `lar-book-templated_snmp`) is an exact match with the hostname in your Zabbix system. We're requesting data specifically from that host. We cannot omit the hostname or use macros such as `{HOST.HOST}` here since the link belongs to two hosts and Zabbix won't understand the context.

19. Let's also click **Add** in the **Link indicators** section and add the following trigger with the color red:

Link indicators	Trigger	Type	Colour	Action
	lar-book-templated_snmp: Interface ens192: Link is down	Line	DD0000	Remove
	Add			

Figure 6.23 – Link indicator filled with a trigger

20. Now, click **Apply** at the bottom of the window and then **Update** in the top-right corner of the page. That's our first map created!

How it works...

After creating and opening our map, we'll see the following:

Figure 6.24 – Our newly created map

The map shows our switch (which is not a monitored host at the moment) and our server (which is a monitored host). This means that when something is wrong with our server, the **OK** status will turn into a **PROBLEM** status on the map.

We can also see our configured label (see *Figure 6.24*), which shows us real-time information on traffic statistics. Now, when we break down the label, we get the following:

Figure 6.25 – Map label breakdown

We can pull real-time statistics into a label by defining which statistics we want to pull into the label between { }. In this case, we collect our values for interface traffic and put them directly in the label, creating a real-time traffic analysis map.

We also put a trigger on this link. The cool thing about putting triggers such as this on our map is that when our link goes down, we can see the following happen:

Figure 6.26 – Map showing problems

Traffic has stopped flowing because the link is now down, and our line has turned red. Also, our host is now showing a **PROBLEM** state under the hostname.

We can even create orange lines with triggers that state 50% traffic utilization like this and trace **Distributed Denial-of-Service (DDoS)** traffic through our network.

Creating dashboards to get the right overview

Now that we've created some graphs and a map, let's continue by not only visualizing our data but also getting the visualization in an overview. In this recipe, we're going to create a dashboard for our Linux-monitored hosts.

Getting ready

Make sure you followed the previous two recipes and that you have your Zabbix server ready. We'll be using our SNMP-monitored host from the previous recipe, as well as some items, triggers, and a map we created earlier.

Feel free to substitute any items you might not have with anything else from your environment. With dashboards, the most important thing is to play around with data, something you can do once you understand the concept of a widget.

How to do it...

Follow these steps:

1. From the sidebar, navigate to **Dashboards** and click **All dashboards** in the left corner of the page.

2. Now, click the **Create dashboard** button in the top-right corner and fill in your dashboard's name, like this:

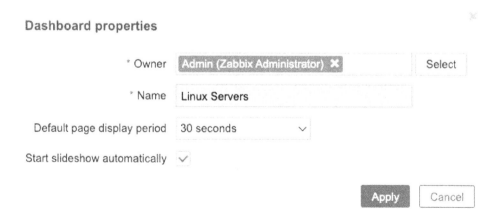

Figure 6.27 – The Dashboard properties area

Start slideshow automatically is enabled here, but it's only useful if you want to use this dashboard in a slideshow, such as on a big screen (TV) in a **Network Operating Center** (**NOC**) room. I always disable it for dashboards that are used on personal computers so that my pages don't jump around while troubleshooting.

3. I've also opened a case to ask Zabbix to change the default behavior: https://support.zabbix.com/browse/ZBXNEXT-7713.

> **Tip**
>
> Keeping Zabbix elements such as maps and dashboards that are meant to be used by entire departments owned by the Zabbix **Admin** user is a good idea. This way, they aren't dependent on a single user who might leave your environment at a later stage, which means we have to change the map owner once we want to delete their account. The elements can be owned by a disabled user as well. If you're not a super admin, don't forget to share the dashboard with yourself before changing the owner.

4. Now, click **Apply**; you'll be taken to your dashboard:

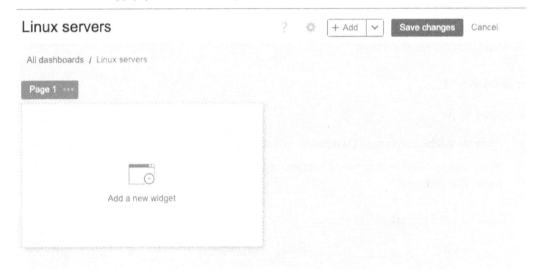

Figure 6.28 – New empty dashboard

After creating our dashboard, we will see that it is empty. We need to fill it with several widgets to create a good overview.

5. Let's start by adding a problem widget. Click + **Add** in the top-right corner. Add the following widget by filling out all the fields:

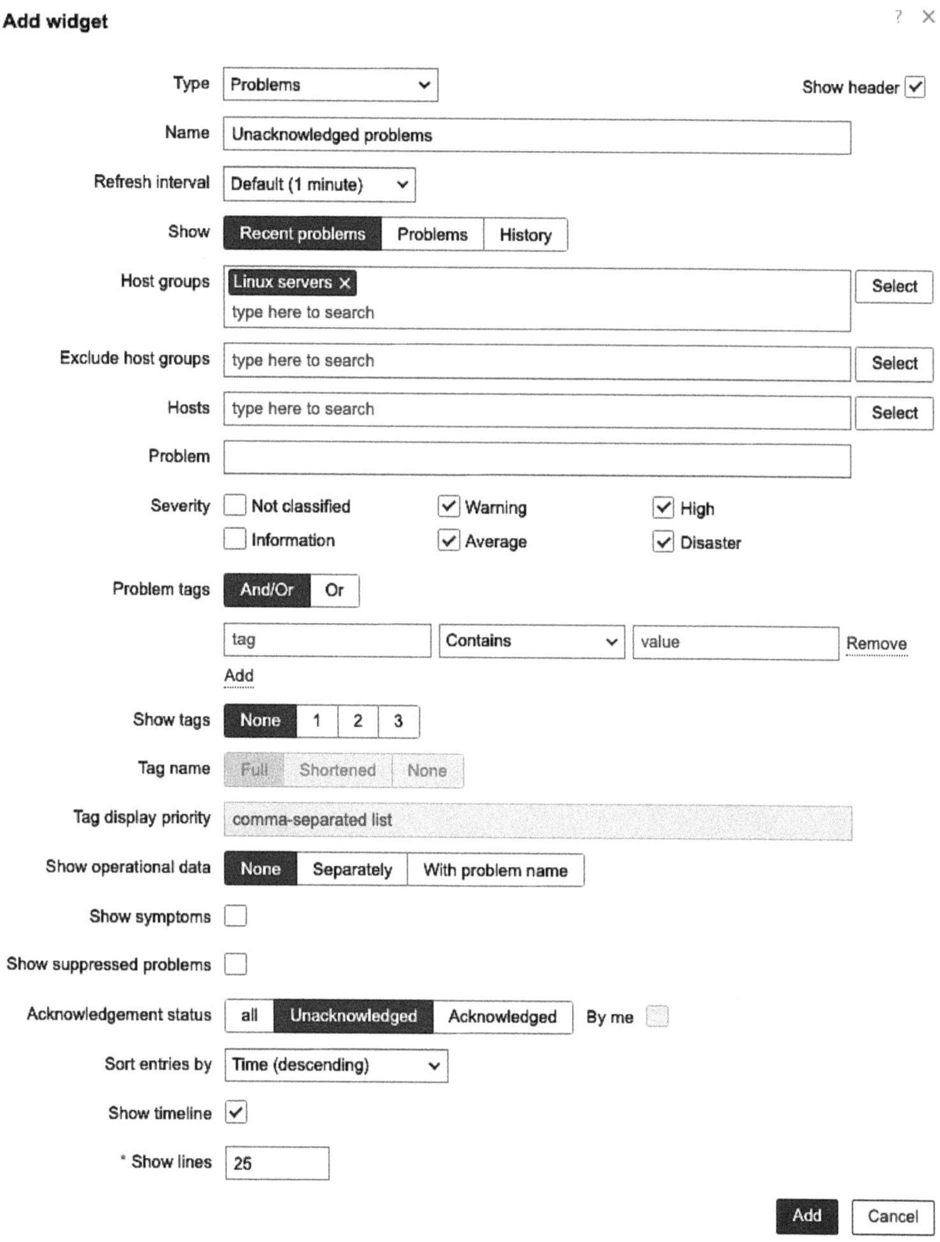

Figure 6.29 – New problem widget creation window

6. Click **Add**. By doing this, we'll have our first widget on our dashboard, displaying all **Unacknowledged problems**. It will only show them for the **Severity** warning and higher on all Linux servers:

Figure 6.30 – The Unacknowledged problems widget

7. Let's immediately add some more widgets, starting with our **Map** widget. Click **+ Add** in the top-right corner and add the following widget:

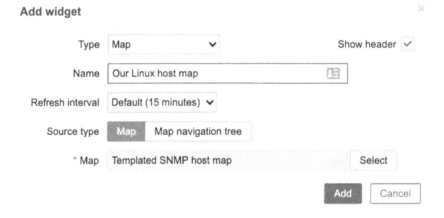

Figure 6.31 – The Add widget page

8. Also, add a **Graph** type widget by clicking **+ Add** in the top-right corner again. This one is a bit more difficult. Let's add our name first:

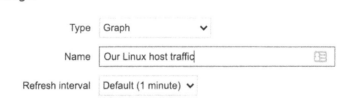

Figure 6.32 – New graph widget creation window

9. Then, we need to add our first **Data set**, like this:

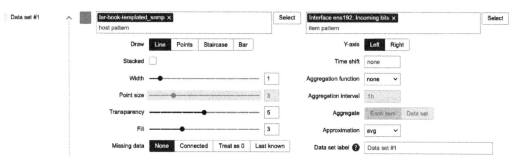

Figure 6.33 – Adding a dataset

10. Then, add a second one by clicking + **Add new data set** and adding the following:

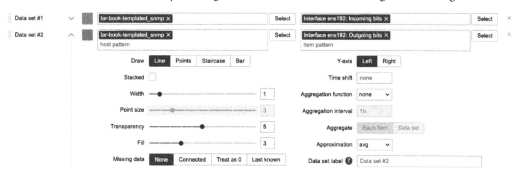

Figure 6.34 – Adding another dataset

11. We can then click **Add**, and our graph will be added to our dashboard.

12. Let's also add the **Item value** widget to the page. Click on + **Add** again. Then, set up the following widget:

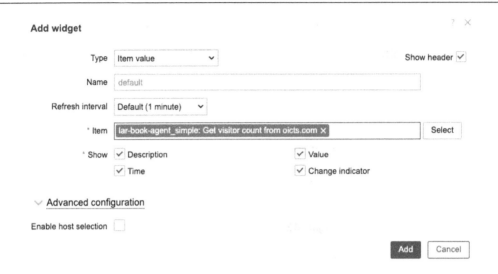

Figure 6.35 – Adding the Item value widget

If you're interested in changing exactly how this widget looks, be sure to use the **Advanced configuration** fields in this widget configuration screen.

13. Another widget we love is the very useful new **Top hosts** widget. Let's add it by using the **+ Add** button again.

14. On the widget configuration screen, set **Host groups** to Linux servers.

15. Next, click on the **Add** button next to **Columns** to add a column with information. Fill out the form like this:

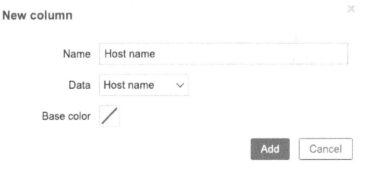

Figure 6.36 – Top hosts widget 1, column 1

16. Click on the **Add** button next to **Columns** again and add the following:

Figure 6.37 – Top hosts widget 1, column 2

17. The result should look like this:

Figure 6.38 – Top hosts widget 1

18. Don't forget to click the blue **Add** button at the bottom of the form to save your changes.

19. Let's create one more **Top hosts** widget by using the + **Add** button again.

20. Set **Host groups** to `Linux servers` again. Then, click on the **Add** button next to **Columns** again. Add the following:

Figure 6.39 – Top hosts widget 2, column 1

21. Click on the **Add** button next to **Columns** again and add the following:

Figure 6.40 – Top hosts widget 2, column 2

22. The result will look like this:

Figure 6.41 – Top hosts widget 2

23. Now, we can freely move around the widgets until we see this:

Figure 6.42 – Our dashboard filled with information

24. Now, let's add another page. Click on the drop-down arrow next to **+ Add** and click **Add page**. This will open the following popup, where we will add a new page called Host data:

Figure 6.43 – Adding a new Host data page

25. Click on **Apply** to add the new page. We can now immediately start adding additional widgets on the first page.

26. Let's click on **+ Add** again to add a new widget and select the widget called **Gauge** first.

27. Let's select an item from the default Zabbix server host. Click **Select** next to the **Item** field and search your host groups for the Zabbix server host. From the list, select the **Available memory in %** item.

28. Make sure the form looks like as follows:

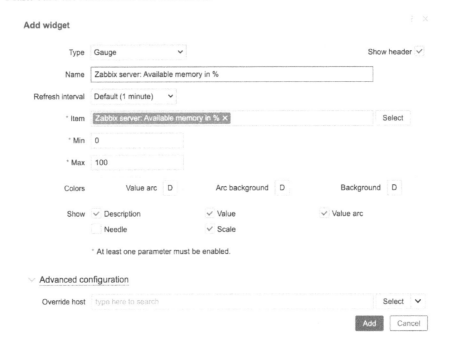

Figure 6.44 – Dashboard Gauge widget creation form

29. Click on the **Add** button at the bottom of this window to save the changes and add this widget.

30. Now, let's add our last widget by clicking + **Add** again. We'll add a new widget called **Pie chart**.

31. Set **Name** to CPU timings. For **host pattern**, set Zabbix server, and for **item pattern**, set CPU * time. It should look like this:

Figure 6.45 – Dashboard Pie chart widget creation form

32. On the **Legend** tab, set **Number of rows** to 3.

33. Now, click on the **Add** button at the bottom of the page to save the changes and add the widget.

34. Your new page should now look like this:

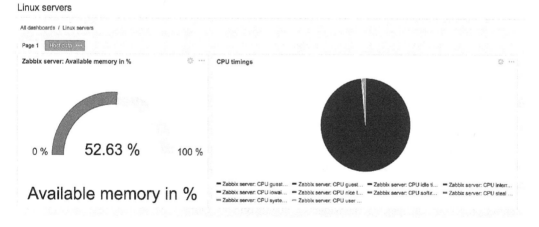

Figure 6.46 – Zabbix global dashboard second page

35. I don't like the name of our first page, so let's click on **Page 1** and then on the three dots next to **Page 1**. It should open a dropdown, where we can select **Properties**.

36. Here, give your page a new name. I will name it **Overview**. I always rename the first page to keep things organized and not have any default non-descriptive names in my dashboard (or anywhere for that matter). It should now show us two pages with different names:

Figure 6.47 – Zabbix Linux servers global dashboard page names

37. Click **Save changes** in the top-right corner and you're done.

How it works...

Creating dashboards is the best way to create overviews for quick access to data during troubleshooting, day-to-day problem monitoring, and – of course – for use with big TV walls. We've probably all seen the big operation centers with TVs displaying data. Zabbix is great for all these purposes and more, as you saw in this recipe.

There's more...

Zabbix has added a lot of new widgets in 7.0 as a big focus for them has been visualization. New widgets will be added in even newer versions of Zabbix as well, so keep your eye on the roadmap if you're still missing something: `https://www.zabbix.com/roadmap`.

We also haven't talked about every single new widget yet, so check out the *what's new* page here: `https://www.zabbix.com/whats_new_7_0`.

Templating dashboards to work at the host level

When Zabbix removed the screens functionality and replaced it with dashboards completely, a lot of people in the Zabbix community got very excited about using the newer widgets on their host-level dashboards. Unfortunately, development time is limited and at the time of writing, the feature is limited to only six widgets.

In Zabbix 7.0, this all changes. All the widgets available on your global dashboards are now available on host-level dashboards, making the whole feature incredibly useful.

Getting ready

Make sure you followed the first two recipes in this chapter and that you have your Zabbix server ready. We'll be using our SNMP-monitored host from the first two recipes, as well as our item, triggers, and map.

Alternatively, anyone can create host-level dashboards, so long as they have some data available on a host that has a template assigned. As such, feel free to apply your own datasets instead.

How to do it...

Let's start building some templated dashboards:

1. Navigate to **Data collection | Templates** and edit the Custom Linux by SNMP template by clicking on **Dashboards** next to it.

2. Now, in the top-right corner, click on **Create dashboard** to start creating your first templated host-level dashboard.

3. We will call this dashboard Host overview as it will contain several different statistics about our host:

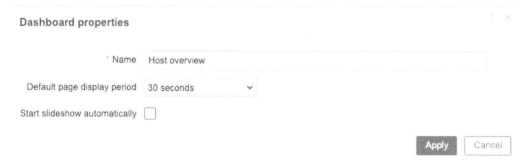

Figure 6.48 – Host dashboard creation

4. **Start slideshow automatically** is enabled here, but this is only useful if you want to use this dashboard in a slideshow, such as on a big screen (TV) in a NOC room. I always disable it for dashboards used on personal computers so that my pages don't jump around while troubleshooting.

5. Click **Apply** to add this new dashboard.

6. Now that the dashboard has been created, we can start adding some widgets. Add your first widget by clicking on + **Add** in the top-right corner.

7. Let's start with a simple **Item value** widget. Add the following information:

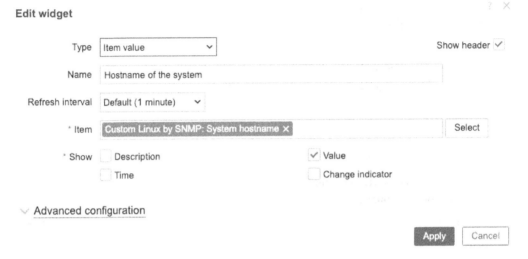

Figure 6.49 – Host dashboard, Item value widget

8. Make sure you apply your changes and add the widget by clicking + **Add**.

9. Next up, we'll add a graph widget. Click + **Add** to create another widget and add the following information:

Figure 6.50 – Host dashboard, Graph widget

10. Still on the widget creation form, we'll also create a dataset. In this case, we can keep it simple by just adding a single-item dataset:

Figure 6.51 – Host dashboard, Graph widget dataset

11. Click **Add** to save your changes and add the widget to the dashboard.

This dataset will simply show the number of incoming ICMP messages in a graph for us. However, it is also possible to use wildcards by using the asterisk symbol (*) to grab multiple items in a single-item pattern.

12. Let's also create a problem overview of the host here. Add the **Problems** widget, as follows:

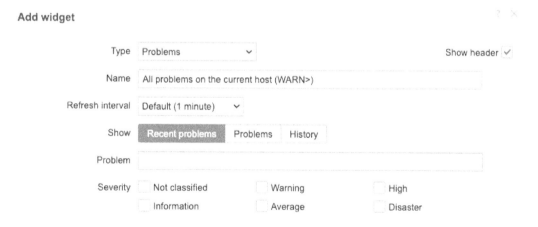

Figure 6.52 – Host dashboard, Problems widget

13. Click **Add** to save your changes and add the widget to the dashboard.

14. Now, make sure you arrange these widgets so that they form a proper dashboard:

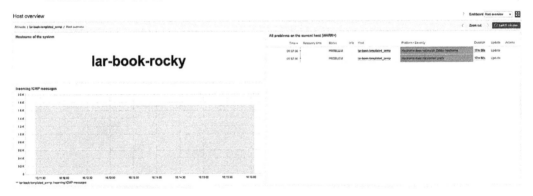

Figure 6.53 – Host dashboard, completed and arranged

At this point, you might be wondering why this dashboard is empty. Templates do not contain data, meaning we have to navigate to a host to look at the actual data.

15. Navigate to **Monitoring | Hosts** and go to the `lar-book-templated_snmp` host. Click **Dashboards** next to it:

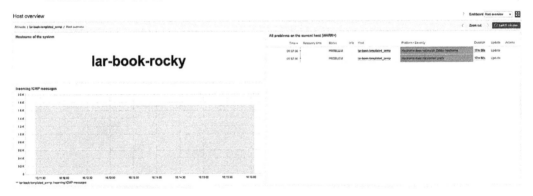

Figure 6.54 – Host dashboard for host lar-book-templated_snmp

As you can see, the dashboard now contains data about our specific host.

How it works...

Host-level dashboards are super useful since they completely rely on templates. We have to set them up at the template level; when we do, all of the hosts that are using that template will have the same dashboard with unique information shown in the widgets. This makes it possible for us to create hundreds or even thousands of dashboards by simply setting it up at the template level.

However, we have to keep in mind that host-level dashboards are different than the global dashboards we set up in the previous recipe. Not only do we access them differently, but they also operate at the host level. This means that they are mainly used to show information about a single host.

With Zabbix 7.0, we can add all of the widgets we have available on the global dashboard level, making host-level dashboards very useful. If we have several templates with one or even multiple dashboards, we can also use the pages at the top to easily navigate several dashboards:

Host dashboards

All hosts / **Zabbix server**   **Network interfaces**   System performance   Zabbix server health   Zabbix server processes

Figure 6.55 – Host dashboard selector for the Zabbix server host

Setting up Zabbix inventory

Zabbix inventory is a feature I love, but it hasn't had a lot of love from the Zabbix development team lately, even though it was on the roadmap for 7.0. Sorry – I still love you Zabbix developers, but if you're reading this, feel free to put some time into the feature!

Specifically, I'm talking about this old gem of a feature request: `https://support.zabbix.com/browse/ZBXNEXT-336`.

The inventory feature makes it possible for us to automatically put collected data in a visual **configuration management database** (**CMDB**) such as inventory in the Zabbix frontend. I think we've all seen CMDB inventory systems not be updated and thus missing out on data we need once we look into the system. Zabbix inventory fixes this by getting the data from the monitored systems. Let's get started.

Getting ready

Make sure that you log in to the Zabbix frontend and keep your SNMP-monitored host from the previous recipes ready.

How to do it...

Follow these steps:

1. Let's start by making sure our Zabbix server puts all of our hosts' inventory information into the fields. I like to do this by going to **Administration** | **General** and then selecting **Other** from the dropdown in the top-left corner.

2. We can then set our **Default host inventory mode** parameter to **Automatic**. Don't forget to click **Update**:

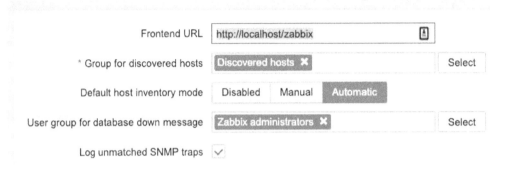

Other configuration parameters ⌄

Frontend URL	http://localhost/zabbix
* Group for discovered hosts	Discovered hosts ✖ Select
Default host inventory mode	Disabled Manual **Automatic**
User group for database down message	Zabbix administrators ✖ Select
Log unmatched SNMP traps	✓

Figure 6.56 – Administration | General | Other configuration parameters

3. Alternatively, we can do this at the host level. Go to **Configuration** | **Hosts** and select our `lar-book-templated_snmp` SNMP-monitored host.

4. Select **Inventory** and set it to **Automatic** here as well. As you may have noticed, the default only applies to newly created hosts from now on.

> **Important note**
> Changing the global setting does not apply it to all existing hosts, only to newly created hosts. It might be a good idea to run a **Mass update** operation for all the hosts or change the inventory mode manually, host by host.

5. Now, let's go to **Data collection** | **Templates** and select **Custom Linux by SNMP**.

6. Go to **Items** and edit **System hostname**. We have to change the **Populates host inventory field** setting, like this:

Populates host inventory field | Name   ⌄ |

Figure 6.57 – Edit item page

7. Click **Update** and navigate to **Inventory** | **Hosts**. You will see the following:

Host ▲	Group	Name
lar-book-templated_snmp	Linux servers	lar-book-agent-t

Figure 6.58 – Inventory | Hosts

How it works...

Zabbix inventory is simple but underdeveloped at the moment. It's not amazing to filter to a point where it shows exactly what we want to see, but it can be very useful nonetheless.

If you're working with a lot of equipment, such as in an MSP environment, it can become overwhelming to log in to every device and get the serial number by hand. If you poll the serial number and populate the **inventory** field, you suddenly have an active list of up-to-date serial numbers.

The same works with anything from hardware information to software versions. We could get the active operating system versions from devices and generate an extensive list of all our operating system versions, which is very useful if you ever have to patch something, for example.

Use Zabbix inventory wisely when creating items, and set the population to **Automatic**, as we did in this chapter – you'll never have to think too much about the feature. You configure it almost automatically this way and have nice lists waiting for you when you need them.

Using the Zabbix Geomap widget

Now that we've seen how to create dashboards, let's set up another dashboard. We'll use this one to create a full-fledged geographical overview of some of our hosts in Zabbix. We'll do this by using the Zabbix inventory functionality we have just learned how to use.

Getting ready

All we need for this recipe is our Zabbix setup with access to the frontend. It is also smart to follow the previous two recipes about dashboards and inventory. If you haven't followed those yet, it is recommended that you follow them first.

How to do it...

Using the Zabbix Geomap functionality is quite easy – we simply need to use our Zabbix inventory on our hosts in combination with a dashboard widget:

1. First, let's navigate to our **Data collection** | **Hosts** page and edit one of our hosts. I'll be using the `lar-book-templated_snmp` host, but it doesn't matter which host you use, so long as it is in the `Linux servers` host group.

2. Go to the **Inventory** tab and make sure that it is set to **Manual** or **Automatic**:

Figure 6.59 – Inventory mode selector on the Zabbix host Inventory tab

3. Now, in the **Location latitude** and **Location longitude** fields, fill in the following:

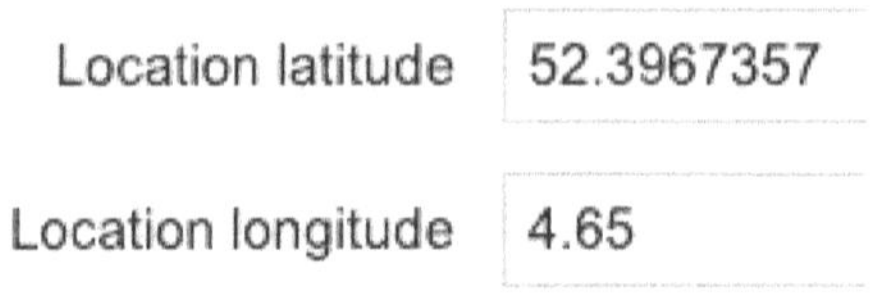

Location latitude   52.3967357

Location longitude   4.65

Figure 6.60 – Inventory tab fields on a Zabbix host

4. Click on the blue **Update** button to save these changes.

5. Back at **Data collection | Hosts**, let's do the same thing for another host. I'll use `lar-book-agent_simple`.

6. Go to the **Inventory** tab and fill in the **Location latitude** and **Location longitude** fields again:

Location latitude   56.9539871637491

Location longitude   24.2207342374544

Figure 6.61 – Inventory tab fields on another Zabbix host

7. Click on the blue **Update** button to save these changes.

8. Now, let's go to **Dashboards** and, at **All dashboards**, create a new dashboard or use our existing **Linux servers** dashboard. This is what I'll do.

9. Click on the blue **Edit dashboard** button in the top-right corner and use the **Add** button dropdown to click on **Add page**:

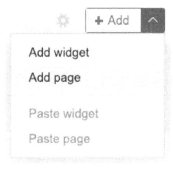

Figure 6.62 – Existing dashboard Add page button

10. We'll add the following new page:

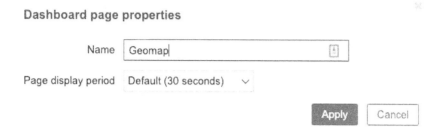

Figure 6.63 – Dashboard page properties

11. Click **Apply** to add this new page.

12. We can now add our **Geomap** widget simply by clicking anywhere on the page. Fill it in as follows:

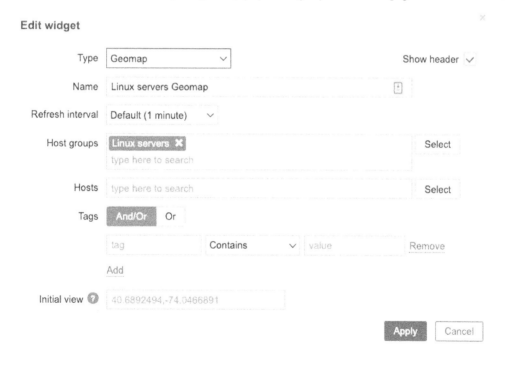

Figure 6.64 – Zabbix Geomap widget properties

13. Click **Apply** to save the widget configuration.

14. We can now click on the blue **Save changes** button in the top-right corner to save our dashboard changes.

15. This will take us back to our dashboard, where we can click the **Geomap** page of this dashboard:

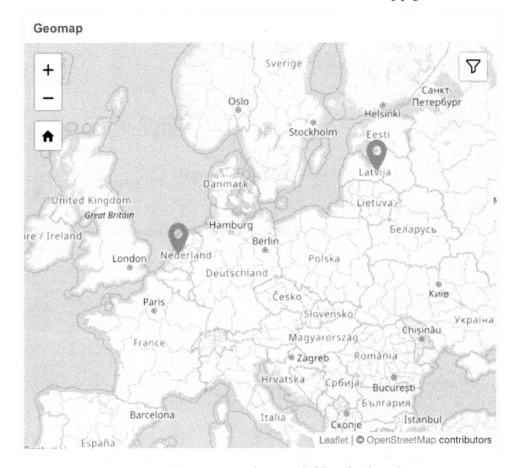

Figure 6.65 – Zabbix Geomap widget on a dashboard with two hosts

We now have a functioning Geomap in our Zabbix dashboard that uses the latitude and longitude that are available in our Zabbix inventory.

How it works...

Instead of creating an entirely new **Monitoring | Geomap** page, Zabbix has chosen to include this new feature via a widget, giving us the option to create even more advanced dashboards. It's important to note here that Zabbix also chose to use existing inventory data. Because it is possible to automatically fill in inventory data, as we saw in the *Setting up Zabbix inventory* recipe, we can also automate our Geomap widget content.

So, whether you go the manual route or the automatic route, the Geomap widget is a valuable extension of our dashboard. In general, Zabbix is extending the dashboard functionality quite a bit by including a bunch of new widgets in Zabbix 7.

We will set up our Zabbix automatic reporting in this chapter, which will also use the dashboard functionality. f you'd like, you can combine a Geomap widget with your automatic report to send out a geographical report. The key takeaway here is that Zabbix is building interoperability between components and giving us flexibility in the way we want to use a new widget like this.

When working with the initial release of the Geomap widget, some people asked us if it was possible to change the kind of map that's used by the Geomap widget. If we navigate to **Administration | General | Geographical maps**, we can choose several built-in map providers:

Geographical maps ⌄

* Tile provider	OpenStreetMap Mapnik ⌄
* Tile URL ?	OpenStreetMap Mapnik
	OpenTopoMap
* Max zoom level ?	Stamen Toner Lite
	Stamen Terrain
	USGS US Topo
	USGS US Imagery
	Other

'g/{z}/{x}/{y}.png

Figure 6.66 – Administration | General | Geographical maps

If that isn't enough, it is also possible to add a custom map provider using the **Other** option under **Tile provider**. Simply fill in the form and you're all set:

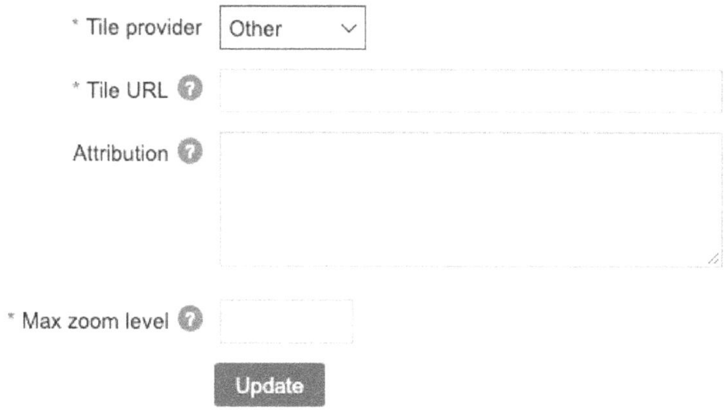

* Tile provider	Other ⌄
* Tile URL ?	
Attribution ?	
* Max zoom level ?	
	Update

Figure 6.67 – Administration | General | Geographical maps – Other

As you can see, a lot of possibilities have been added through this single widget. One of the most requested features from the Zabbix community, we can now set it up and use it in the latest Zabbix releases.

Working through Zabbix reporting

Zabbix reporting got some well-deserved love from Zabbix development, especially concerning getting reports out of the system and improving the audit log. First, let's take a look at some powerful features to show you exactly what's going on with your statistics right from the Zabbix frontend. Then, in the next recipe, we'll take a look at how to create automatic PDF reports, a new and much-anticipated feature.

Getting ready

For this recipe, all you'll need is the Zabbix frontend and a monitored host. I'll be using the SNMP-monitored host from the previous recipes.

How to do it...

There isn't anything to configure really as reporting is present in Zabbix from the start. So, let's dive into what each page of reporting offers us.

System information

If you navigate to **Reports | System information**, you will find the following table:

Parameter	Value	Details
Zabbix server is running	Yes	localhost:10051
Zabbix server version	7.0.0	
Zabbix frontend version	7.0.0	
Software update last checked	2024-06-01	
Latest release		
Number of hosts (enabled/disabled)	21	16 / 5
Number of templates	284	
Number of items (enabled/disabled/not supported)	1023	565 / 241 / 217
Number of triggers (enabled/disabled [problem/ok])	411	373 / 38 [12 / 361]
Number of users (online)	10	1
Required server performance, new values per second	8.64	
High availability cluster	Disabled	

Figure 6.68 – Reports | System information

You might have seen this table before as it can also be configured as a dashboard widget. This page gives us all of the information we need about our Zabbix server, such as the following:

- **Zabbix server is running**: This informs us whether the Zabbix server backend is running and where it is running. In this case, it's running, and it's running on `localhost:10051`.

- **Zabbix server version**: The version of the Zabbix server daemon installed on our instance.

- **Zabbix frontend version**: The version of the Zabbix frontend currently running on our web server.

- **Software update last checked**: Details when Zabbix last checked for a new available version.

- **Latest release**: This shows us if there is a new version of Zabbix available.

- **Number of hosts**: This will detail the number of hosts enabled (`16`) and the number of hosts disabled (`5`). It gives us a quick overview of our Zabbix server host information.

- **Number of templates**: A simple counter showing the number of templates currently available on this Zabbix system.

- **Number of items**: Here, we can see details of our Zabbix server's items – in this case, enabled (`565`), disabled (`241`), and not supported (`217`).

- **Number of triggers**: This details the number of triggers we have. We can see how many are enabled (`373`) and disabled (`38`), but also how many are in a problem state (`12`) and how many are in an OK state (`361`).

- **Number of users (online)**: The first value details the total number of users. The second value details the number of users currently logged in to the Zabbix frontend.

- **Required server performance, new values per second**: Perhaps I'm introducing you to a completely new concept here, which is **new values per second** (**NVPS**). A server receives or requests values through items and writes this to our Zabbix database. The information detailed here shows the estimated number of NVPS received by the Zabbix server.

You might also see two additional values here, depending on your setup:

- **Database name**: If you see the name of your database with the value of your version, it might indicate you are running a non-supported database version. You could see a message like `Warning! Unsupported <DATABASE NAME> database server version. Should be at least <DATABASE VERSION>`.

- **High availability cluster**: If you are running a Zabbix server high availability cluster, you will see if it is enabled here and what the failover delay is. Additionally, the **Reports | System information** page will display additional high-availability information.

Availability report

Navigating to **Reports | Availability report** will give us some useful information about how long a trigger has been in a **Problems** state versus an **Ok** state for a certain period:

Figure 6.69 – Reports | Availability report

Looking at one of our hosts, we can see that in the last 30 days, the **Zabbix agent is not available (for 3m)** trigger has been in a **Problems** state for **10.0000%** of the time. This might be useful for us to know so that we can determine how often a certain problem arises.

Trigger top 100

Upon navigating to **Reports | Trigger top 100**, we will find the top 100 triggers that have been firing in a certain amount of time:

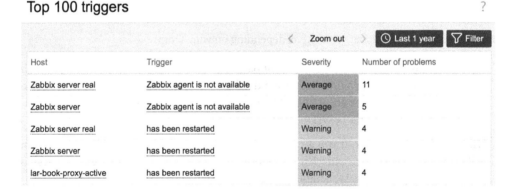

Figure 6.70 – Reports | Trigger top 100

For my Zabbix server, the busiest trigger was a **Zabbix agent is not available** trigger on a server. It's a useful page to see what we are putting most of our time into, problem-wise.

Audit

The audit log, which is a handy addition to Zabbix, can be found by going to **Reports | Audit**:

Time	User	IP	Resource	ID	Action	Recordset ID	Details
2024-06-26 09:23:27	Admin	192.168.2.254	Trigger	24020	Update	clxvie5270000hd3yq07lxxwl	Description: Zabbix agent is not available
							trigger.expression: max(/lar-book-agent_passive/zabbix[host,agent,available],30s)=0 => last(/lar-book-agent_passive/zabbix[host,agent,available])=0
2024-06-26 09:22:48	Admin	192.168.2.254	Trigger	24020	Update	clxvidbfh0000hh3yhtmda8ug	Description: Zabbix agent is not available
							trigger.expression: max(/lar-book-agent_passive/zabbix[host,agent,available],{$AGENT.TIMEOUT})=0 => max(/lar-book-agent_passive/zabbix[host,agent,available],30s)=0
2024-06-26 09:21:49	Admin	192.168.2.254	LLD rule	46900	Delete	clxvic1e50000hq3yuo0jpxs4	Description: Mounted filesystem discovery

Figure 6.71 – Reports | Audit

Here, we can see which user has done what on our Zabbix server – identifying a culprit for something that shouldn't have been done, for instance.

Action log

When we go to **Reports | Action log**, we land on a page that shows which actions have been fired. If you've configured **Actions**, then you can get a list here, like this:

Action log

			Zoom out >	Last 30 days	Filter

Time	Action	Type	Recipient	Message	Status	Info
2020-09-07 14:27:42	Action to notify our book reader of a problem		Admin (Zabbix Administrator)	**Subject:**	Failed	ℹ
				Message:		
2020-09-05 12:30:12	Action to notify our book reader of a problem		Admin (Zabbix Administrator)	**Subject:**	Failed	ℹ
				Message:		

Figure 6.72 – Reports | Action log

If you're not sure if your action succeeded, then look at this list. It is very useful to troubleshoot your actions to a point where you get them up and running as you want.

When you hover over the **Info** box, you also get to see what went wrong. For example, for the **Failed** items on my Zabbix instance, I must define the appropriate media type for the **Admin** user:

Recipient	Message	Status	Info
Admin (Zabbix Administrator)	**Subject:**	Failed	ℹ
	Message:	No media defined for user.	

Figure 6.73 – Reports | Action log – Info

Notifications

Last, but not least, navigating to **Reports** | **Notifications** will show us the number of notifications that have been sent to a certain user over a certain period:

Notifications

From	Till	MS Teams	brian@oicts.nl (Brian OICTS SAML)	bvbaekel (bvbaekel Administrator)
2021-12-27 00:00	2022-01-03 00:00	10		30
2022-01-03 00:00	2022-01-10 00:00	10		30
2022-01-10 00:00	2022-01-17 00:00	15		45
2022-01-17 00:00	2022-01-24 00:00	17		51
2022-01-24 00:00	2022-01-31 00:00	22		66
2022-01-31 00:00	2022-02-07 00:00	28		96
2022-02-07 00:00	2022-02-14 00:00	85		258
2022-02-14 00:00	2022-02-21 00:00	42		129
2022-02-21 00:00	2022-02-28 00:00	38		123
2022-02-28 00:00	2022-03-07 00:00	34		102

Figure 6.74 – Reports | Notifications

In my case, 50 notifications have been sent to the **Admin** user this year, and 0 to other users.

Setting up scheduled PDF reports

A much-wanted feature was added in Zabbix 5.4: sending automatic PDF reports through email. Let me start by stating that this implementation might not fully cover every Zabbix user's situation yet. What this feature does is take a screenshot of any Zabbix dashboard and send it through email. It's not just a screenshot, though – data is converted into text in the PDF file and the resolution is very high. It's the first setup from the Zabbix developers and I think we should appreciate it for what it is.

On top of that, it is a very flexible way of implementing this as we can choose any kind of widget available, along with its filters, and send it in an automatic report. On top of that, it gives the Zabbix development team the flexibility to add new widgets on the fly that immediately work with your PDF reports.

Getting ready

We will need an existing Zabbix installation with access to the frontend and the CLI. You can use the server we have been using throughout this book for this or you can use your own installation.

In the case of a multi-host setup, the easiest method is to install this where the Zabbix server is also running, but it is possible to run this on any host. In this example, we've used our single-host installation.

You will also need to set up a user with an email media type.

How to do it...

To get started with Zabbix scheduled reports, we need to install some things on our Zabbix server:

1. Let's log in to our Zabbix server CLI and execute the following command to install the Google Chrome browser.

   On RHEL-based systems, run the following command:

   ```
   vim /etc/yum.repos.d/google-chrome.repo
   ```

   Then, add the following to the file:

   ```
   [google-chrome]
   name=google-chrome
   baseurl=http://dl.google.com/linux/chrome/rpm/stable/$basearch
   enabled=1
   gpgcheck=1
   gpgkey=https://dl-ssl.google.com/linux/linux_signing_key.pub
   ```

Finally, install Google Chrome:

```
dnf install -y google-chrome-stable
```

On Ubuntu systems, run the following command:

```
wget -q -O - https://dl-ssl.google.com/linux/linux_signing_key.
pub | sudo apt-key add -
sudo sh -c 'echo "deb http://dl.google.com/linux/chrome/deb/
stable main" >> /etc/apt/sources.list.d/google.list'
sudo apt update
sudo apt-get install google-chrome-stable
```

2. Now, let's install our required Zabbix web services package with the following commands.

    Here's the command for RHEL-based systems:

    ```
    dnf install zabbix-web-service
    ```

    Here's the command for Ubuntu systems:

    ```
    apt install install zabbix-web-service
    ```

3. Now, let's edit our new Zabbix web service configuration file:

    ```
    vim /etc/zabbix/zabbix_web_service.conf
    ```

4. We can find a bunch of Zabbix web-service-specific parameters here, including encryption. Make sure the following line is set up to match your Zabbix server(s) IP(s):

    ```
    AllowedIP=127.0.0.1,::1
    ```

5. Now, let's edit our Zabbix server configuration file:

    ```
    vim /etc/zabbix/zabbix_server.conf
    ```

6. Edit the `WebServiceURL` parameter so that it matches your Zabbix web service IP and `StartReportWriters` to make sure we have a reporting subprocess:

    ```
    WebServiceURL=https://localhost:10053/report
    StartReportWriters=3
    ```

Important note

For scheduled reporting to work, you will need to set up SSL encryption for your Zabbix frontend; we recommend using Let's Encrypt. Alternatively, set the `IgnoreURLCertErrors=1` parameter in `/etc/zabbix/zabbix_web_service.conf`.

---

7. That's it for the CLI part. Let's log in to our frontend and navigate to **Administration | General | Other**.

8. Make sure you fill out the **Frontend URL** parameter on this page with your frontend URL, like this:

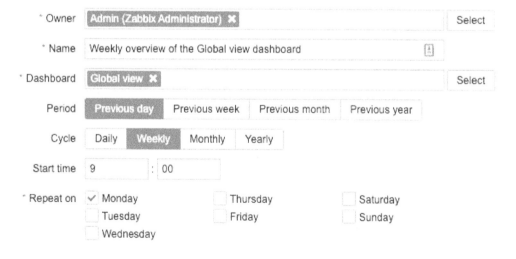

Figure 6.75 – Administration | General | Others with Frontend URL filled in

9. Click the blue **Update** button at the bottom of the page and navigate to **Reports | Scheduled reports**.

10. Now, we are on the page where we can set up and maintain our scheduled reports. So, let's create a new one using the blue **Create report** button in the top-right corner.

11. This will take us to a new page where we can set up a new report. Let's set up a weekly report using our existing dashboard's **Global view**. First, we'll name this report Weekly overview of the Global view dashboard.

12. Select the dashboard's **Global view** by clicking the **Select** button next to **Dashboard**.

13. Set **Cycle** to **Weekly** with a start time of 9 : 00 and set **Repeat on** to **Monday**:

Figure 6.76 – Reports | Scheduled reports – creating a new report, part I

14. Also, make sure to fill in **Subject** and **Message** and set up **Subscriptions** so that they match users that have media with the type of email set on their user profile:

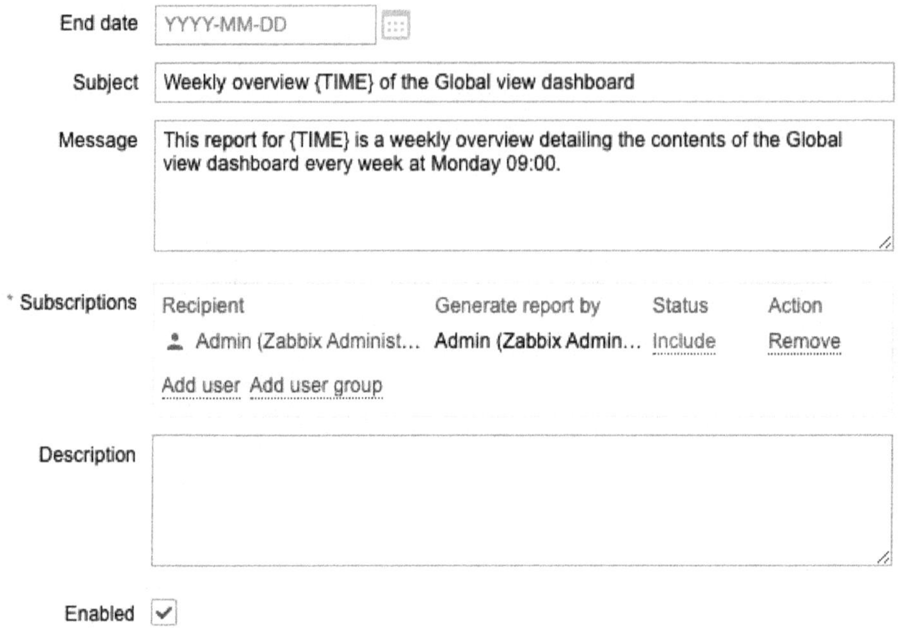

End date	YYYY-MM-DD

Subject	Weekly overview {TIME} of the Global view dashboard

Message	This report for {TIME} is a weekly overview detailing the contents of the Global view dashboard every week at Monday 09:00.

* Subscriptions	Recipient	Generate report by	Status	Action
	👤 Admin (Zabbix Administ...	Admin (Zabbix Admin...	Include	Remove
	Add user Add user group			

Description	

Enabled	✔

Figure 6.77 – Reports | Scheduled reports – creating a new report page, part II

15. You can now click the **Test** button to see if the report is working. Once it is, use the blue **Add** button to finish setting up this scheduled report.

How it works...

This feature is long-awaited and is finally here, but it's not finished and is simply still a building block for more advanced scheduled reports coming later. There are some key things to keep in mind with this new reporting functionality. I always state that Zabbix development tries to keep everything as customizable as possible by adding features and interconnecting them to make sure we can use existing functionality in new ways.

The Zabbix development team could have decided to create a fully fledged PDF reporting engine for Zabbix. But by going the way of using Zabbix dashboards as building blocks for all your PDF reports, they have created versatility and customizability. Every single new dashboard widget that is added is now available for you to use in your PDF reports, and more and more reporting-focused widgets will be added in the near future.

Zabbix simply grabs the information from your dashboard and sends it to you in a PDF form using the new Zabbix web services module and the Google Chrome browser. Once we get these prerequisites out of the way, we are provided with a way to send PDF reports to any of our Zabbix users, provided they have an email media type set up.

Setting up improved business service monitoring

Business service monitoring is a way to monitor the services that we, as a business, offer to our customers or internally. Think of a CRM system, email, and our website. It all has to work and we'd like to know if it does for the people using them. It also allows us to monitor the SLA of those services, if we want to define them.

Starting from Zabbix 6.0, business service monitoring has had an entire overhaul. If you've set it up in versions before 6.0, it might be wise to spend some time rediscovering the basics using this recipe. If you're starting with 7.0 here and you're entirely new to business service monitoring, don't worry as we will go through setting it up step by step in this recipe.

Getting ready

We will need our Zabbix server and access to its frontend. I'll be using my `lar-book-centos` host with the configuration we have done so far. We will also need a monitored host, for which I will use the Zabbix server itself.

How to do it...

I'll be using the Zabbix frontend as an example to set up business service monitoring, for which we will create a new host called `lar-book-zabbix-frontend` with some items and triggers.

Setting up items and triggers

If you have followed the previous recipes, you should have a good understanding of setting up items and triggers. Let's go through it again and set up some for our business service monitoring example:

1. First, let's create a new template by logging in to our Zabbix frontend and navigating to **Data collection | Templates**.

2. Click on the blue **Create template** button in the top-right corner and fill in the page, as follows:

Template
Template
Tags
Macros
Value mapping
* Template name
Zabbix frontend by Zabbix agent
Visible name
Zabbix frontend by Zabbix agent
Templates
type here to search
Select
* Template groups
Templates/Applications
type here to search
Select
Description

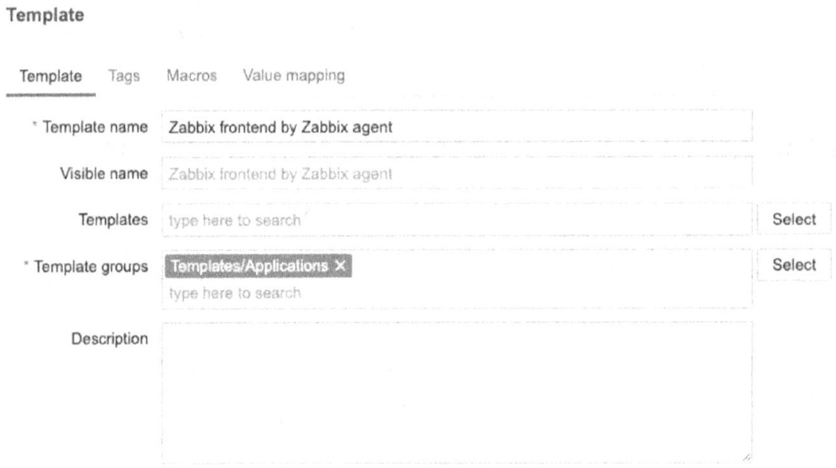

Figure 6.78 – New Zabbix frontend template configuration page

3. Make sure you save this new template by clicking the blue **Add** button.
4. Now, let's set up our new host by navigating to **Data collection | Hosts**.
5. Click on the blue **Create host** button in the top-right corner and fill in the page, as follows:

Host
IPMI
Tags
Macros
Inventory
Encryption
Value mapping
* Host name
Zabbix frontend
Visible name
Zabbix frontend
Templates
Zabbix frontend by Zabbix agent
type here to search
Select
* Host groups
Linux servers
Zabbix servers
type here to search
Select
Interfaces
Type
IP address
DNS name
Connect to
Port
Default
Agent
127.0.0.1
IP
DNS
10050
Remove
Add
Description
Monitored by
Server
Proxy
Proxy group
Enabled

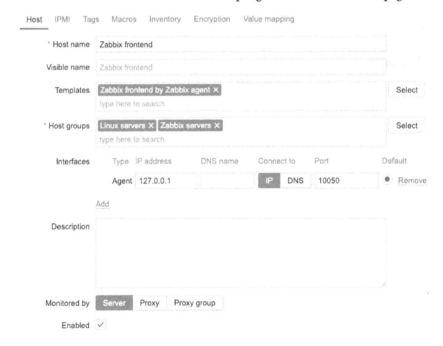

Figure 6.79 – New Zabbix frontend host configuration page

6. Then, add the following tag by navigating to the **Tags** tab:

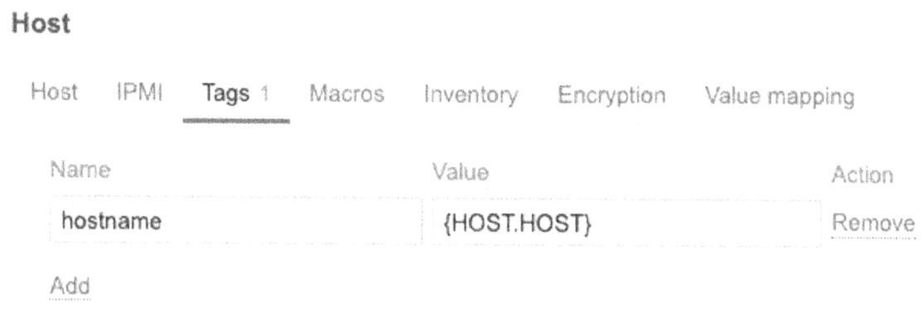

Figure 6.80 – New Zabbix frontend host configuration page – the Tags tab

7. Click the blue **Add** button to save this new host configuration and navigate to **Data collection | Templates**.

8. Edit the **Zabbix frontend by Zabbix agent** template and go to **Value mapping**.

9. Click on the small **Add** button with the blue dotted line under it and add the following value mapping:

Figure 6.81 – Template Zabbix frontend by Zabbix agent, Service state value mapping

10. Make sure you click the blue **Update** button. Then, back on the template, go to **Items**.

11. Click on the blue **Create item** button and add the following:

Figure 6.82 – ICMP ping item

12. Before adding the item, make sure you also add the **Value mapping** value, as follows:

Figure 6.83 – ICMP ping item value mapping

13. We must also go to the **Tags** tab to add some tags to this item:

Figure 6.84 – ICMP ping item – the Tags tab

14. Now, click the blue **Add** button at the bottom of the page.

15. Back at **Items**, click on the blue **Create item** button to create another item. Fill it in, as follows:

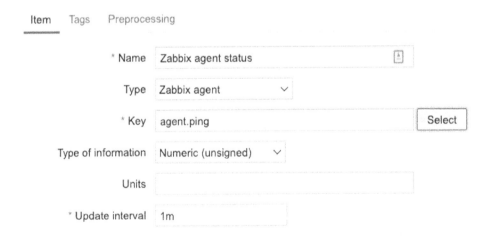

Figure 6.85 – Agent ping item

16. We must also go to the **Tags** tab to add some tags to this item:

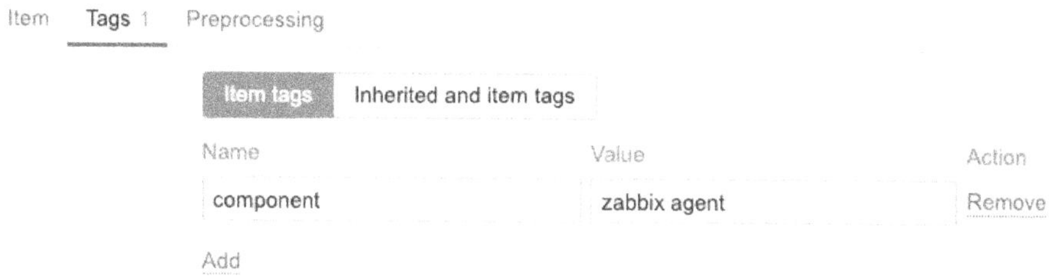

Figure 6.86 – HTTP service state item – the Tags tab

17. Now, save the new item by clicking the blue **Add** button at the bottom of the page.

18. Now that we have two new items, let's navigate to the **Triggers** page for this template.

19. Click the **Create trigger** button in the top-right corner and add the following trigger:

Figure 6.87 – ICMP down trigger configuration

20. On the **Tags** tab, we need to add a new tag, indicating that this trigger will be used in our SLA monitoring:

Figure 6.88 – ICMP down trigger – the Tags tab

21. Now, let's click the blue **Add** button to add this trigger. Then, create another trigger using the **Create trigger** button in the top-right corner.

22. Let's add the following trigger:

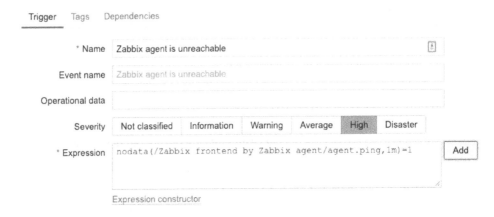

Figure 6.89 – Zabbix agent is unreachable trigger configuration

23. Make sure you add a tag for the SLA on this trigger as well by going to the **Tags** tab:

Figure 6.90 – Zabbix agent is unreachable trigger – the Tags tab

24. Click the blue **Add** button to finish setting up this trigger.

Adding the business service monitoring configuration

That concludes our item and trigger configuration. We can now continue with setting up our business service monitoring:

1. First, let's define our SLA period by going to **Services | SLA** and clicking on the blue **Create SLA** button in the top-right corner. We'll define the following SLA:

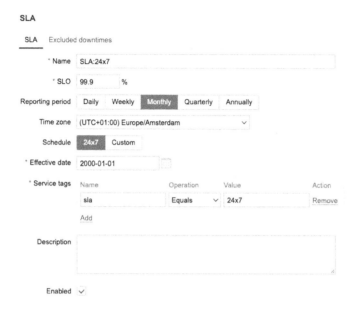

Figure 6.91 – Services | SLA – Zabbix SLA setup

2. Click **Add** at the bottom of the window to save this SLA.

3. Next, go to **Service | Service** and select **Edit** using the slider in the top-right corner.

4. Now, click **Create service** in the top-right corner to add a new service. Here, we will add a new service for our Zabbix setup:

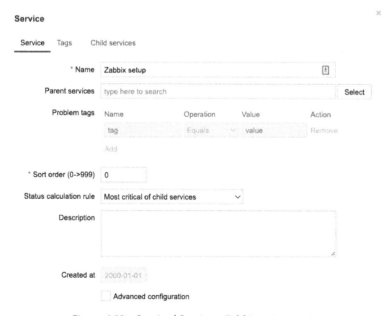

Figure 6.92 – Service | Service – Zabbix setup service

5.　On the **Tags** tab, make sure to add the following:

Figure 6.93 – Service | Service – Zabbix setup service – the Tags tab

6.　Click on the blue **Add** button at the bottom of the window to add this new service. Then, click **Create service** in the top-right corner again to add the following:

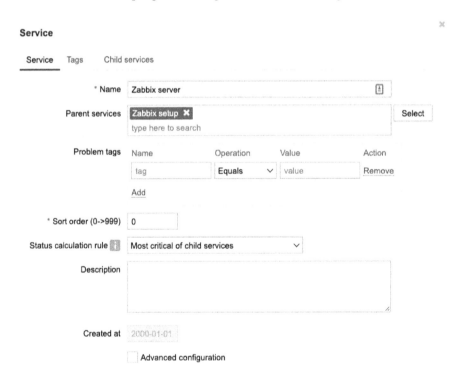

Figure 6.94 – Service | Service – Zabbix server service

7.   On the **Tags** tab, make sure to add the following:

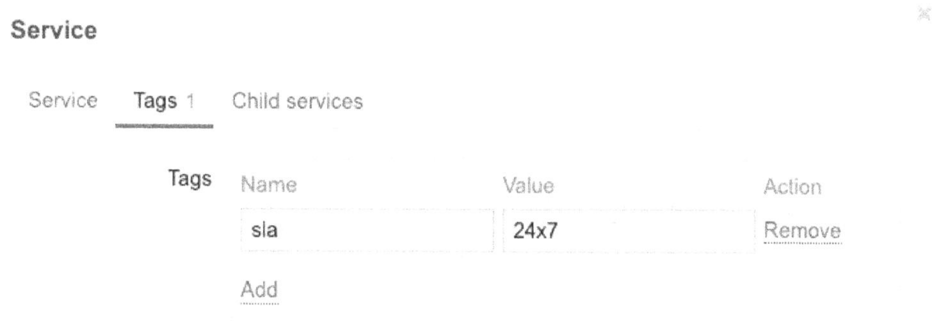

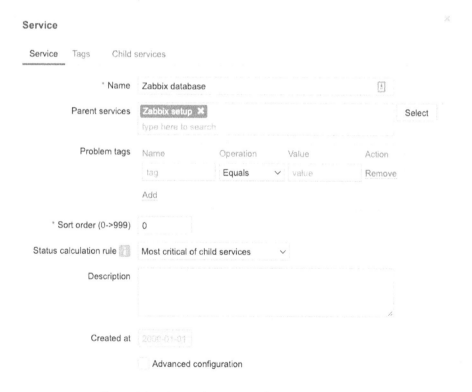

Figure 6.95 – Service | Service – Zabbix server service – the Tags tab

8.   Click the blue **Add** button again and then **Create service** in the top-right corner. Add another service at the same level, as follows:

Figure 6.96 – Service | Service – Zabbix database service

9. On the **Tags** tab, make sure to add the following:

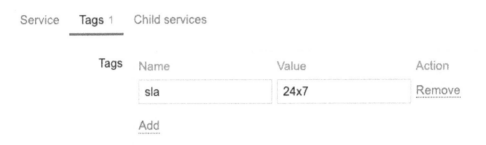

Figure 6.97 – Service | Service – Zabbix database service – the Tags tab

10. Click **Add** again to add this service.

11. Finally, we'll add the last child of the Zabbix setup by clicking the **Create service** button again:

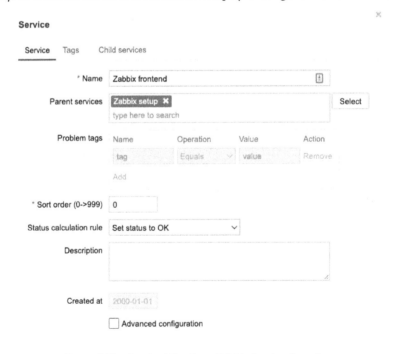

Figure 6.98 – Service | Service – Zabbix frontend service

12. Select **Advanced configuration** and click **Add** under **New additional rule**. We will add the following calculation here:

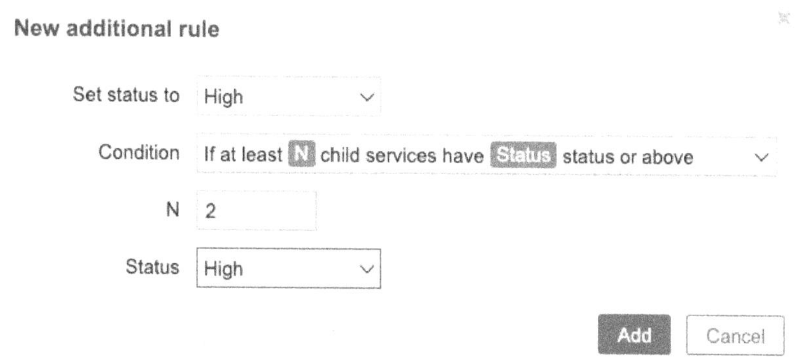

Figure 6.99 – Service | Service – Zabbix frontend service, additional rules

13. On the **Tags** tab, make sure to add the following:

Service

Service   **Tags** 1   Child services

Tags	Name	Value	Action
	sla	24x7	Remove
	Add		

Figure 6.100 – Service | Service – Zabbix database service – the Tags tab

14. Finish setting up this service by clicking the **Add** button at the bottom of the window.

15. Now, we'll have to add two more services, but this time under the Zabbix frontend. Click on **Zabbix frontend** and then **Create service** again and add the following:

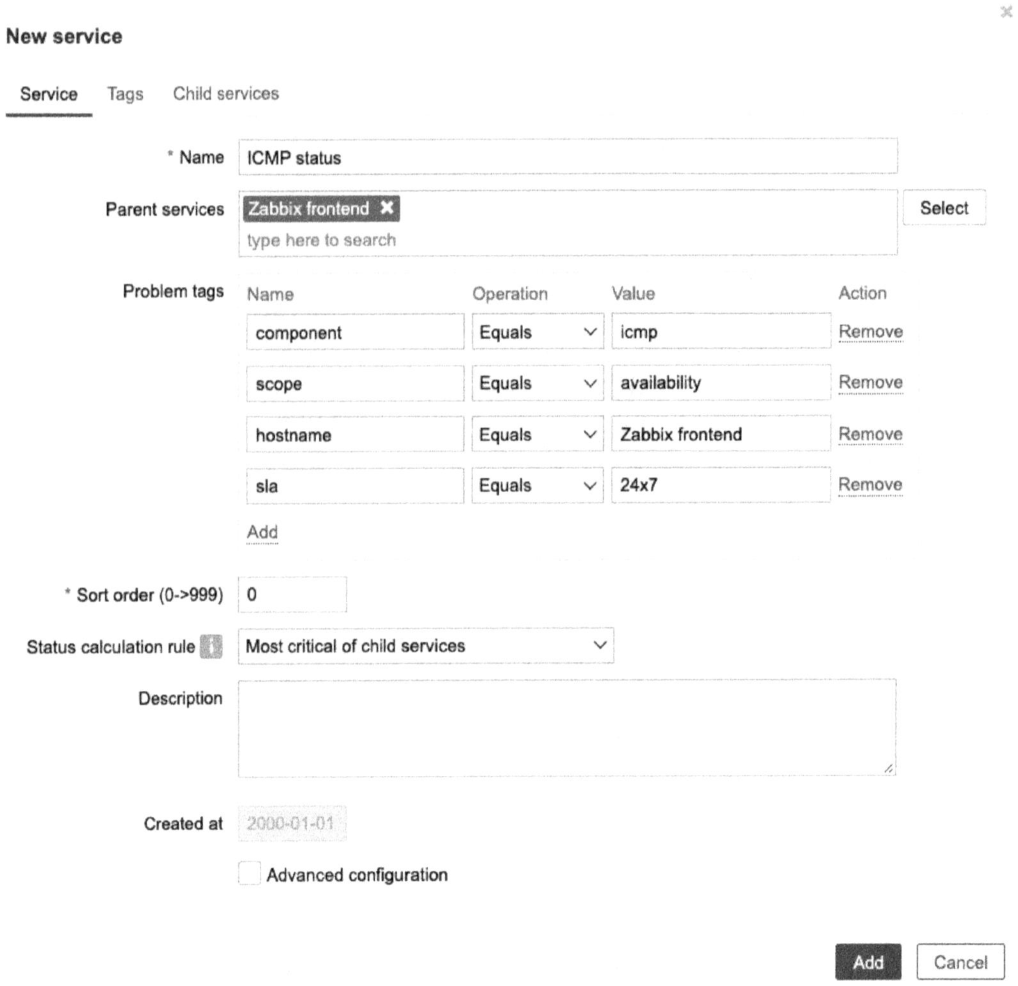

New service

| Service | Tags | Child services |

* Name   ICMP status

Parent services   Zabbix frontend ✖   Select
               type here to search

Problem tags	Name	Operation	Value	Action
	component	Equals	icmp	Remove
	scope	Equals	availability	Remove
	hostname	Equals	Zabbix frontend	Remove
	sla	Equals	24x7	Remove

Add

* Sort order (0->999)   0

Status calculation rule   Most critical of child services

Description

Created at   2000-01-01

☐ Advanced configuration

Add   Cancel

Figure 6.101 – Service | Service – Zabbix frontend, ICMP status child service

16. Click the blue **Add** button and then **Create service** again to add the last service.

17. Add the final service:

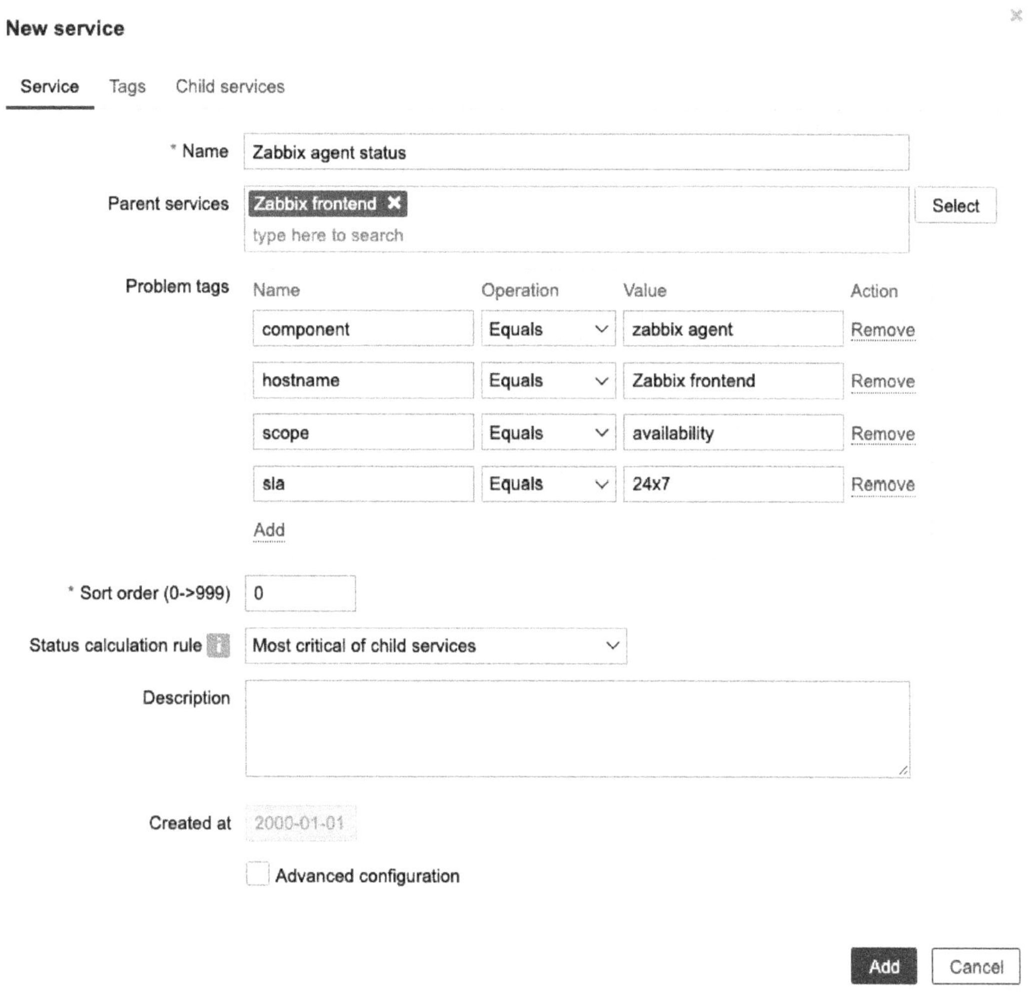

New service

Service   Tags   Child services

* Name	Zabbix agent status
Parent services	Zabbix frontend ✖
	type here to search

Problem tags	Name	Operation	Value	Action
	component	Equals ⌄	zabbix agent	Remove
	hostname	Equals ⌄	Zabbix frontend	Remove
	scope	Equals ⌄	availability	Remove
	sla	Equals ⌄	24x7	Remove
	Add			

* Sort order (0->999)	0
Status calculation rule	Most critical of child services ⌄
Description	
Created at	2000-01-01
	☐ Advanced configuration

Add   Cancel

Figure 6.102 – Service | Service – Zabbix frontend, Zabbix agent status child service

18. Click the blue **Add** button to add this service. Let's see if it works as expected.

How it works...

Let's take a look at what we have set up in our current configuration. We've used business service monitoring to monitor part of our Zabbix stack. Look at business service monitoring as a tree, where we just created two levels. Our initial level is the Zabbix setup, which consists of our Zabbix server, Zabbix database, and Zabbix frontend.

Beneath the Zabbix frontend level, we have one more level where we have defined two more services that represent the status of ICMP and the Zabbix agent. We only want to calculate the SLA if both the ICMP and Zabbix agent are in a problem state:

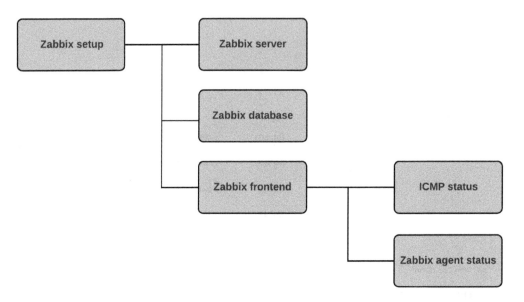

Figure 6.103 – Business service monitoring tree structure example

As you can see, we have a distinct tree structure forming once we start to visualize this. The part where the magic happens in this case is the Zabbix frontend service because this is where we defined what should happen to our SLA once something goes wrong with services.

Let's take another look at that level:

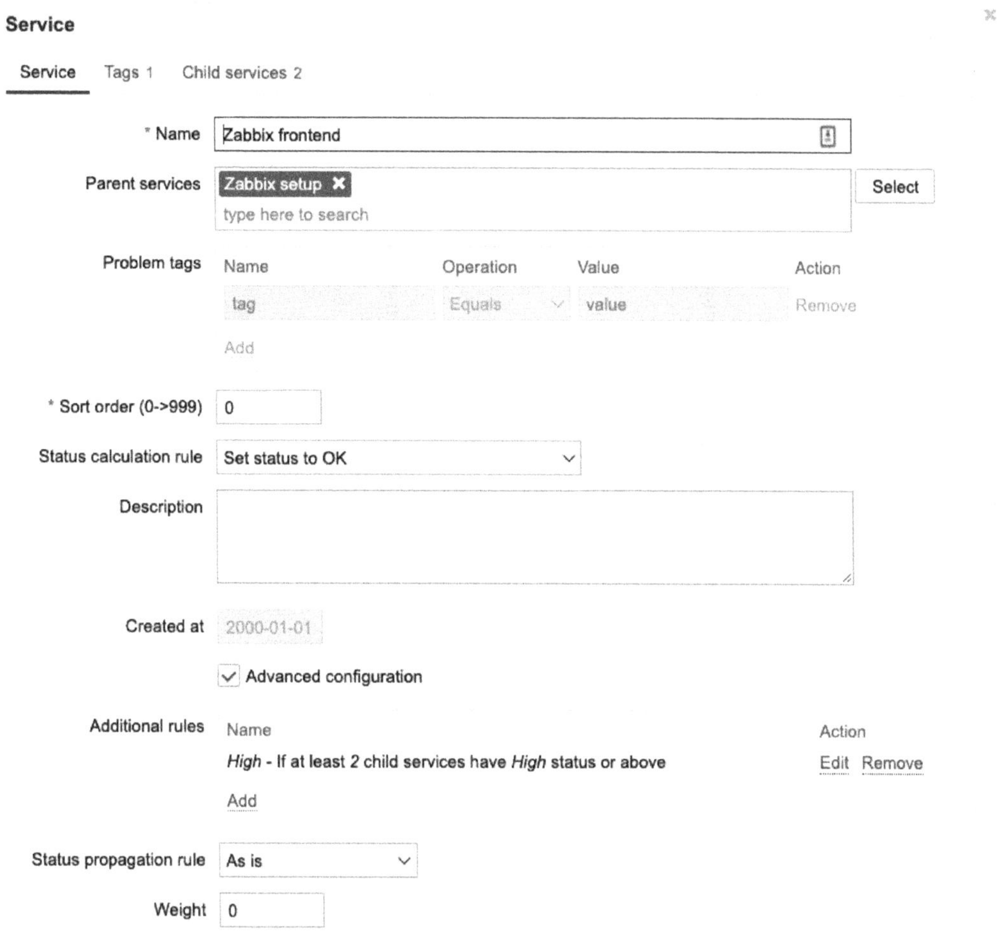

Figure 6.104 – Zabbix frontend service completed

Because we defined that the service should always **Set status to OK**, it will only use what we defined in our **Additional rules** section. This is where we specified that we only want to affect our SLA calculation: **If at least 2 child services have High status or above**. Effectively, this means that our SLA is only going down if the Zabbix agent can't be reached and ICMP is down.

We've built in a security measure for ourselves here, making sure that if someone stops the Zabbix agent but the server can still be reached by ICMP, the SLA won't be affected.

Now, let's take a look at the result, which we can use to monitor these SLAs. Over at **Services | SLA report**, we can find all we need to know about whether our SLA is being met. We can set the filter to the period for which we want to find the SLA. We'll see the following output:

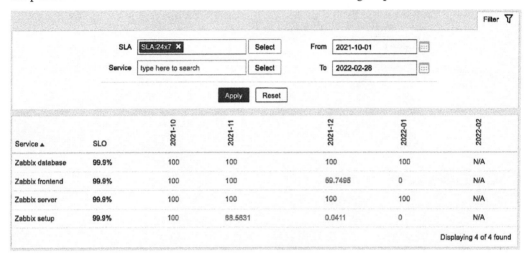

Figure 6.105 – Service | SLA report, all services with our SLA:24x7 tag

Here, we can see our monthly 24/7 SLA, where a SLA of 99.9% is expected. For our Zabbix setup back in October 2021, the SLA was 100, so we met our required SLA. However, in November 2021, we noticed that the SLA dropped below 100, clearly indicating in red that our SLA was not met.

Drilling down even further and selecting our specific service Zabbix setup, we can create a more detailed overview:

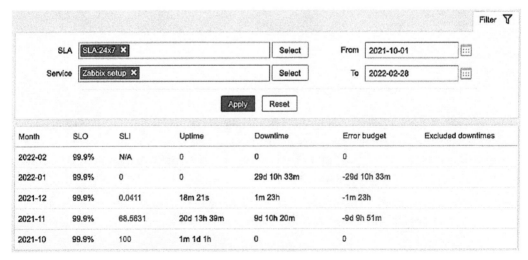

Figure 6.106 – Service | SLA report, Zabbix setup services with our SLA:24x7 tag

Here, we can see all the details regarding the uptime and downtime of our service and what our leftover error budget is like.

Using business service monitoring calculations like this, we can narrow down where weak points in our services might be while attaching useful statistics to that measure. In this case, we used a simple example with ICMP and the Zabbix agent trigger, but the possibilities are endless when using services in combination with tags.

There's more...

One of the main concerns with the old way of monitoring services through business service monitoring was the inability to automate and customize it. This automation has been mostly resolved through the use of tags as we can now define tags at the host, template, or trigger level to define what's used in the business service monitoring configuration.

In terms of customization, Zabbix has given us a lot more options to do calculations:

Figure 6.107 – Zabbix service, additional rule options

Looking at the numerous options here, we can see that we have a lot more to play around with. Not only can we specify the exact number of child services we'd like to use in our calculations, but we can also work with weights and percentages, giving us the options we might need to build more complex setups.

7
Using Discovery for Automatic Creation

This chapter is going to be all about making sure that you as a Zabbix administrator are doing as little work as possible on host and item creation. We are going to learn how to perform (or perfect, maybe) automatic host, item trigger, and graph creation. Check out the recipes featured here to see just what we are going to discover.

In this chapter, we will first learn how to set up Zabbix network discovery with Zabbix agent and **Simple Network Management Protocol** (**SNMP**). We will then set up active agent autoregistration. Later, we will also cover the automatic creation of Windows performance counters, **Java Management Extensions** (**JMX**), and SNMP items using **low-level discovery** (**LLD**).

In this chapter, we will cover the following recipes:

- Setting up Zabbix agent host discovery
- Working with Zabbix SNMP network discovery
- Automating host creation with active agent autoregistration
- Using Windows performance counter discovery
- Discovering JMX objects
- Setting up Zabbix SNMP discovery the new way
- Creating hosts with LLD and custom JSON

Technical requirements

As this chapter is all about host and item discovery, besides our Zabbix server, we will need one new Linux host and a Windows host. Both these hosts will need Zabbix agent 2 installed, but not configured just yet.

Furthermore, we are going to need our JMX host, as configured in *Chapter 3*, *Setting Up Zabbix Monitoring*, and a new host with SNMP set up. To learn more about setting up an SNMP-monitored host, check out the *Working with SNMP monitoring the old way* recipe in *Chapter 3*, *Setting Up Zabbix Monitoring*.

Setting up Zabbix agent network discovery

A lot of Zabbix administrators use Zabbix agent extensively and thus spend a lot of time creating Zabbix agent hosts by hand. Maybe they don't know how to set up Zabbix agent discovery, maybe they didn't have time to set it up yet, or maybe they just prefer it this way. If you are ready to get started with Zabbix agent discovery, in this recipe we will learn just how easy it is to set it up.

Getting ready

Besides our Zabbix server, in this chapter's introduction, I mentioned that we will need two (empty) hosts with Zabbix agent 2 installed: one Windows host and one Linux host. If you don't know how to install Zabbix agent 2, check out *Chapter 3*, *Setting Up Zabbix Monitoring*, or see the Zabbix documentation at https://www.zabbix.com/documentation/current/en/manual/concepts/agent2.

Let's give the servers the following hostnames:

- `lar-book-disc-lnx`: For the Linux host (use Zabbix agent 2)
- `lar-book-disc-win`: For the Windows host (use Zabbix agent 2)

How to do it...

Follow these steps:

1. Let's get started by logging in to our `lar-book-disc-lnx` Linux host and editing the following file:

    `vim /etc/zabbix/zabbix_agent2.conf`

2. Now, make sure your Zabbix agent 2 configuration file contains at least the following line:

    `Hostname=lar-book-disc-lnx`

3. For your Windows Zabbix agent, it's important to do the same. Edit the following file:

    `C:\Program Files\Zabbix Agent 2\zabbix_agent2.conf`

4. Now, change the hostname by editing the following line:

    `Hostname=lar-book-disc-win`

5. Next up, in our Zabbix frontend, navigate to **Data collection | Discovery**, and on this page, we click on **Create discovery rule** to create a rule with the following settings:

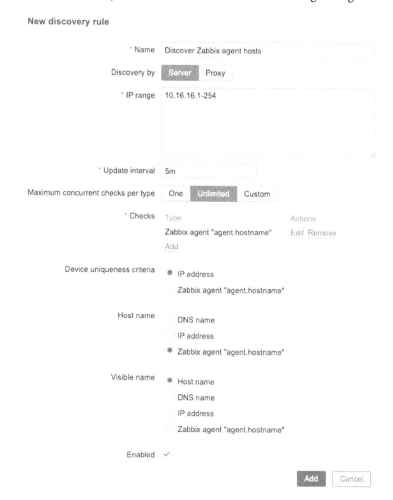

Figure 7.1 – Discovery rules page for Zabbix agent hosts

Important note

We are using an update interval of 5 minutes in this example. As this might take up a lot of resources on your server, make sure to adjust this value for your production environment. For example, one hour might be a better production value to make sure we put less load onto our Zabbix processes. Depending on the size of the IP range we are scanning and how fast you want to discover things, we can adjust this value.

6. Click the blue **Add** button to move on.

7. After setting up the discovery rule, we will also need to set up an action to actually create the host with the right template. Navigate to **Alerts | Actions | Discovery actions**:

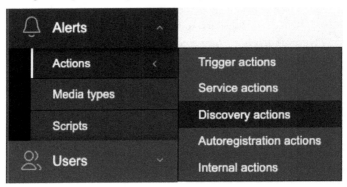

Figure 7.2 – Alerts | Actions | Discovery actions

8. Here, we will click the **Create action** button in the top-right corner and fill out the next page with the following information:

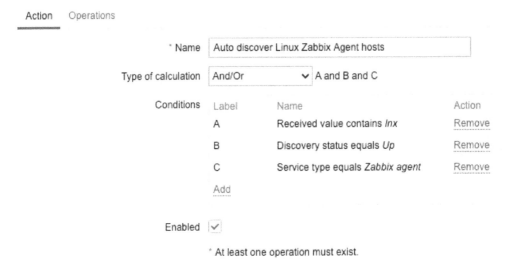

Figure 7.3 – The discovery action creation page for Zabbix agent hosts

> **Tip**
>
> When creating Zabbix actions, it's important to keep the order of creation for **Conditions** in mind. The labels seen in the preceding screenshot will be added in order of creation. This means that it's easier to keep track of your Zabbix actions if you keep the order of creation the same for all actions.

9. Next up, click the **Operations** tab. This is where we will add the following:

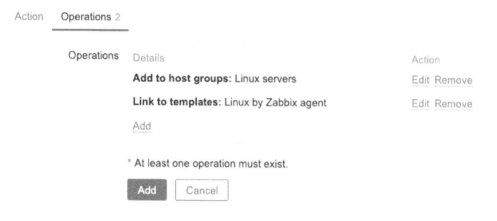

Figure 7.4 – The Operations page for Zabbix agent hosts

10. That's it for the Linux agent. Click the blue **Add** button, and let's continue with our Windows discovery rule.

11. Navigate to **Data collection | Host groups**. Create a host group for our Windows hosts by clicking **Create host group** in the top-right corner and filling out the group name:

Figure 7.5 – The Create host group page for Windows server hosts

12. Click the blue **Add** button and navigate to **Alerts | Actions**.

13. Go to **Discovery actions** again and click **Create action**. We will fill out the same thing but for our Windows hosts this time:

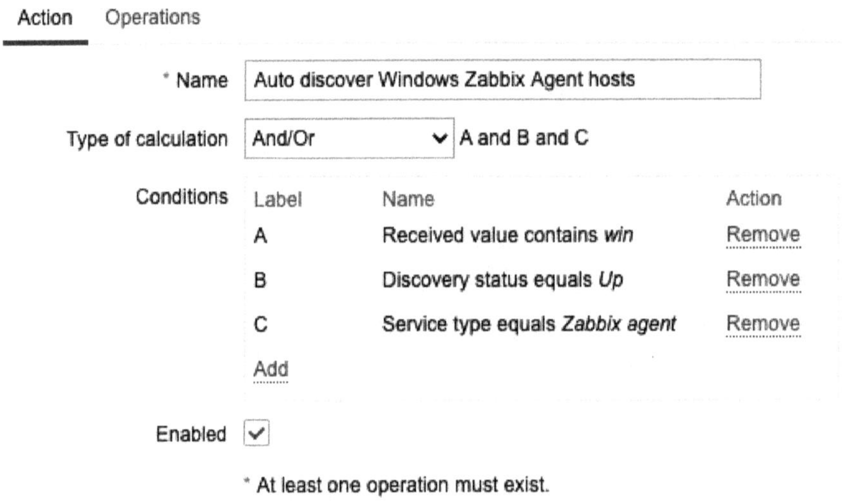

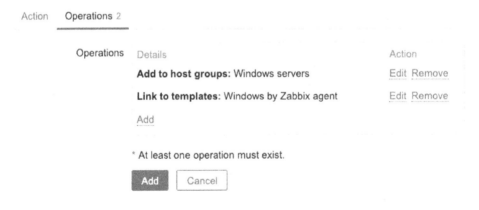

Figure 7.6 – Discovery action creation page for Windows Zabbix agents

14. Before clicking **Add**, let's also fill out the **Operations** page with the operations shown in the following screenshot:

Figure 7.7 – The Operations page for Windows Zabbix agents

15. Now, we can click the blue **Add** button, and our second discovery action is present.

16. Move on to **Monitoring | Discovery**. This is where we can see whether and when our hosts are discovered:

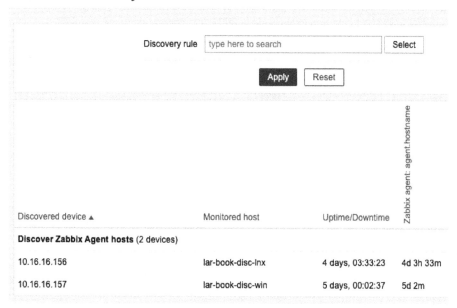

Figure 7.8 – The Monitoring | Discovery page

> **Tip**
>
> Use the **Monitoring | Discovery** page to keep a close eye on the hosts you expect to show up in your Zabbix setup. It's very useful to track new hosts coming in and see which Zabbix discovery rule was used to create the host.

How it works...

Network discovery might not be very hard to set up initially, but there are loads of options to configure. For this example, we chose to use the `agent.hostname` key as our check. We create the Zabbix hostname based on what's configured in the Zabbix agent configuration file.

What happens is that Zabbix network discovery finds our hosts and performs our check. In this case, the check is *What is the hostname used by Zabbix agent?* This information, plus our IP address, is then triggering the action. Our action then performs our configured checks:

- Does the hostname contain `lnx` or `win`?

- Is the discovery status UP?

- Is the service type `Zabbix Agent`?

If all of those checks are true, our action will then create our newly discovered host with the following:

- Our configured host group plus the default `Discovered hosts` host group

- Our template as configured in our action

We will end up with two newly created hosts, with all the right settings:

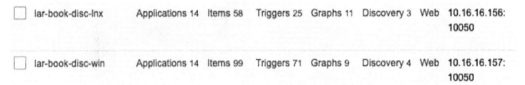

Figure 7.9 – The Data collection | Hosts page with our new hosts, Windows and Linux

There's more...

Creating the host by using the configuration file settings isn't always the right way to go, but it's a solid start to working with network discovery.

If you want a more flexible environment where you don't have to even touch the Zabbix agent configuration file, then you might want to use different checks on the discovery rule. Check out which keys we can use to build different discovery rules in the Zabbix documentation at `https://www.zabbix.com/documentation/current/en/manual/config/items/itemtypes/zabbix_agent`.

Working with Zabbix SNMP network discovery

If you work with a lot of SNMP devices but don't always want to set up monitoring manually, network discovery is the way to go. Zabbix network discovery uses the same functionality as Zabbix agent discovery but with a different configuration approach.

Getting ready

To get started with network discovery, we are going to need a host that we can monitor with SNMP. If you don't know how to set up a host such as this, check out the *Working with SNMP monitoring the old way* recipe in *Chapter 3, Setting Up Zabbix Monitoring*. We'll also need our Zabbix server.

How to do it...

Follow these steps:

1. First, log in to your new SNMP-monitored host and change the hostname to the following:

    ```
    hostnamectl set-hostname lar-book-disc-snmp
    exec bash
    ```

2. Then, restart the SNMP daemon using the following command:

    ```
    systemctl restart snmpd
    ```

3. Now, navigate to **Data collection | Discovery** and click on **Create discovery rule** in the top-right corner.

4. We are going to create a new SNMP discovery rule, with an SNMP **object identifier** (**OID**) check type. Fill out the **Name** and **IP range** fields first, like this:

Figure 7.10 – Data collection | Discovery, discovery rule creation page for SNMPv2

5. Make sure to fill out your own IP range in the **IP range** field.

6. Now, we are going to create our SNMP check. Click on **Add** next to **Checks**, and you'll see the following pop-up screen:

Figure 7.11 – Data collection | Discovery, discovery check creation pop-up window

7. We want **Check type** to be **SNMPv2 agent** and we want to fill it with our community and a useful OID, which in this case will be the OID for the system name. Fill it out like this:

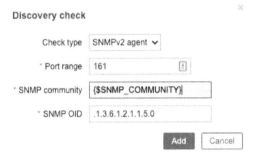

Figure 7.12 – Data collection | Discovery, discovery check creation
pop-up window filled with an SNMPv2 check

Important note

Please note that our check type is *not* SNMP version independent. We have three SNMP versions and thus three different check types to choose from, unlike our new SNMP interface selection on the Zabbix 7 host screen.

---

8. After clicking **Add** again, fill out the rest of the page, as follows:

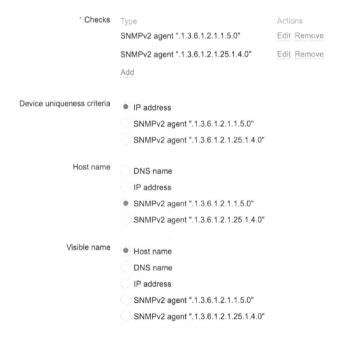

Figure 7.13 – The Data collection | Discovery page for SNMPv2 agents

9. Last, but not least, click the **Add** button at the bottom of the page. This concludes creating our Zabbix discovery rule.

10. We will also need an action for creating our hosts from the discovery rule. Navigate to **Alerts | Actions**, and after using the dropdown to select **Discovery actions**, click on **Create action**.

11. We will fill out the page with the following information:

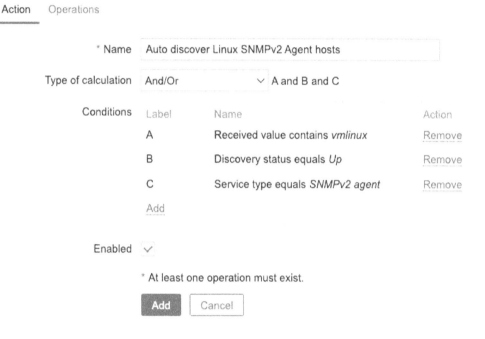

Figure 7.14 – Alerts | Actions, action creation page for SNMPv2 agents

12. Before clicking **Add**, navigate to **Operations** and fill out this page with the following details:

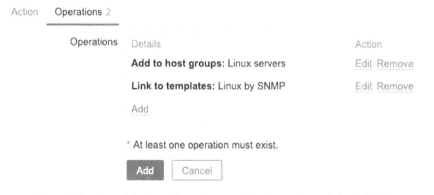

Figure 7.15 – Alerts | Actions, the action creation Operations tab for SNMPv2 agents

13. Now, click **Add** and navigate to **Monitoring | Discovery** to see whether our host gets created:

Discovered device ▲		Monitored host	Uptime/Downtime	SNMPv2 agent: .1.3.6.1.2.1.1.5.0	SNMPv2 agent: .1.3.6.1.2.1.25.1.4.0
Discover Linux SNMPv2 Agent hosts (3 devices)					
10.16.16.152 (lar-book-centos)		lar-book-centos_2	5 days, 18:25:21	5d 18h 25m	
10.16.16.153		lar-book-agent_passive	5 days, 18:25:19	5d 18h 25m	5d 18h 25m
10.16.16.158		lar-book-disc-snmp	23:48:48	23h 48m 48s	3m 28s

Figure 7.16 – The Monitoring | Discovery page for SNMPv2 agents

How it works...

In this recipe, we've created another discovery rule, but this time for SNMP. As you've noticed, the principle remains the same, but the application is a bit different.

When we created this Zabbix discovery rule, we gave it two checks instead of the one check we did in the previous recipe. We created one check on the `.1.3.6.1.2.1.1.5.0` SNMP OID to retrieve the hostname of the device through SNMP. We then put the hostname retrieved from the system into Zabbix as the Zabbix hostname of the system.

We also created a check on the `.1.3.6.1.2.1.25.1.4.0` SNMP OID. This check will retrieve the following string, if present:

```
"BOOT_IMAGE=(hd0,gpt2)/vmlinuz-4.18.0-193.6.3.el8_2.x86_64 root=/dev/
mapper/cl-root ro crashkernel=auto resume=/dev/mapper/cl-swa"
```

If the string is present, it means that the boot image is Linux on this host. This is a perfect example of how we can retrieve multiple OIDs to do multiple checks in our Zabbix discovery rules. If we'd been monitoring a networking device, for instance, we could have picked an OID to see whether it was a Cisco or a Juniper device. We would replace `.1.3.6.1.2.1.25.1.4.0` with any OID and poll it. Then, we would create our action based on what we received (Juniper or Cisco) and add our templates accordingly.

> **Important note**
> General knowledge of SNMP structure is very important when creating Zabbix discovery rules. We want to make sure we use the right SNMP OIDs as checks. Make sure to do your research well, utilize SNMP walks, and plan out what OIDs you want to use to discover SNMP agents. This way, you'll end up with a solid monitoring infrastructure.

Automating host creation with active agent autoregistration

Using discovery to set up your Zabbix agents is a very useful method to automate your host creation. But what if we want to be even more upfront with our environment and automate further? That's when we use a Zabbix feature called **active agent autoregistration**.

Getting ready

For this recipe, we will need a new Linux host. We will call this host `lar-book-lnx-agent-auto`. Make sure to install Zabbix agent 2 on this host. Besides this new host, we'll also need our Zabbix server.

How to do it...

1. Let's start by logging in to our new `lar-book-lnx-agent-auto` host and changing the following file:

   ```
   vim /etc/zabbix/zabbix_agent2.conf
   ```

2. We will then edit the following line in the file. Make sure to enter your Zabbix server IP on this line:

   ```
   ServerActive=10.16.16.152
   ```

3. We can also change the following line in the file if we want to set our hostname in the file manually:

   ```
   Hostname=lar-book—lnx-agent-auto
   ```

   This is not a requirement though, as Zabbix agent will use the system hostname if it is not filled out.

4. Next up, we will navigate to our Zabbix frontend, where we'll go to **Alerts | Actions**.

5. Use the drop-down menu to go to **Autoregistration actions**, as in the following screenshot:

Figure 7.17 – The Alerts | Actions page drop-down menu

6. Now, we will click the blue **Create action** button in the top-right corner to create a new action.

7. Fill out the **Name** field and then click on the **Add** text link:

Figure 7.18 – Alerts | Actions, create new action page

8. We can set up a condition here to only register hosts with a certain hostname. Let's do this by filling out the window like this:

Figure 7.19 – Create action | New condition for the lar-book-lnx host

Your page should now look like this:

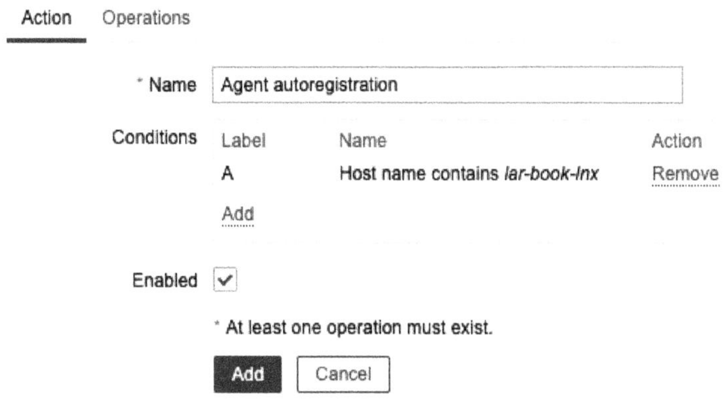

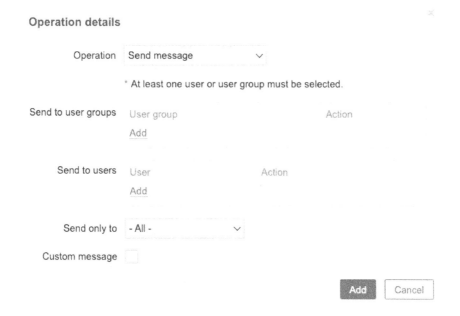

Figure 7.20 – The Create action page, filled with our information for the lar-book-lnx host

> **Tip**
> We can set up conditions for different types of hosts. For instance, if we want to add Windows
> hosts, we set up a new action with a different hostname filter. This way, it is easy to maintain
> the right groups and templates, even with autoregistration.

9.   Before clicking the blue **Add** button, let's go to the **Operations** tab.

10.   Click on the **Add** text link, and you will see the following window:

Figure 7.21 – The Send message operation for the lar-book-lnx host

11. Create an action to add the host to the **Linux servers** host group:

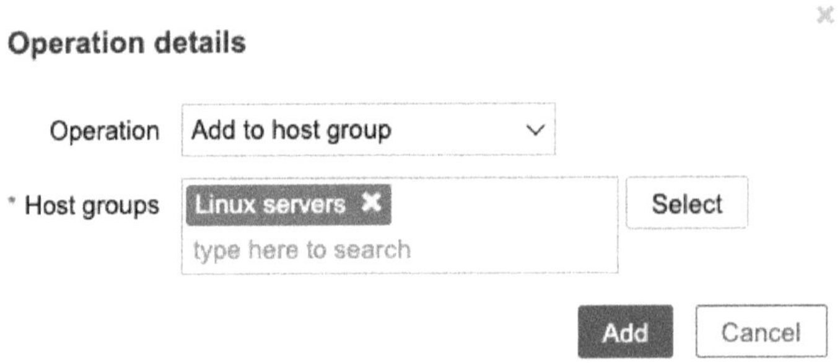

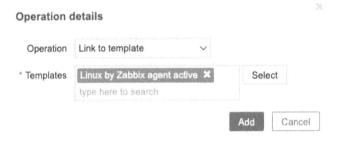

Figure 7.22 – The Add to host group operation creation

12. Create an action to add the host to the **Linux by Zabbix agent active** template:

Figure 7.23 – The Link to template operation creation

13. Your finalized **Operations** page should now look like this, and we can click the blue **Add** button:

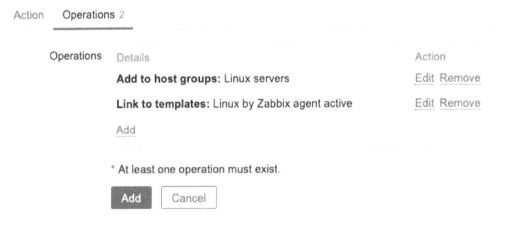

Figure 7.24 – The Operations page, filled with our information

14. Navigate to **Data collection | Hosts**, and we can see our new active autoregistered host:

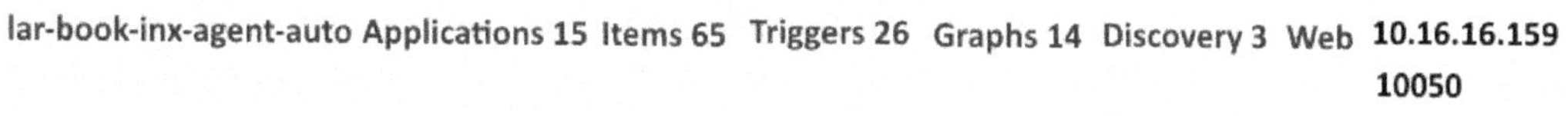

lar-book-inx-agent-auto Applications **15** Items **65** Triggers **26** Graphs **14** Discovery **3** Web **10.16.16.159**
10050

Figure 7.25 – The Data collection | Hosts page with host lar-book-lnx-agent-auto

How it works...

Active agent autoregistration is a solid method to let a host register itself. Once the `ServerActive=` line is set up with the Zabbix server or proxy IP, the host agent will start requesting configuration data from the Zabbix server or Proxy. The Zabbix server will receive these requests, and if there is an action set up in Zabbix (as we just did in this recipe), the host autoregisters to Zabbix:

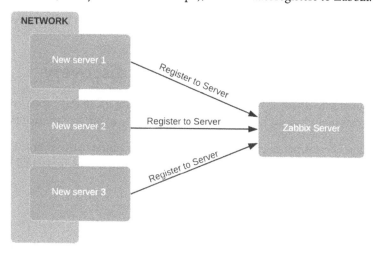

Figure 7.26 – Host autoregistration process

We can do a bunch of cool automation with this functionality. We could create a script to fill up our Zabbix agent configuration file with the right `ServerActive=` line on our hosts in a certain IP pool.

It would also be super easy to set up new hosts with Ansible. We can automate the Zabbix agent installation with Ansible and we can add the `ServerActive=` line in the `/etc/zabbix/ zabbix_agent2.conf` file using Ansible as well. Our Zabbix server autoregistration action will take care of the rest from here.

Zabbix agent autoregistration is a perfect way to get a zero-touch monitoring environment that's always up to date with our latest new hosts.

There's more...

Not every company uses hostnames that reflect the machine's OS or other attributes. This is when Zabbix HostMetadata can come in very useful. We can add this field to the active Zabbix agent configuration to reflect the attributes of the machine.

Afterward, we can use HostMetadata in our Zabbix discovery action to do the same kind of filtering we did on the hostname.

We also have the HostInterface and HostInterfaceItem parameters in the Zabbix agent configuration file, which are used for autoregistration. The host will use the specified IP or DNS name as its Zabbix agent interface IP or DNS, as seen in the Zabbix frontend. We can also use this functionality to enable passive agent monitoring while using autoregistration to create the host.

Check out this link for more information:

```
https://www.zabbix.com/documentation/current/manual/discovery/auto_
registration#using_host_metadata
```

Using Windows performance counter discovery

In Zabbix 7, it is possible to discover Windows performance counters. In this recipe, we will go over the process of discovering Windows performance counters to use in our environments.

Discovering Windows performance counters might seem to be a little tricky at first, as it uses both Windows- and Zabbix-specific concepts. However, once we finish this recipe, you'll know exactly how to set it up.

Getting ready

In this chapter, we added the lar-book-disc-win host to our setup, which is the host used in our Zabbix agent discovery process. We can reuse this host to discover Windows performance counters easily.

Of course, we'll also need our Zabbix server.

How to do it...

1. Let's start by navigating to **Data collection** | **Templates** and creating a new template by clicking **Create template** in the top-right corner.

2. Create the following template:

Templates   Tags   Macros   Value mapping

* Template name	Windows performance by Zabbix agent
Visible name	Windows performance by Zabbix agent
Templates	type here to search
* Groups	Templates/Operating systems ✗
	type here to search
Description	

Select · Select

Add   Cancel

Figure 7.27 – The Windows performance by Zabbix agent template creation

3. Click on the blue **Add** button, which will bring you back to **Data collection | Templates**. Select the new template.

4. Now, before continuing with our template, navigate to your Windows frontend and open `perfmon.exe`:

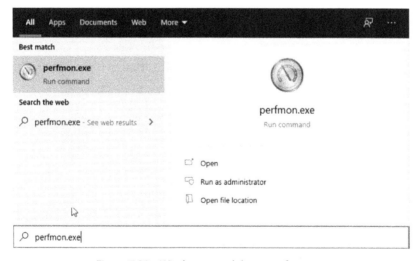

Figure 7.28 – Windows search bar – perfmon.exe

5. Doing so will open the following window:

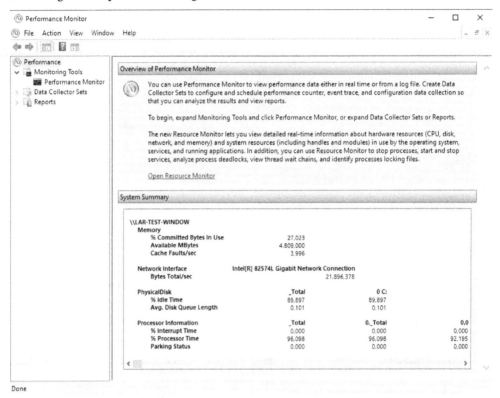

Figure 7.29 – Windows perfmon.exe

6. Let's click on `Performance Monitor` and then on the green + icon. This will show you all the available Windows performance counters.

7. Let's start by using the **Processor** counter.

8. Go back to the **Data collection | Templates** page in Zabbix and edit our new **Windows performance by Zabbix agent** template.

9. When you are at the **Edit template** page, click on **Discovery rules** in the bar next to your template name.

10. Click on **Create new discovery rule** in the top-right corner and add the following rule:

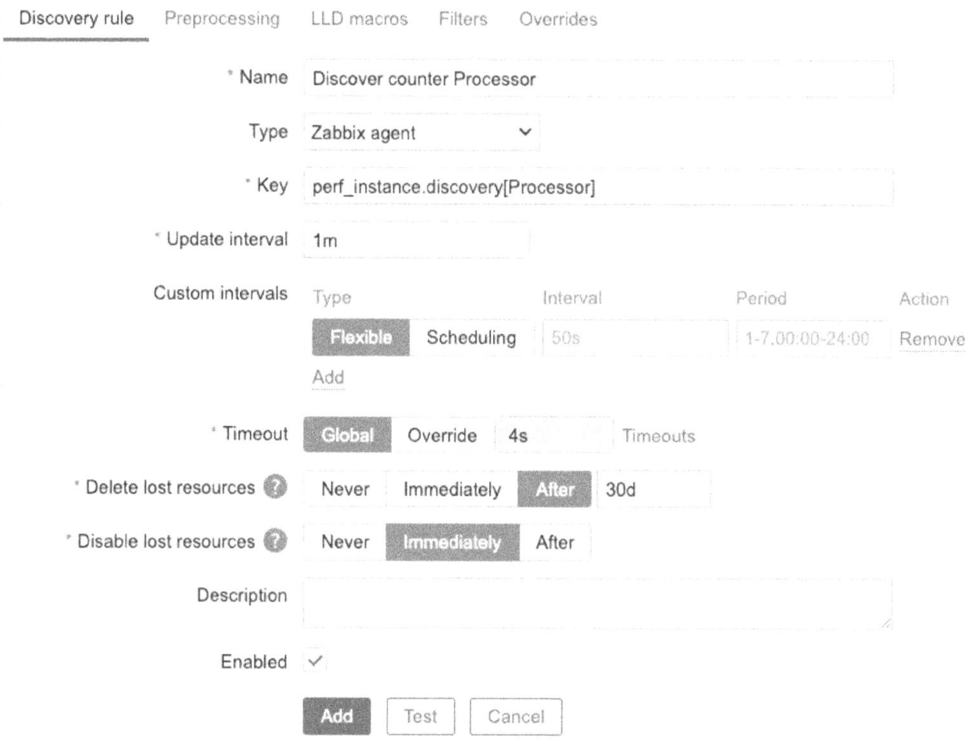

Discovery rule	Preprocessing	LLD macros	Filters	Overrides

* Name	Discover counter Processor
Type	Zabbix agent ⌄
* Key	perf_instance.discovery[Processor]
* Update interval	1m

Custom intervals	Type	Interval	Period	Action
	Flexible Scheduling	50s	1-7,00:00-24:00	Remove
	Add			

* Timeout	Global Override 4s Timeouts
* Delete lost resources ❓	Never Immediately After 30d
* Disable lost resources ❓	Never Immediately After
Description	
Enabled	✓

Add   Test   Cancel

Figure 7.30 – Create an LLD rule page – Discover counter Processor

---

Important note

We are using an update interval of 1 minute in this example. As this might take up a lot of resources on your server, make sure to adjust this value to your production environment. For example, one hour is a much better production value.

---

11. Click the blue **Add** button at the bottom and click our new **Discover counter Processor** discovery rule.

12. Click on **Item prototypes**, and in the top-right corner click on **Create item prototype**. We will then create the following item prototype:

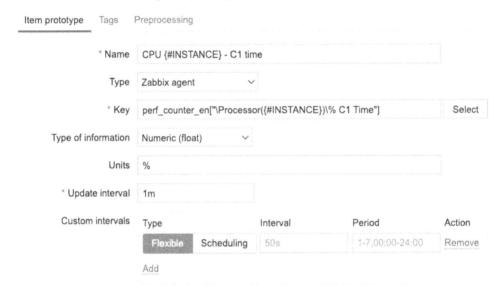

Figure 7.31 – The CPU instance C1 time item prototype creation

13. On the **Tags** tab, do not forget to add some new tags as follows:

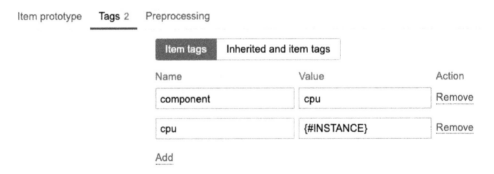

Figure 7.32 – The Tags tab

14. Save the new **Item prototype**, go to **Data collection | Hosts**, and click on `lar-book-disc-win`.

15. Add our **Windows performance by Zabbix agent** template:

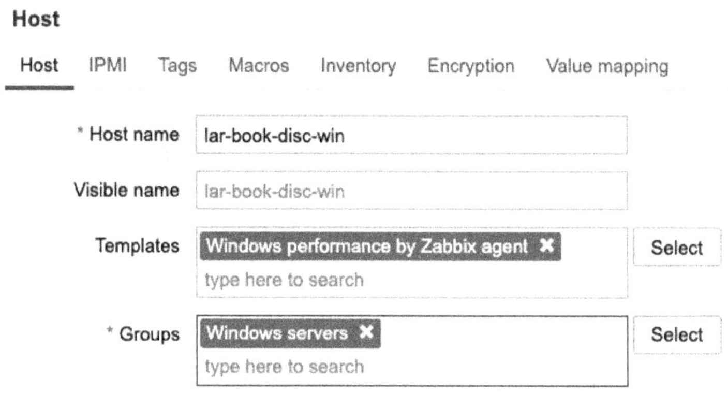

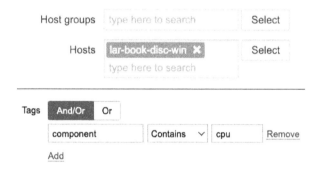

Figure 7.33 – Add Windows performance by Zabbix agent template to lar-book-disc-win

16. After clicking on the blue **Update** button, we can navigate to **Monitoring | Latest data**. Add the following filters:

Figure 7.34 – Latest data filter on host lar-book-disc-win

17. We can now see our three newly created items:

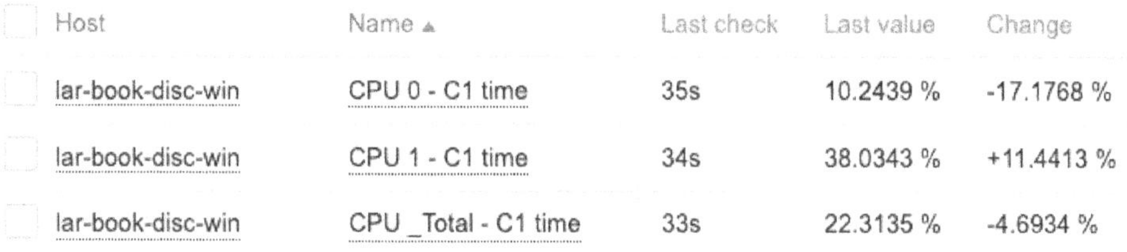

Host	Name ▲	Last check	Last value	Change
lar-book-disc-win	CPU 0 - C1 time	35s	10.2439 %	-17.1768 %
lar-book-disc-win	CPU 1 - C1 time	34s	38.0343 %	+11.4413 %
lar-book-disc-win	CPU _Total - C1 time	33s	22.3135 %	-4.6934 %

Figure 7.35 – The Monitoring | Latest data page for our host lar-book-disc-win

How it works...

Windows performance counters have been around for a long time and they are very important to anyone who wants to monitor Windows machines with Zabbix. Using LLD in combination with Windows performance counters makes it a lot easier and more flexible to build solid Windows monitoring.

In this recipe, we created a very simple but effective Windows performance counter discovery rule by adding the discovery rule with the perf_instance.discovery[Processor] item key. The [Processor] part of this item key directly correlates to the perfmon.exe window we saw. If we look at the following screenshot, we already see **Processor** listed:

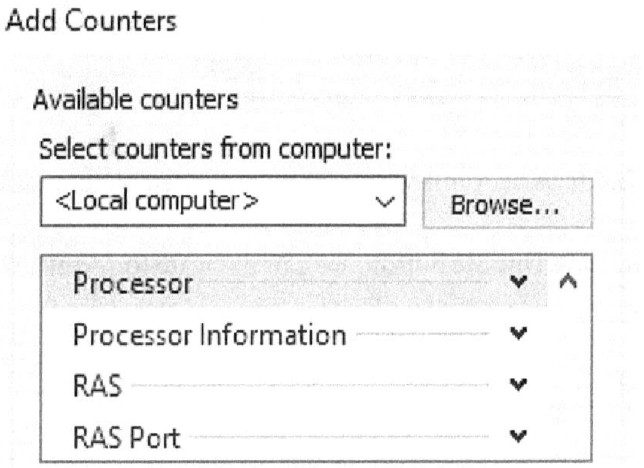

Add Counters

Available counters

Select counters from computer:

<Local computer>   ∨   Browse...

Processor   ∨   ∧

Processor Information   ∨

RAS   ∨

RAS Port   ∨

Figure 7.36 – perfmon.exe | Add Counters – Processor

When our discovery rule polls this item key, Zabbix agent will return the following value for our host:

```
[
    {
        "{#INSTANCE}":"0"
    },
    {
        "{#INSTANCE}":"1"
    },
    {
        "{#INSTANCE}":"_Total"
    }
]
```

This value means that Zabbix will fill the {#INSTANCE} macro with three values:

- 0
- 1
- _Total

We can then use these three values by using the `{#INSTANCE}` macro in **Item prototype**, as we did here:

Name ▲
Key
CPU {#INSTANCE} - C1 time
perf_counter_en["\Processor({#INSTANCE})\% C1 Time"]

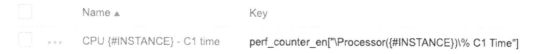

Figure 7.37 – Our created item prototype, CPU C1 time

It will then create three items with our macro values, with the right keys to monitor the second part of our counter – `% C1 time`. If you expand the window in your `perfmon.exe` file, you can see all the different counters we could add to our item prototypes to monitor more Windows performance counters:

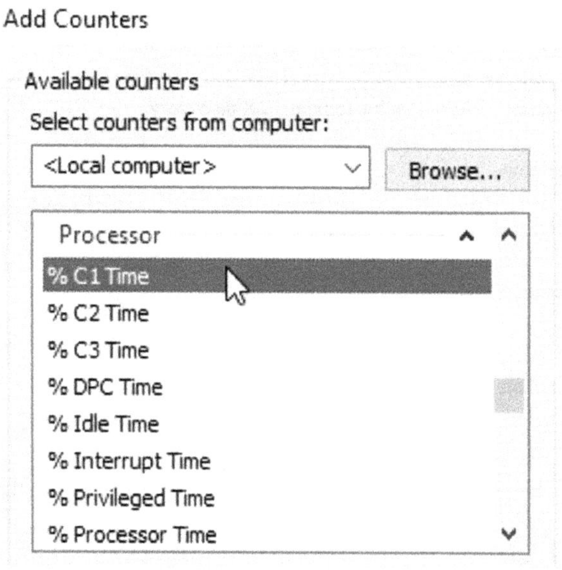
Add Counters
Available counters
Select counters from computer:
<Local computer>
Browse...
Processor
% C1 Time
% C2 Time
% C3 Time
% DPC Time
% Idle Time
% Interrupt Time
% Privileged Time
% Processor Time

Figure 7.38 – Perfmon.exe | Add Counters – Processor expanded

Discovering JMX objects

In *Chapter 3*, *Setting Up Zabbix Monitoring*, we went over setting up JMX monitoring in the recipe titled *Setting up JMX monitoring*. What we didn't cover yet though was discovering JMX objects.

In this recipe, we will go over how to set up JMX objects with LLD, and after you've finished this recipe, you'll know just how to set it up.

Getting ready

For this recipe, we will need the JMX host that you set up for the *Setting up JMX monitoring* recipe in *Chapter 3, Setting Up Zabbix Monitoring*. Make sure to finish that recipe before working on this one.

We will also need our Zabbix server with our Zabbix JMX host titled `lar-book-jmx`.

How to do it...

1. Let's start this recipe off by logging in to our Zabbix frontend and navigating to **Data collection | Templates**.

2. Create a new template by clicking on **Create template** in the top-right corner. Fill in the following fields:

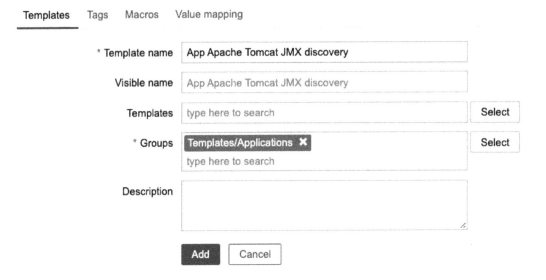

Figure 7.39 – The App Apache Tomcat JMX discovery template creation

3. After clicking the blue **Add** button, you will be taken back to **Data collection | Templates**. Click on your new **App Apache Tomcat JMX discovery** template.

4. We will now add our JMX discovery rule. Click on **Discovery rules** next to our template name.

5. Now, click on **Create discovery rule** and fill in the following fields:

Figure 7.40 – The Discover JMX object MemoryPool discovery rule creation

6. Click on the blue **Add** button at the bottom of the page. Then, click on **Item prototypes** next to your newly created **Discover JMX object MemoryPool** discovery rule.

7. We will now click on the **Create item prototype** button in the top-right corner and create the following item prototype:

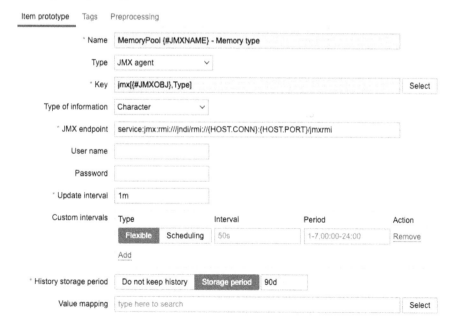

Figure 7.41 – The Item prototype creation page – MemoryPool Memory type

8. Also, make sure that on the **Tags** tab, you add a new tag with the name of `component` and a value of `memory pool`:

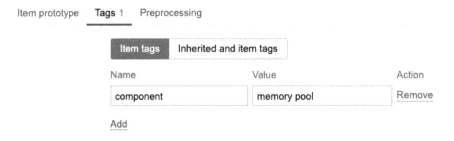

Figure 7.42 – The Tags tab – the MemoryPool Memory type

9. Let's click on the blue **Add** button and move on.

10. Go to **Data collection | Hosts** and click on `lar-book-jmx`. We will add our template to this host.

11. Click on **Templates** and add the template, like this:

Figure 7.43 – Data collection | Host – add a template to the lar-book-jmx host

12. Click on the blue **Update** button.

13. When we navigate to **Monitoring | Latest data** now, we will select `lar-book-jmx` for **Hosts** and `component` for **Tags** with `memory pool` as its value, like this:

Figure 7.44 – The Monitoring | Latest data page filters – the lar-book-jmx host

14. We will then see the following results:

	Host	Name ▲	Last check	Last value	Change	Tags
	lar-book-jmx	MemoryPool: Code Cache - Memory type	13s	NON_HEAP		component: memory p...
	lar-book-jmx	MemoryPool: Compressed Class Space - Memory type	13s	NON_HEAP		component: memory p...
	lar-book-jmx	MemoryPool: Eden Space - Memory type	13s	HEAP		component: memory p...
	lar-book-jmx	MemoryPool: Metaspace - Memory type	13s	NON_HEAP		component: memory p...
	lar-book-jmx	MemoryPool: Survivor Space - Memory type	13s	HEAP		component: memory p...
	lar-book-jmx	MemoryPool: Tenured Gen - Memory type	13s	HEAP		component: memory p...

Figure 7.45 – The Monitoring | Latest data page for the lar-book-jmx host with our results

How it works...

Monitoring JMX applications can be quite daunting at first, as there is a lot of work to figure out while building your own LLD rules. But now that you've built your first LLD rule for JMX, there is a clear structure in it.

First, for our discovery rule, we've picked the item key:

```
jmx.discovery[beans,"*:type=MemoryPool,name=*"]
```

`MemoryPool` is what we call an **MBean** in Java. We poll this MBean object for several JMX objects and fill the macros accordingly.

We picked the `name=*` object to fill the `{#JMXNAME}` macro in this discovery rule. Our macro is then used in our item prototype to create our items.

Our items are then created, like this:

Name	Triggers	Key
Discover JMX object MemoryPool: MemoryPool Metaspace - Memory type		jmx["java.lang:type=MemoryPool,name=Metaspace",Type]
Discover JMX object MemoryPool: MemoryPool Tenured Gen - Memory type		jmx["java.lang:type=MemoryPool,name=Tenured Gen",Type]
Discover JMX object MemoryPool: MemoryPool Eden Space - Memory type		jmx["java.lang:type=MemoryPool,name=Eden Space",Type]
Discover JMX object MemoryPool: MemoryPool Survivor Space - Memory type		jmx["java.lang:type=MemoryPool,name=Survivor Space",Type]
Discover JMX object MemoryPool: MemoryPool Compressed Class Space - Memory type		jmx["java.lang:type=MemoryPool,name=Compressed Class Space",Type]
Discover JMX object MemoryPool: MemoryPool Code Cache - Memory type		jmx["java.lang:type=MemoryPool,name=Code Cache",Type]

Figure 7.46 – Items on our JMX-monitored host

If we look at the keys of the items, we can see that we poll the `Type` JMX attribute on every `MemoryPool` with different names.

That's how we create JMX LLD rules with ease.

There's more...

If you are not familiar with MBeans, then make sure to check out the Java documentation. This will explain to you a lot about what MBeans are and how they can be used for monitoring JMX attributes: `https://docs.oracle.com/javase/tutorial/jmx/mbeans/index.html`.

> **Tip**
> Before diving deeper into using JMX object discovery, dive deeper into the preceding JMX object documentation. There's a lot of information in it and it will greatly improve your skills in creating these LLD rules.

Setting up Zabbix SNMP LLD the new way

Zabbix 6.4 introduced an overhaul to using SNMP in our Zabbix environments. Although the old way is still available (and explained in this book) as an option, it might be better to use the new way to build your SNMP monitoring as it will actually use the `GetBulk` requests. This makes SNMP monitoring a lot more efficient and less strenuous on the SNMP device we are collecting data from.

Getting ready

Before starting with the recipe, please make sure to read *Chapter 3, Setting Up SNMP Monitoring the New Way*, first. We will need the knowledge from that chapter to set up SNMP LLD discovery as well as we will use some hosts and items from that chapter. Make sure you have the following:

- Your Zabbix environment
- The `lar-book-snmp_bulk` host as set up in *Chapter 3, Setting Up Zabbix Monitoring*

How to do it...

As we have already set up the SNMP server to start monitoring in *Chapter 3, Setting Up Zabbix Monitoring*, we can start immediately on the frontend. In *Chapter 5, Building your own Structured Templates*, we also learned about creating templates for all our monitoring, so let's start by doing that. Follow these steps:

1. In the Zabbix frontend, navigate to **Data collection | Templates** and click on **Create template** in the top-right corner. We will create a new template as follows.

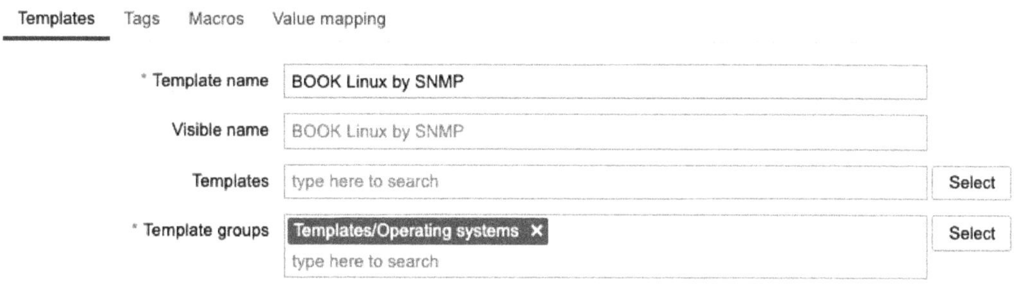

Figure 7.47 – The BOOK Linux by SNMP template

2. Also, make sure to switch to the **Tags** tab to add some tags according to the new tag policy:

Name	Value	
class	os	Remove
target	linux	Remove

Add

Figure 7.48 – Template BOOK Linux by SNMP Macros tab

3. At the bottom of the window, click on the big **Add** button to finish setting up this new template.

    We already created a value mapping and some items on the `lar-book-snmp_bulk` host we set up earlier. Let's start by using the mass update functionally to copy the value mapping to our new template.

4. Select your template in the list with the checkbox and click on the big **Mass update** button at the bottom of the window.

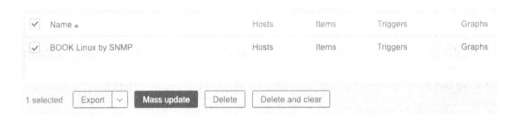

Figure 7.49 – Mass update

5. At **Mass update**, switch to the **Value mapping** tab, check the box, and click on the small dotted underlined **Add from host** button.

6. Find your `lar-book-snmp_bulk` host and select **Interface Up/Down** from the list. It should look like this.

Figure 7.50 – Mass update – add value mapping from the host

7. You can now press the big **Update** button at the bottom of the window to add this value mapping.

8. Now, let's copy over our existing items from the template. Go to **Data collection | Hosts** and go to **Items** for `lar-book-snmp_bulk`. Select the two items we created earlier and click on **Copy** at the bottom of the window.

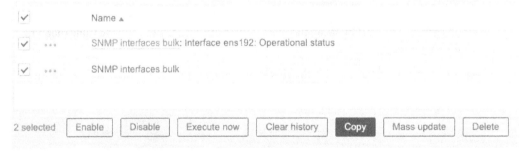

Figure 7.51 – The lar-book-snmp_bulk items to copy

9. Set **Target type** to **Templates** and type in BOOK Linux by SNMP. Select it and then press **Copy**.

Figure 7.52 – The lar-book-snmp_bulk items copy window

10. Now, go back to **Data collection | Templates** and click on **Discovery** for your **BOOK Linux by SNMP** template.

11. In the top-right corner, click on the **Create discovery rule** button. We will create the following LLD rule here.

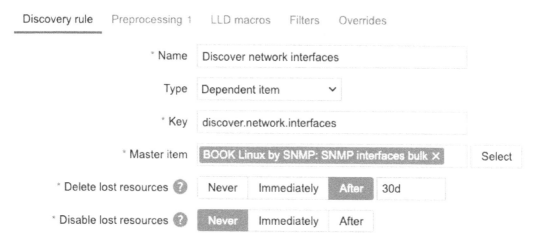

Figure 7.53 – Discovery rule

We will make the LLD rule of the **Dependent item** type to make sure we use the data collected in bulk earlier on the **SNMP interfaces bulk** item. However, all LLD data has to be presented in the JSON data format, so let's make sure to convert the data first.

12. Switch to the **Preprocessing** tab and add the following.

Figure 7.54 – Preprocessing

13. Now, press the big **Add** button at the bottom of the window to add the LLD rule. Then, go to **Item prototypes** to add our first item in an automated manner.

14. In the top-right corner, press **Create item prototype** and create the following item prototype:

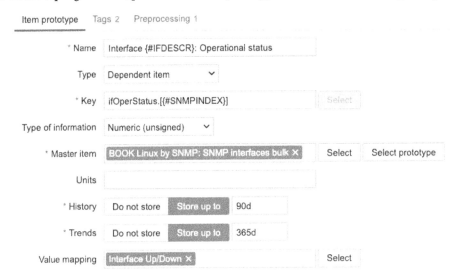

Figure 7.55 – Item prototype

15. Don't forget to add your tags at the **Tags** tab before adding the item prototype:

Figure 7.56 – Item tags

16. We will need a preprocessing step to extract the right information as well, so let's add that too by going to the **Preprocessing** tab:

Figure 7.57 – Preprocessing

17. Now, press the big **Add** button at the bottom of the page to finish setting up this new item prototype.

18. We aren't using the template on our host yet, so let's navigate to **Data collection | Hosts** and click on our `lar-book-snmp_bulk` host. Then, add the template.

Figure 7.58 – Adding the BOOK Linux by SNMP template to lar-book-snmp_bulk

19. Click on **Update** at the bottom of the window to add the template.

20. The LLD rule should now be added and executed. Let's see whether the items are created by navigating to **Monitoring | Latest data** and filtering on the `lar-book-snmp_bulk` host. Please keep in mind it can take around one minute for the item to show up and another minute for it to collect data.

Figure 7.59 – lar-book-snmp_bulk – latest data after LLD rule

How it works...

So, how does this new LLD discovery work? As you might have noticed, we are still using the same item as we used in *Chapter 3, Setting Up Zabbix Monitoring*. The values are for everything under OID `.1.3.6.1.2.1.2.2.1` are being collected in bulk still. As a remember of the bulk metric collection, let's have another look at the data:

Timestamp	Value
2023-03-18 10:31:30	`.1.3.6.1.2.1.2.2.1.1.1 = INTEGER: 1`
	`.1.3.6.1.2.1.2.2.1.1.2 = INTEGER: 2`
	`.1.3.6.1.2.1.2.2.1.2.1 = STRING: lo`
	`.1.3.6.1.2.1.2.2.1.2.2 = STRING: ens192`
	`.1.3.6.1.2.1.2.2.1.3.1 = INTEGER: 24`
	`.1.3.6.1.2.1.2.2.1.3.2 = INTEGER: 6`
	`.1.3.6.1.2.1.2.2.1.4.1 = INTEGER: 65536`
	`.1.3.6.1.2.1.2.2.1.4.2 = INTEGER: 1500`
	`.1.3.6.1.2.1.2.2.1.5.1 = Gauge32: 10000000`
	`.1.3.6.1.2.1.2.2.1.5.2 = Gauge32: 4294967295`
	`.1.3.6.1.2.1.2.2.1.6.1 = STRING:`
	`.1.3.6.1.2.1.2.2.1.6.2 = STRING: 0:50:56:9a:25:79`
	`.1.3.6.1.2.1.2.2.1.7.1 = INTEGER: 1`
	`.1.3.6.1.2.1.2.2.1.7.2 = INTEGER: 1`
	`.1.3.6.1.2.1.2.2.1.8.1 = INTEGER: 1`
	`.1.3.6.1.2.1.2.2.1.8.2 = INTEGER: 1`
	`.1.3.6.1.2.1.2.2.1.9.1 = 0`
	`.1.3.6.1.2.1.2.2.1.9.2 = 0`
	`.1.3.6.1.2.1.2.2.1.10.1 = Counter32: 0`
	`.1.3.6.1.2.1.2.2.1.10.2 = Counter32: 22457056`
	`.1.3.6.1.2.1.2.2.1.11.1 = Counter32: 0`
	`.1.3.6.1.2.1.2.2.1.11.2 = Counter32: 30150`
	`.1.3.6.1.2.1.2.2.1.12.1 = Counter32: 0`
	`.1.3.6.1.2.1.2.2.1.12.2 = Counter32: 455`
	`.1.3.6.1.2.1.2.2.1.13.1 = Counter32: 0`
	`.1.3.6.1.2.1.2.2.1.13.2 = Counter32: 6316`
	`.1.3.6.1.2.1.2.2.1.14.1 = Counter32: 0`
	`.1.3.6.1.2.1.2.2.1.14.2 = Counter32: 0`
	`.1.3.6.1.2.1.2.2.1.15.1 = Counter32: 0`
	`.1.3.6.1.2.1.2.2.1.15.2 = Counter32: 0`
	`.1.3.6.1.2.1.2.2.1.16.1 = Counter32: 0`
	`.1.3.6.1.2.1.2.2.1.16.2 = Counter32: 2153620`
	`.1.3.6.1.2.1.2.2.1.17.1 = Counter32: 0`
	`.1.3.6.1.2.1.2.2.1.17.2 = Counter32: 13936`
	`.1.3.6.1.2.1.2.2.1.18.1 = Counter32: 0`
	`.1.3.6.1.2.1.2.2.1.18.2 = Counter32: 0`
	`.1.3.6.1.2.1.2.2.1.19.1 = Counter32: 0`
	`.1.3.6.1.2.1.2.2.1.19.2 = Counter32: 0`
	`.1.3.6.1.2.1.2.2.1.20.1 = Counter32: 0`
	`.1.3.6.1.2.1.2.2.1.20.2 = Counter32: 0`
	`.1.3.6.1.2.1.2.2.1.21.1 = Gauge32: 0`
	`.1.3.6.1.2.1.2.2.1.21.2 = Gauge32: 0`
	`.1.3.6.1.2.1.2.2.1.22.1 = OID: .0.0`
	`.1.3.6.1.2.1.2.2.1.22.2 = OID: .0.0`

Figure 7.60 – The lar-book-snmp_bulk raw bulk metrics

We have all the data we need right there in the SNMP walk item. All of the items and discovery rules we then added afterward are using that data and parsing it internally using the Zabbix server (or proxy) preprocessing processes.

In the case of LLD, we have to add the SNMP walk to JSON preprocessing step, as you can see in *Figure 7.54*, which is what will convert the normal SNMP walk data to a JSON data format. It will look like this afterward:

```
[
    {
        "{#SNMPINDEX}":"2",
        "{#IFDESCR}":"ens192"
    },
    {
        "{#SNMPINDEX}":"1",
        "{#IFDESCR}":"lo"
    }
]
```

Figure 7.61 – The lar-book-snmp_bulk raw bulk metrics converted to JSON

It collects the values we want by finding the OID `.1.3.6.1.2.1.2.2.1.2` and adding its values to the `{#IFDESCR}` discovery macro. It also retains the SNMP index and it's it to the `{#SNMPINDEX}` macro.

Now, all that's left to do is set up our item prototypes and use the same raw SNMP walk item to extract data with the preprocessing step SNMP walk value as we see in *Figure 7.57*. We also have to make sure it's unique for each item, so we add the `{#SNMPINDEX}` macro to find the correct value for every item that will be created.

Just like that, we did a single SNMP `GetBulk` call in the **SNMP interfaces bulk** item and used the power of Zabbix-dependent items and preprocessing to split it up further.

Creating hosts with LLD and custom JSON

Creating hosts from LLD works the same as creating anything else from LLD rules. We will simply feed our Zabbix installation with a compatible JSON formatted dataset and use that data to create new hosts. However, starting with Zabbix 6.2, something has changed. Hosts created by LLD are now customizable after creation, so, let's have a look at how to do it and how it works.

Getting ready

For this recipe, we are going to need two things: any Zabbix 7 installation and a compatible JSON-formatted dataset containing hosts and their data. Some good default template examples to create hosts from LLD are as follows:

- VMware host and hypervisors
- Kubernetes
- Azure and AWS

For the example, however, we will be using a custom dataset, which you can find on GitHub here: `https://github.com/PacktPublishing/Zabbix-7-IT-Infrastructure-Monitoring-Cookbook/blob/main/chapter07/lldhosts.json`.

It's also important to have a basic understanding of Zabbix sender, dependent items, and preprocessing. I recommend reading the following recipes from *Chapter 3, Setting Up Zabbix Monitoring*, first:

- *Creating Zabbix simple checks and the Zabbix trapper*
- *Working with calculated and dependent items*
- *Using Zabbix preprocessing to alter item values*

How to do it...

Let's get started on building this new LLD rule with a custom JSON dataset. To do that, we will first need to build a JSON file or get it from some of our own data sources. We have one prepared for you:

1. First, let's have a look at the JSON file located here:

   `https://raw.githubusercontent.com/PacktPublishing/Zabbix-7-IT-Infrastructure-Monitoring-Cookbook/main/chapter07/lldhosts.json`

2. Next up, let's log in to the Zabbix server CLI and make sure we have our Zabbix sender application installed. We are going to use this to send the file to our Zabbix environment:

   For RHEL-based systems, use the following:

   ```
   dnf install zabbix-sender
   ```

   For Ubuntu systems, use the following:

   ```
   apt install zabbix-sender
   ```

3. Let's set up the `zabbix-sender` application to send this JSON file to our system every minute. We will use a CronJob to make things easy for now:

   ```
   crontab -e
   ```

4. Add the following to the CronJob file:

```
* * * * * zabbix_sender -z 127.0.0.1 -s lar-lldhost-creation -k
lldhosts.raw -o '[{"vmname":"lar-lld-host1","vmip":"10.16.16.
200","vmlocation":"Amsterdam"},{"vmname":"lar-lld-host2","vmip
":"10.16.16.201","vmlocation":"London"},{"vmname":"lar-lld-
host3","vmip":"10.16.16.202","vmlocation":"Chicago"},{"vmname":"
lar-lld-host4","vmip":"10.16.16.203","vmlocation":"Tirana"}]' >/
dev/null 2>&1
```

The preceding is just the CronJob + Zabbix sender command built up as follows:

```
* * * * *
zabbix_sender -z 127.0.0.1 -s lar-lldhost-creation -k lldhosts.
raw -o
FILE CONTENT FROM GITHUB
>/dev/null 2>&1
```

5. Now, let's switch to our Zabbix frontend where we will add a host to receive the JSON file from.

6. To add the host, navigate to **Data collection | Hosts** and click on the **Create host** button in the top-right corner. Add the following host:

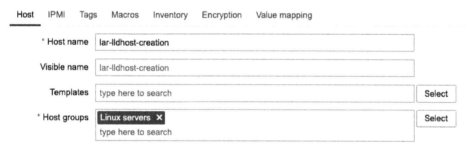

Figure 7.62 – The Zabbix host creation window for the lar-lldhost-creation host

7. Click on the **Add** button at the bottom of the window to finish creating this host. You will be brought back to **Data collection | Hosts**.

8. On this page, click on **Items** next to the `lar-lldhost-creation` host we just created. In the top-right corner, click on **Create item** and create the following item:

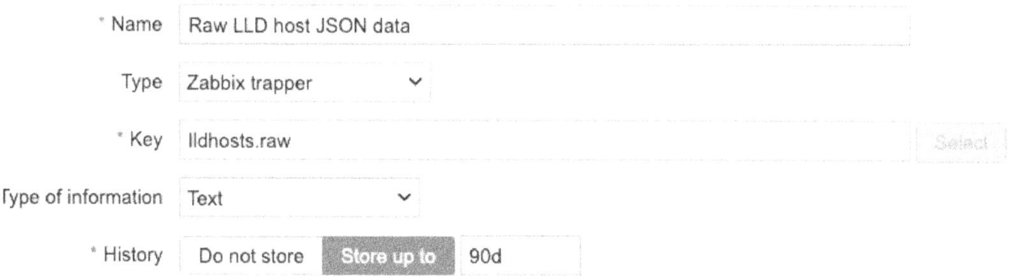

Figure 7.63 – The Zabbix item creation page for the lar-lldhost-creation host

9. Make sure to also add a tag to the item:

Item   **Tags** 1   Preprocessing

Item tags	Inherited and item tags

Name		Value	
component		raw	Remove

Add

Figure 7.64 – The Zabbix tag creation page for the lldhosts.raw item

10. Now, press the big **Add** button at the bottom of the page. After the Zabbix server reloads its configuration cache, we should see data coming into this item within a minute or two.

11. Meanwhile, we can start building the LLD rule. While still on the host edit page, navigate to **Discovery rules**. Click on **Create discovery rule** in the top-right corner. We will add the following:

Figure 7.65 – The Zabbix LLD rule creation page for hosts.from.json

12. Next, click on **LLD macros**, and let's define some macros to use from the file:

```
{
    "vmname":"lar-lld-host1",
    "vmip":"10.16.16.200",
    "vmlocation":"Amsterdam"
}
```

We are going to use JSONPath to convert the preceding blocks of data to the following:

```
{
    "{#VMNAME}":"lar-lld-host1",
    "{#VMIP}":"10.16.16.200",
    "{#VMLOCATION}":"Amsterdam"
}
```

13. To do so, switch to the tab called **LLD macros** and define the following:

Figure 7.66 – The Zabbix LLD macros tab for the hosts.from.json LLD rule

14. Now, we can click on the big **Add** button at the bottom of this page and the LLD rule is created.

15. To do something with the LLD rule, however, we will have to create a new host prototype. To do so, click on **Host prototypes** and then **Create host prototype** in the top-right corner. We will add the following:

Figure 7.67 – The Zabbix LLD host prototype creation page

As you can see, we use our macros to create the name of the **virtual machine** (**VM**) as well as make a unique host group.

16. We will also define an interface using our macro, by switching interface to **Custom**, pressing the small underlined **Add** button, and adding the following:

Figure 7.68 – The Zabbix LLD host prototype creation page with a custom interface

17. You can now press the big **Add** button at the bottom of the page. This will add the host prototype.

18. Now, navigate to **Data collection | Hosts** where, after waiting for a few minutes, we should see our new hosts:

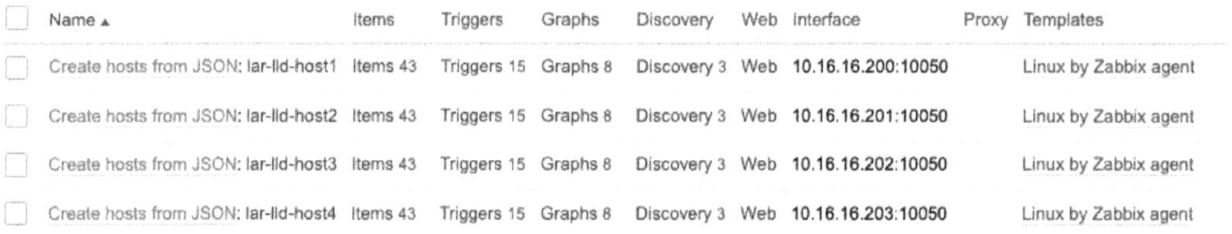

Name ▲	Items	Triggers	Graphs	Discovery	Web	Interface	Proxy	Templates
Create hosts from JSON: lar-lld-host1	Items 43	Triggers 15	Graphs 8	Discovery 3	Web	10.16.16.200:10050		Linux by Zabbix agent
Create hosts from JSON: lar-lld-host2	Items 43	Triggers 15	Graphs 8	Discovery 3	Web	10.16.16.201:10050		Linux by Zabbix agent
Create hosts from JSON: lar-lld-host3	Items 43	Triggers 15	Graphs 8	Discovery 3	Web	10.16.16.202:10050		Linux by Zabbix agent
Create hosts from JSON: lar-lld-host4	Items 43	Triggers 15	Graphs 8	Discovery 3	Web	10.16.16.203:10050		Linux by Zabbix agent

Figure 7.69 – Zabbix hosts created from LLD

How it works...

LLD is an extensive topic in the Zabbix world and, as such, it can become quite complicated. As we saw, we used a completely custom JSON file in this recipe.

Before we dive deeper into what the results are, keep in mind that custom JSON can be created from anywhere. This could be a custom (Python, Perl, PowerShell) script, some API checks, or anything else. Also, sometimes, JSON is already provided by the Zabbix environment itself. As long as we follow the following format (with or without macros straight in the file), anything can be parsed to Zabbix LLD rules:

```
[
    {
        "vmname":"lar-lld-host1",
        "vmip":"10.16.16.200",
        "vmlocation":"Amsterdam"
    },
    {
        "vmname":"lar-lld-host2",
        "vmip":"10.16.16.201",
        "vmlocation":"London"
```

```
    },
    {
        "vmname":"lar-lld-host3",
        "vmip":"10.16.16.202",
        "vmlocation":"Chicago"
    },
    {
        "vmname":"lar-lld-host4",
        "vmip":"10.16.16.203",
        "vmlocation":"Tirana"
    }
]
```

Check out the following link for more examples:

`https://www.zabbix.com/documentation/current/en/manual/discovery/ low_level_discovery`

Now, what did we actually create with the preceding JSON file? Well, we used the vmname, vmip, and vmlocation JSON keys and their values to create some custom hosts, fully automated. To do that first, we have to use **JSONPath** to parse the JSON keys to the LLD macros every LLD rule needs:

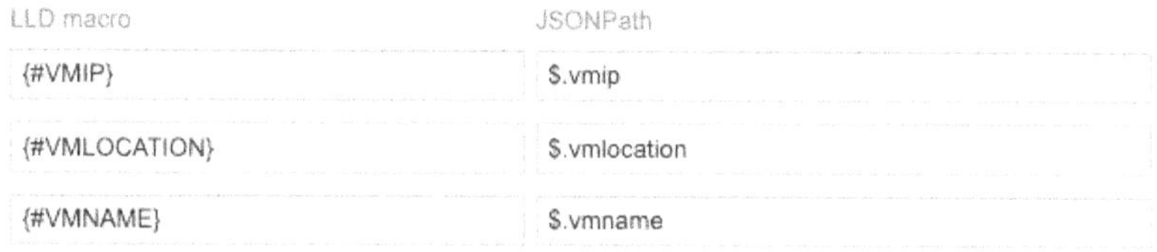

LLD macro	JSONPath
{#VMIP}	$.vmip
{#VMLOCATION}	$.vmlocation
{#VMNAME}	$.vmname

Figure 7.70 – JSONPath usage to convert keys to LLD macros

This converts vmip to {#VMIP}, vmlocation to {#VMLOCATION}, and vmname to {#VMNAME}. **JSONPath** is searching for the keys and Zabbix is converting them to macros for us.

We then use these macros in the LLD host prototype to define what the values are going to be for each host:

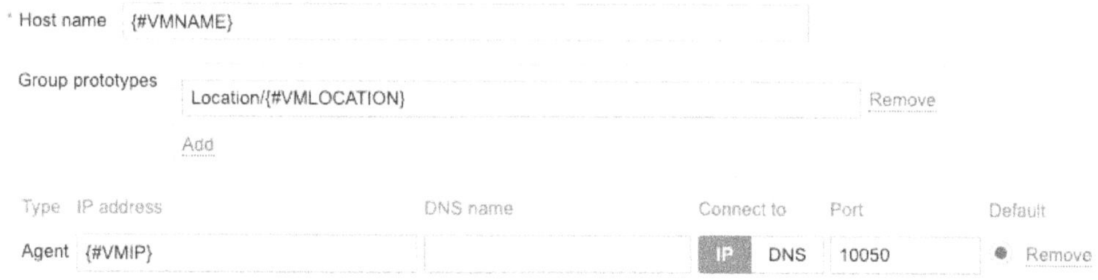

* Host name	{#VMNAME}				
Group prototypes	Location/{#VMLOCATION}				Remove
	Add				

Type	IP address	DNS name	Connect to	Port	Default
Agent	{#VMIP}		IP DNS	10050	● Remove

Figure 7.71 – Our LLD macros in use

The result will be that our hosts are created with the correct settings:

Figure 7.72 – lar-lld-host4 created by our LLD rule

As you can see, our hostname is filled in, as is the custom host group. We also have an interface defined with the correct IP. We will now use that IP on the interface to start monitoring Zabbix agent on the host with the template that we hooked up to the host.

There's more...

But wait – there's more! Starting from Zabbix 6.2, host prototypes are actually practically useful, by introducing some major changes. It is now possible to fully customize the host settings such as macros, tags, and even templates.

This means that even though a host is created from a prototype, we can still customize our monitoring. For example, if you discover your VMs in VMware, we could customize the LLD rule to also automatically start monitoring with the Zabbix agent. If you then want to override some macros for the host, to change, let's say, a trigger threshold, you actually can.

Do keep in mind, though, that if you remove the host or LLD rule that discovered the hosts, all the discovered hosts will also be removed. Be careful with removing things and always have backups at the ready!

8
Setting Up Zabbix Proxies

You can't preach about Zabbix without actually preaching about the use of Zabbix **proxies** – a nice addition at first, but a no-brainer by now. Anyone who's expecting to set up a Zabbix environment of a medium/larger size will need proxies. The main reason to use proxies is scalability, as Zabbix proxies offload the data collection and preprocessing load from the Zabbix server. This way, we can scale up our Zabbix environment further and with greater ease. Furthermore, proxies can provide an additional layer of security by collecting data in a local network and then sending it to the Zabbix server, splitting your data collection access requirements between hosts instead of consolidating it on a single Zabbix server.

In this chapter, we will first learn how to set up a Zabbix proxy. We will then learn how to work with passive and active Zabbix proxies, and also how to monitor hosts with either form of the Zabbix proxy. Zabbix 7 also introduces the possibility of adding proxies in a load-balancing pool together, which we will also discover in this chapter. We will also cover some Zabbix network discovery using the proxies, and we'll learn how to monitor Zabbix proxies to keep them healthy. After these recipes, you'll have no more excuses for not setting up the proxies, as we'll cover most of the possible forms of proxy use in this chapter.

So, let's go through the following recipes and check out how to work with Zabbix proxies:

- Setting up a Zabbix proxy
- Working with passive Zabbix proxies
- Working with active Zabbix proxies
- Monitoring hosts with a Zabbix proxy
- Encrypting your Zabbix proxy connection with pre-shared keys
- Setting up Zabbix proxy load balancing
- Using discovery with Zabbix proxies
- Monitoring your Zabbix proxies

Technical requirements

We are going to need several new Linux hosts for this chapter, as we'll be building them as Zabbix proxies.

Set up two proxies by installing your preferred RHEL-based or Ubuntu Linux distribution on the following two new hosts:

- `lar-book-proxy-passive`
- `lar-book-proxy-active`

You'll also need the Zabbix server, with at least one monitored host. We'll be using the following new host to monitor with a Zabbix agent installed:

`lar-book-agent-by-proxy`

Setting up a Zabbix proxy

Setting up a Zabbix proxy can be quite daunting if you don't have a lot of experience with Linux, but the task is quite simple once you get the hang of it. We will install a Zabbix proxy on our `lar-book-proxy-passive` server; you can repeat the task on `lar-book-proxy-active`.

Getting ready

Make sure to have a new empty Linux host, with your distribution of choice ready and installed. We won't need our Zabbix server in this recipe yet.

How to do it...

1. Let's start by logging in to the **command-line interface (CLI)** of our new `lar-book-proxy-passive` host.

2. Now, execute the following command to add the Zabbix repository. For RHEL-based systems, use the following:

    ```
    rpm -Uvh https://repo.zabbix.com/zabbix/7.0/rhel/8/ x86_64/
    zabbix-release-7.0-1.el8.noarch.rpm
    dnf clean all
    ```

    For Ubuntu systems, use the following:

    ```
    wget https://repo.zabbix.com/zabbix/7.0/ubuntu/pool/ main/z/
    zabbix-release/zabbix-release_7.0-1+ubuntu22.04_all.deb
    dpkg -i zabbix-release_7.0-1+ubuntu22.04_all.deb
    apt update
    ```

3. Now, install the Zabbix proxy by executing the following command.

    RHEL-based systems, use the following:

    ```
    dnf install zabbix-proxy-sqlite3 zabbix-selinux-policy
    ```

    Ubuntu systems, use the following:

    ```
    apt install zabbix-proxy-sqlite3
    ```

> **Tip**
>
> On RHEL-based servers, don't forget to set **Security-Enhanced Linux** (**SELinux**) to permissive or allow Zabbix proxy in SELinux for production. For lab environments, it is fine to set SELinux to permissive, but in production, I would recommend leaving it enabled. For Ubuntu systems, in a lab environment, we can disable AppArmor.

4. Now, edit the Zabbix proxy configuration by executing the following command:

    ```
    vim /etc/zabbix/zabbix_proxy.conf
    ```

    Let's start by setting the proxy mode on the `passive` proxy. The mode will be `1` on this proxy. On the `active` proxy, this will be `0`:

    ```
    ProxyMode=1
    ```

5. Change the following line to your Zabbix server address:

    ```
    Server=10.16.16.152
    ```

> **Important note**
>
> When working with a Zabbix server in **high availability** (**HA**), make sure to add the Zabbix server IP addresses here for every single node in your cluster. The proxy will only be sending data to the active node. Keep in mind that HA nodes are delimited by a semi-colon (;) instead of a comma (,).

6. Change the following line to your proxy hostname:

    ```
    Hostname=lar-book-proxy-passive
    ```

    As we'll be using the `sqlite` version of the proxy for the example, change the `DBName` parameter to the following:

    ```
    DBName=/tmp/zabbix_proxy.sqlite3
    ```

7. You can now enable the Zabbix proxy and start it with the following two commands:

    ```
    systemctl enable zabbix-proxy
    systemctl start zabbix-proxy
    ```

8. You should want to check that the Zabbix proxy logs are not showing any errors, with the following command:

```
tail -f /var/log/zabbix/zabbix_proxy.log
```

How it works...

There are three versions of Zabbix proxy to work with:

- `zabbix-proxy-mysql`
- `zabbix-proxy-pgsql`
- `zabbix-proxy-sqlite3`

We've just done the setup for the `zabbix-proxy-sqlite3` package, which is the easiest method if you ask me. The `sqlite3` version of Zabbix proxy makes it possible for us to set up a Zabbix proxy with great ease as we don't actually need to worry too much about database setup and maintenance.

Please do note that the `sqlite3` versions might not be suited to Zabbix proxies with a very high load as `sqlite3` cannot be further scaled up. You get more options to scale a `mysql` or `postgresql` database when using Zabbix proxy by the fine-tuning mechanisms available in those database types. However, since the idea behind proxies in most cases is dividing the load between proxies, I find it easier to add the SQLite3 proxy once my database has reached its capacity.

> **Tip**
> Always pick the right type of Zabbix proxy for what you expect to need in the future. Although it is easy to switch proxies later, don't go too easy on this choice as you might save yourself hours in the future.

The amazing part about the `sqlite3` version is that if we run into database issues, it's very easy to just remove the database by running the following:

```
rm /tmp/zabbix_proxy.sqlite3
```

With this, the `sqlite3` Zabbix proxy then automatically creates a new database on startup, and we're all ready to start collecting again. Do note that we might lose some information that is in the proxy database though, which functions as a cache that might contain data still to be sent to Zabbix. If the data hasn't been sent to Zabbix yet and you delete the database, that's when you will lose the metrics still in the database.

In Zabbix 7, however, the proxy database concept has been changed slightly, by the addition of an in-memory cache. By default, proxies in Zabbix 7 will now run in one of three modes:

- Hybrid (default for new installations)
- Memory
- Disk (default for existing installations)

We can adjust this setting by editing `ProxyBufferMode=` in the Zabbix proxy configuration file.

This means that a Zabbix proxy will now buffer metrics to be sent in memory instead of in the database by default. This is only in the case that the Zabbix proxy memory cache is full with the proxy start buffering data to the database.

One last thing to keep in mind is that, in hybrid mode, we can still restart the Zabbix proxy safely; however, upon restarting, the Zabbix proxy process data will be flushed into the database. In memory mode, however, this is not the cache and data will be lost upon restarting the Zabbix proxy process.

There's more...

More information about installing Zabbix proxies can be found here:

```
https://www.zabbix.com/documentation/current/manual/installation/
install_from_packages
```

Choose the distribution you are using, and you can find the guides for all the different variants of proxy installations.

Working with passive Zabbix proxies

Now that we have installed our Zabbix proxy in the previous recipe, we can start working with it. Let's start by setting up our passive Zabbix proxy in the frontend and see what we can do with it from the start.

Getting ready

You will need the `lar-book-proxy-passive` host for this recipe ready and installed with Zabbix proxy. We will also be using our Zabbix server in this recipe again.

How to do it...

1. Let's start by logging in to our Zabbix frontend and navigating to **Administration | Proxies**:

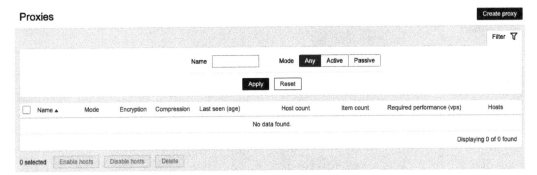

Figure 8.1 – Administration | Proxies page, without proxies

Our **Proxies** page is where we do all frontend proxy-related configuration.

2. Let's add a new proxy with the blue **Create proxy** button in the top-right corner.

    This will show us to the **Create proxy** popup, where we will fill out the following information:

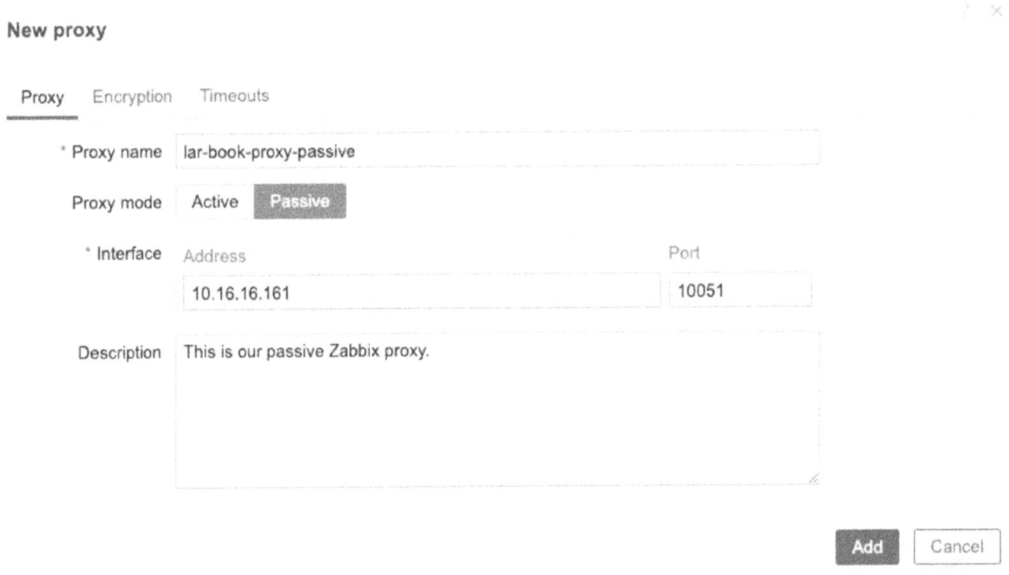

Figure 8.2 – Administration | Proxies, Create proxy popup, lar-book-proxy-passive

Before clicking the blue **Add** button, let's take a look at the **Encryption** tab:

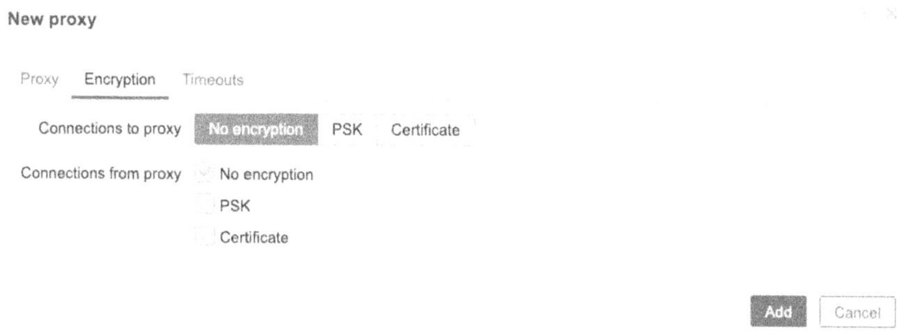

Figure 8.3 – Administration | Proxies, Create proxy popup Encryption tab, lar-book-proxy-passive

By default, **No encryption** is checked here, which we'll address in another recipe. For now, leave it set to **No encryption**.

> **Important note**
>
> A lot of valuable information is exchanged between Zabbix servers and Zabbix proxies. If you are working with insecure networks or just need an extra layer of security, use Zabbix proxy encryption. You can find more information on Zabbix encryption here: `https://www.zabbix.com/documentation/current/en/manual/encryption`.

Before we move on, there's one more tab in our Zabbix proxy creation popup that we need to have a look at – **Timeouts**:

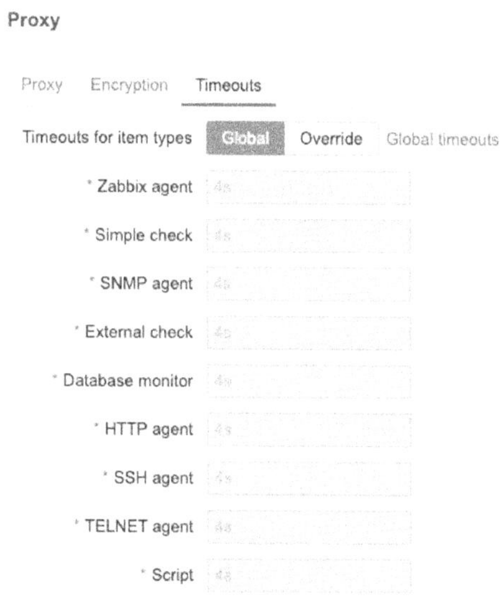

Figure 8.4 – Administration | Proxies, Create proxy popup Timeouts tab, lar-book-proxy-passive

We won't change anything on the **Timeouts** tab here, but since Zabbix 7, it is possible to override global timeouts for various types of item checks, allowing us to significantly tweak our Zabbix setup to avoid timeout performance issues.

1. Now, click the blue **Add** button, which will take us back to our proxy overview page.

2. The **Last seen (age)** part of your newly added proxy should now show a time value, instead of **Never**:

Name ▾	Mode	Encryption	Version	Last seen (age)
lar-book-proxy-passive	Passive	None	7.0.0	4s

Figure 8.5 – Administration | Proxies page, Last seen (age)

How it works...

Adding proxies isn't the hardest task after we've already done the installation part. After the steps we took in this recipe, we are ready to start monitoring with this proxy.

The proxy we just added is a **passive proxy**. These proxies work by receiving configuration from the Zabbix server, which the Zabbix server sends to the Zabbix proxy on port 10051:

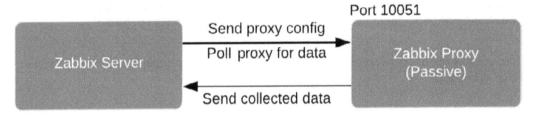

Figure 8.6 – Diagram showing an active proxy connection

Once the passive proxy knows what to monitor, every time the Zabbix server polls for data, data is sent back within the same TCP connection. This means that the connection is always initiated from the Zabbix server side. Once it's set up, the Zabbix server will keep sending configuration changes and it will keep polling for new data.

Working with active Zabbix proxies

We now know how to install and add proxies. Let's set up our active proxy, like we did with the passive proxy in the previous recipe, and see how it works.

Getting ready

You will need the lar-book-proxy-active host for this recipe, ready and installed with Zabbix proxy, as set up in the first recipe of this chapter. We will also be using our Zabbix server in this recipe.

How to do it...

1. Let's start by logging in to our Zabbix frontend and navigating to **Administration | Proxies**:

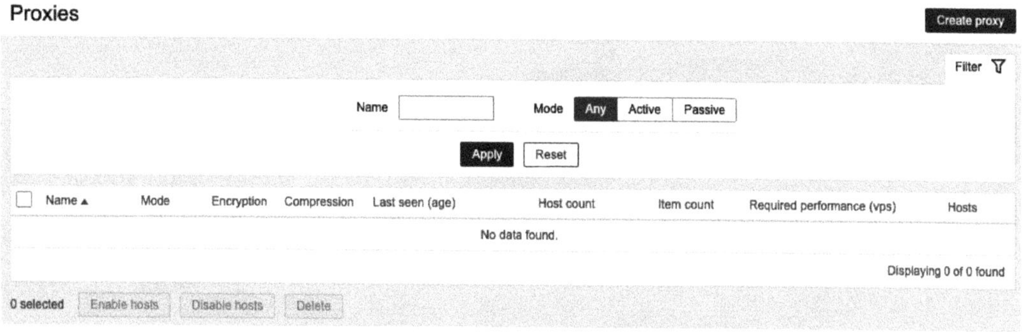

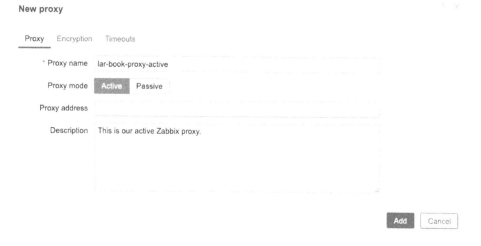

Figure 8.7 – Administration | Proxies page, no active proxies

Our **Proxies** page is where we do all configurations that are proxy related.

Let's add a new proxy by clicking the blue **Create proxy** button in the top-right corner.

2. This will open the **Create proxy** popup, where we will fill out the following information:

Figure 8.8 – Administration | Proxies, Create proxy popup, lar-book-proxy-active

---

Tip

The **Proxy address** field is actually optional for our active proxy. You do not have to add this for the Zabbix proxy to function, but it is recommended to keep things clear. Adding the **Proxy address** field also functions as a sort of whitelist in this case, as only the IP address listed will be allowed to connect.

Before clicking the blue **Add** button, let's take a look at the **Encryption** tab:

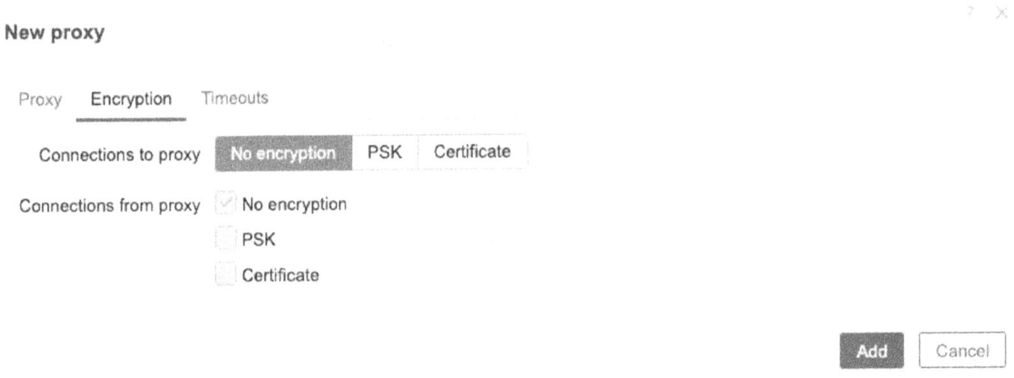

New proxy

Proxy Encryption Timeouts

Connections to proxy | No encryption | PSK | Certificate |

Connections from proxy   ✓ No encryption

  ☐ PSK

  ☐ Certificate

Add   Cancel

Figure 8.9 – Administration | Proxies, Create proxy popup Encryption tab, lar-book-proxy-active

By default, **No encryption** is checked here, which we'll leave be for this recipe.

Before we move on, there's one more tab in our Zabbix proxy creation popup that we need to have a look at – **Timeouts**:

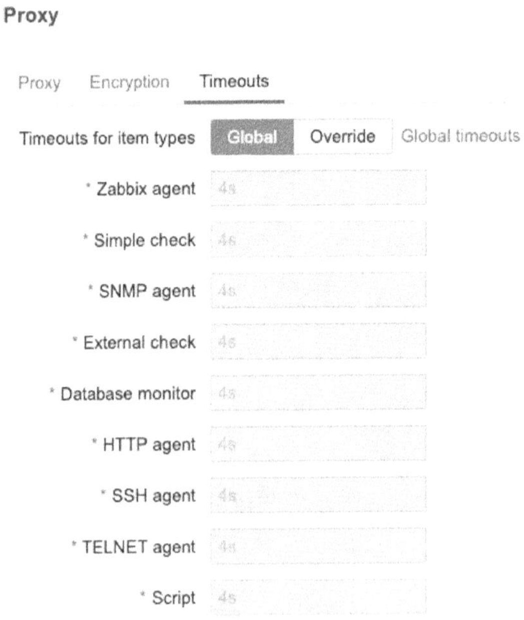

Proxy

Proxy Encryption Timeouts

Timeouts for item types | Global | Override | Global timeouts |

* Zabbix agent   4s

* Simple check   4s

* SNMP agent   4s

* External check   4s

* Database monitor   4s

* HTTP agent   4s

* SSH agent   4s

* TELNET agent   4s

* Script   4s

Figure 8.10 – Administration | Proxies, Create proxy popup Timeouts tab, lar-book-proxy-passive

We won't change anything on the **Timeouts** tab here, but since Zabbix 7, it is possible to override global timeouts for various types of item checks, allowing us to significantly tweak our Zabbix setup to avoid timeout performance issues.

1. Now, click the blue **Add** button, which will take us back to the proxy overview page.

2. Log in to the CLI and check the configuration with the following command:

   ```
   vim /etc/zabbix/zabbix_proxy.conf
   ```

   Let's change the proxy mode on the `active` proxy. The mode will be 0 on this proxy instead of 1:

   ```
   ProxyMode=0
   ```

3. Change the following line to your proxy hostname:

   ```
   Hostname=lar-book-proxy-active
   ```

4. Then, restart the proxy:

   ```
   systemctl restart Zabbix-proxy
   ```

5. The **Last seen (age)** part of your newly added proxy should now show a time value, instead of **Never**.

Name ▲	Mode	Encryption	Version	Last seen (age)
lar-book-proxy-active	Active	None	7.0.0	5s

Figure 8.11 – Administration | proxies page, Last seen (age)

Depending on your setting in the proxy configuration file, the **Last seen (age)** part may take a while to change.

How it works...

If you followed the *Working with passive Zabbix proxies* recipe from this chapter, the steps are about the same, except for the part where we add the proxy mode.

The proxy we just added is an active proxy that works by requesting a configuration from the Zabbix server on port `10051`.

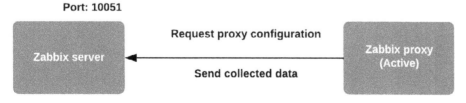

Figure 8.12 – Diagram showing an active proxy connection

The Zabbix proxy keeps requesting configuration changes, and it keeps sending any newly collected data to the Zabbix server every second or it sends out a heartbeat if no data is available.

> **Important note**
>
> It is recommended to use active Zabbix proxies, as we can use them to reduce the load on our Zabbix server. Use the passive proxy only when you have a good reason to do so.

Monitoring hosts with Zabbix proxy

We have our `active` and `passive` Zabbix proxies ready to use, so it's now time to add some hosts to them. Setting up the Zabbix frontend to monitor hosts with Zabbix proxies works in about the same way as monitoring directly from the Zabbix server. The backend and design change completely though, which I'll explain in the *How it works…* section of this recipe.

Getting ready

Make sure you have your `lar-book-proxy-passive` passive proxy and your `lar-book-proxy-active` active proxy ready by following all of the previous recipes in this chapter.

You will also need your Zabbix server and at least two hosts to monitor. We will be using `lar-book-agent_snmp` and `lar-book-agent` in the example, but any host with an active and passive Zabbix agent will work.

How to do it...

We'll configure a host on both our active and our passive proxies to show you what the difference is between these two. Let's start with the passive proxy.

Passive proxy

1. Let's start this recipe by logging in to our Zabbix frontend and navigating to **Data collection | Hosts**.

Let's add the host with the passive agent to our passive proxy. In my case, this is the `lar-book-agent_snmp` host.

Click on the `lar-book-agent_snmp` host and change the **Monitored by proxy** field to `lar-book-proxy-passive`, as in the following screenshot:

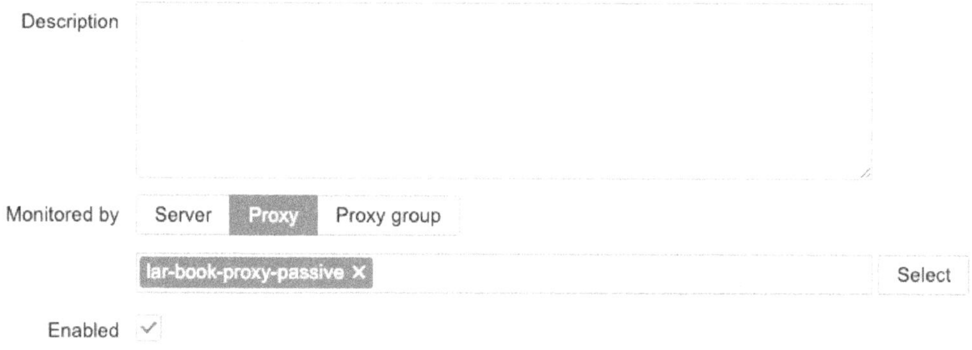

Figure 8.13 – Configuration | Hosts, Edit host page for host lar-book-agent_snmp

Now, click on the blue **Update** button. Our host will now be monitored by the Zabbix proxy.

> **Important note**
> Due to the configuration update interval and the fact we just switched the monitoring source to a proxy, we can see the SNMP icon turn gray temporarily.

Active proxy

1. Let's do the same for our other `lar-book-agent` host by navigating back to **Data collection| Hosts**.

2. Click on the `lar-book-agent` host and change the **Monitored by proxy** field to `lar-book-proxy-active`, as in the following screenshot:

Figure 8.14 – Configuration | Hosts, Edit host page for host lar-book-agent

3. Now, click on the blue **Update** button.

4. On the CLI of our monitored Linux host, the `lar-book-agent` host, execute the following command:

   ```
   vim /etc/zabbix/zabbix_agent2.conf
   ```

5. When working with an active Zabbix agent, we need to make sure to add the proxy IP address to the following line:

   ```
   ServerActive=
   ```

Our host will now be monitored by the Zabbix proxy once the proxy has received its new configuration data and the agent has executed the checks.

How it works...

Monitoring hosts with a Zabbix proxy in passive or active mode works in the same way from the frontend. We merely configure which host is monitored by which proxy in our Zabbix frontend, and it will be done.

Let's take a look at how our **Simple Network Management Protocol** (**SNMP**) agent is now monitored by the passive proxy:

Figure 8.15 – A completely passive Zabbix setup with proxy

Our passive Zabbix proxy now collects data from our SNMP agent, and after this is collected, the Zabbix server collects this data from our Zabbix proxy. Sounds like a whole process already, right?

Let's look at our active Zabbix proxy setup:

Figure 8.16 – A completely active Zabbix setup with proxy

Our active Zabbix proxy receives data from our active Zabbix agent and then sends this data to our Zabbix server. We've eliminated the part where the passive proxy is waiting to be polled in this proxy setup altogether. Furthermore, we only have to allow firewall connections going toward the proxy or server, meaning this could provide additional security.

There are some of the reasons I would always recommend working with active proxies – and even active agents – as much as possible. If we look at the following screenshot, we can see a setup that you might see at a company:

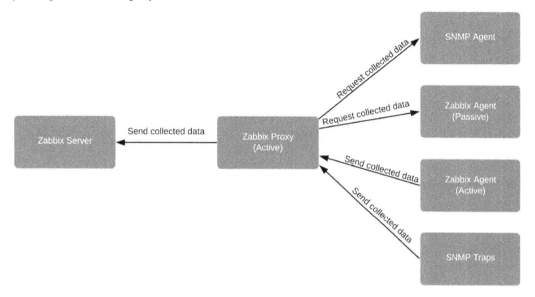

Figure 8.17 – An active Zabbix proxy setup with different monitored types

Fortunately, we have the option of using a lot of different combinations of setups. It is perfectly possible – and even logical – to combine your checks from a proxy, just as much as it would be with a Zabbix server. We can monitor all types from our proxy, whether it's a Zabbix agent, SNMP, or even **Java Management Extensions (JMX)** and the **Intelligent Platform Management Interface (IPMI)**.

> Tip
> When designing a new Zabbix hosting infrastructure, start with adding proxies to your design if possible. This way, you don't have to change a lot later. It's easy to add and change proxies, but it's harder to go from just using the Zabbix server to using Zabbix proxies in your design.

There's more...

We now have a solid setup with some proxies up and running. We've figured out the difference between active and passive proxies and how they affect monitoring. But why would we build a setup like this? Well, Zabbix proxies are great for many environments – not just the big ones, but even sometimes in the smallest ones.

We can use Zabbix proxies to offload polling and preprocessing from our main Zabbix server, thus keeping the server clear for handling data such as when writing to the Zabbix database.

We can use Zabbix proxies to monitor offsite locations, such as when you're a **managed service provider** (**MSP**) and want to monitor a customer network. We simply place a proxy on-site and monitor it. We can use industry-standard techniques such as monitoring through a **virtual private network** (**VPN**) or simply set up a connection using built-in Zabbix proxy encryption.

When the connection to the remote site goes down, our proxy will keep collecting data on-site and send this to our Zabbix server when the connection comes back up. By default, Zabbix will keep the data on disk for one hour, which is specified by the `ProxyOfflineBuffer=` parameter in the Zabbix proxy configuration file.

We can also use the Zabbix proxy to bypass firewall complications. When we place a proxy behind a firewall in a monitored network, we only need one firewall rule between the Zabbix server and the Zabbix proxy. Our Zabbix proxy then monitors the different hosts and sends the collected data in one stream to the Zabbix server. This means we don't have to poke holes for ports `161`, `162`, `10050`, and `10051`, Java, API ports, and many more through our firewall.

With this, you have loads of options to use Zabbix proxies already.

See also

Check out this interesting blog post about some more cool hidden benefits of Zabbix proxies: `https://blog.zabbix.com/hidden-benefits-of-zabbix-proxy/9359/`.

Encrypting your Zabbix proxy connection with pre-shared keys

For additional security, it's recommended to make sure your Zabbix proxy is connecting over the network encrypted. The simple reason for this is to make sure that any possible intruder on your network cannot see all the data sent between the Zabbix server and Zabbix proxy from your network in plain text.

Image you have macros configured with important passwords. These macros will flow over the network in plain text if you do not encrypt the connection between the Zabbix server and proxy.

Getting ready

For this recipe, we will need our Zabbix server and a connected proxy of either the passive or active type.

How to do it...

Let's get started on the CLI of our proxy, where we need to make some configuration changes:

1. First things first, let's edit the Zabbix proxy configuration file:

    ```
    vim /etc/zabbix/zabbix_proxy.conf
    ```

2. We will find the following and edit the following variables:

```
TLSConnect=psk
TLSAccept=psk
TLSPSKIdentity=lar-book-proxy-active
TLSPSKFile=/etc/zabbix/proxy_psk.key
```

`TLSConnect` is used for active proxy connections and `TLSAccept` is used for passive proxy connections. It is smart, however, to set both parameters at all times, as then we will always ensure an encrypted connection.

> **Important note**
>
> In the example, I'm setting the identity to the hostname of the proxy for simplicity's sake. If the PSK identity and PSK itself are a unique combination every time, you can use anything though. *Do not* use the same PSK with a different identity, as this will result in errors and the proxy won't connect.

1. Save your proxy configuration file and exit.

    We will create a new file, the `/etc/zabbix` folder, which contains the PSK. To do that, execute the following command:

    ```
    openssl rand -hex 128 > /etc/zabbix/proxy_psk.key
    ```

    This will create the file with the new pre-shared key. You can make sure it was created correctly with the following command:

    ```
    cat /etc/zabbix/proxy_psk.key
    ```

2. It should look as follows:

```
[root@lar-book-proxy-active ~]# cat /etc/zabbix/proxy_psk.key
b1fd5e5fddf3334da7e35afbae39a9f2c1dc22bdecb5708e6f3ae2c8b9f319cecc89739f34bdb255ff5
cf38b6b68109ee626cdcbc405cc9e5d49c700345006d8c6b595be49f30af5d777d965f799de583a2cab
bce25ede46583efad40ab1a04b77f62f550a9a9ab519a6a76c1f37e8ac8e2126656bfc94c4cd66ad459
8ed6d3f
```

Figure 8.18 – Created proxy PSK file

Now, let's make sure only the Zabbix proxy can read this PSK file:

```
chmod 400/etc/zabbix/proxy_psk.key
chown zabbix:zabbix /etc/zabbix/proxy_psk.key
```

3. Restart your Zabbix proxy to make the changes take effect:

    ```
    Systemctl restart zabbix-proxy
    ```

4. Now, it's time to move on to the frontend. Navigate to **Administration | Proxies**.

    You will probably see that the proxy is no longer connected and the **Last seen (age)** value is getting higher. Edit the proxy on which you are adding encryption and go to the encryption tab. For an active proxy, it will look as follows:

Proxy

Proxy	Encryption ●	Timeouts

Connections to proxy	No encryption PSK Certificate
Connections from proxy	☐ No encryption
	☑ PSK
	☐ Certificate
* PSK identity	lar-book-proxy-active
* PSK	e7f6005f1e287016e41fbf3b52dfbce5134a03b309a2be0563d33bbb7b619b9e0be37

Figure 8.19 – lar-book-proxy-active PSK settings

For a passive proxy, make sure **Connections to proxy** is set to **PSK**, and for the active proxy, select **PSK** as the option we'd like to use for **Connections from proxy**.

Fill in the **PSK identity** and **PSK** as set up in *Steps 2* and *6* of this recipe.

5. Now, click on **Update** and your proxy should connect again.

How it works...

After setting up your active or passive Zabbix proxy encryption, not much will change. Your proxy will still connect to or be connected from the Zabbix server. The thing that changes is that the Zabbix server and proxy will now use encryption when communicating with each other.

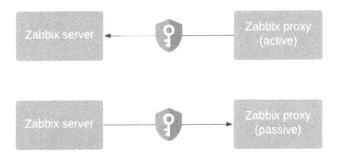

Figure 8.20 – Encrypted proxies connections

The method we have used utilizes a pre-shared key-based encryption method. Although this provides a relatively safe method of adding encryption to these connections, there's always the chance of a pre-shared key leaking somehow. Since a pre-shared key never expires, that would create a permanent hole in your security that only patches whenever you are going to change the pre-shared key. In practice, this is usually never.

One of the benefits of utilizing certificate-based encryption methods is that certificates expire. This does mean for an added layer of security, but at the same time it means that you will be regularly forced to update your encryption settings or risk losing the connection between the Zabbix proxy and server.

Setting up Zabbix proxy load balancing

Long awaited and finally implemented, Zabbix has finally introduced proxy high availability and load balancing. This completes all the required functionality to truly make Zabbix a product that is highly reliable even in cases of outages.

It also means that Zabbix is now a lot easier to scale, utilizing the load balancing on proxies to divide the load between available proxies.

Getting ready

For this recipe, we will utilize the active and passive proxy we've set up in earlier recipes in this chapter. Besides that, all we need is the Zabbix setup and a host to monitor for which we will use a Zabbix agent in active mode.

How to do it...

Let's get started on the frontend, where we should already have two (or more) Zabbix proxies available:

1. Navigate to **Administration | Proxies** and make sure you have two proxies available. It does not matter what the proxy mode is, as we can combine active and passive proxies in proxy load balancing.

Name ▾	Mode	Encryption	State	Version	Last seen (age)
lar-book-proxy-passive	Passive	None	Online	7.0.0	2s
lar-book-proxy-active	Active	PSK	Online	7.0.0	1s

Figure 8.21 – Administration | Proxies page with the two proxies we will use for load balancing

2. Next, let's navigate to **Administration | Proxy groups** to define our first proxy load balancing and high availability group.

3. In the top-right corner, click on the **Create proxy group** button. This will open the following pop-up window:

New proxy group

* Name	
* Failover period	1m
* Minimum number of proxies	1
Description	

Add   Cancel

Figure 8.22 – Administration | Proxy groups, New proxy group popup

4. Here, we basically just have to give the proxy group a name and we will be done for now. Let's name this group `lar-proxy-group`. Then, we can press the **Add** button to finish adding this proxy group.

5. Now, navigate back to **Administration | Proxies**, and let's add our two proxies to the new proxy group.

6. Click on a proxy and add it to the proxy group by filling in the **Proxy group** parameter or by using the **Select** button.

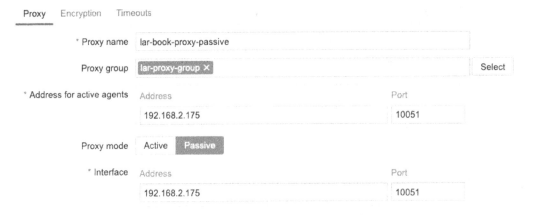

Figure 8.23 – Administration | Proxies, add proxy group to a proxy

Do this for all proxies we want to add to the group.

7. Also, make sure to fill in the **Address for active agents** parameter as when we work with active agents, we'll need to know where to connect to on each proxy.

8. We'll do the same for our second proxy:

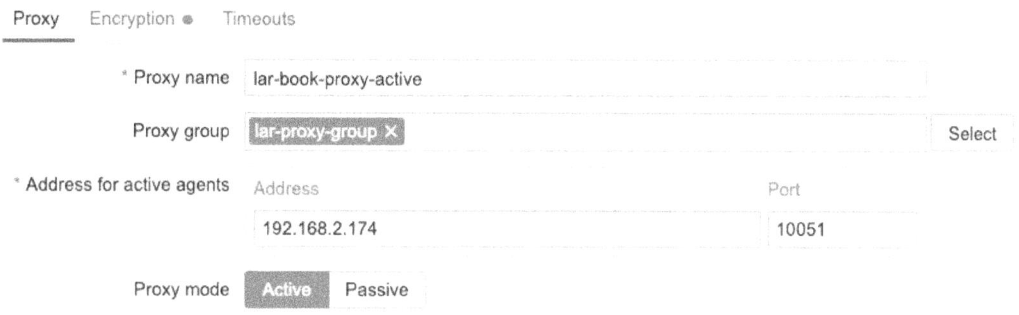

Proxy   Encryption ●   Timeouts

* Proxy name	lar-book-proxy-active
Proxy group	lar-proxy-group ×

* Address for active agents   Address 192.168.2.174   Port 10051

Proxy mode   **Active**   Passive

Figure 8.24 – Administration | Proxies, add proxy group to a proxy

9. Now, let's add a new host to be monitored by this proxy group. I will use a new host called `lar-book-proxyha-test`. When creating the host, make sure to add the **Proxy group** parameter like this:

Monitored by   Server   Proxy   **Proxy group**

lar-proxy-group ×   Select

Figure 8.25 – Data collection | Hosts, Edit or create host popup window

10. In the agent configuration file for the host we are adding, let's change some parameters. If it's a Linux host, edit the configuration file:

```
vim /etc/zabbix/zabbix_agent2.conf
```

11. Now, edit the following two parameters:

```
ServerActive=192.168.2.175,192.168.2.174
Hostname=lar-book-proxyha-test
```

12. With two proxies active and configured as part of a proxy group and a host monitored by that proxy, we are now done and ready to load balance our hosts between the proxies

How it works...

As you can see, proxy load balancing and high availability is a pretty straightforward setup and there's not much complexity to work with. We simply add our proxies to a group and Zabbix will take care of the rest for us.

In our case, we've added two proxies to the group, an active and a passive proxy. These two proxies will now work together, load balancing the load of the hosts between them.

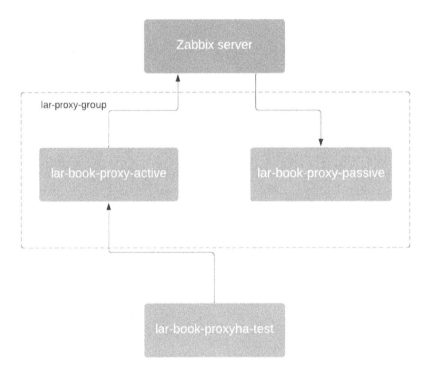

Figure 8.26 – Our proxy group working normally

Our active agent `lar-book-proxyha-test` host will connect to one of the two proxies on the defined IP address under the `ServerActive=` parameter. Because we entered **Address for active agents** under the proxy configuration in the Zabbix frontend, all the proxies in the proxy group also know the address for redirecting checks to in case load balancing needs to be done.

Our Zabbix proxy group will now make sure to load balance the hosts, even if they are in active mode. In case there is an outage, the proxy group will need to recalculate the load balancing.

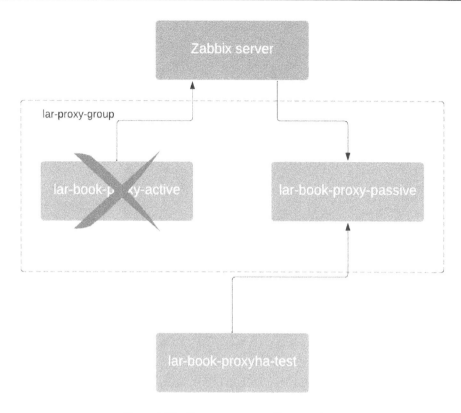

Figure 8.27 – Our proxy group after an outage

The host will now be redirected to a different proxy in the proxy group for further monitoring. This is why we do not even need any floating IPs or virtual IPs to redirect active agent checks, as our proxies can take care of it for us.

Do keep the proxy parameters in mind, however, to make sure our proxy group is functioning optimally. If you have 4 proxies in a proxy group and set **Minimum number of proxies** to 1, make sure to configure every single proxy in the group to be able to handle the full load of all hosts in the proxy group. Otherwise, your monitoring performance will still be compromised.

This is also true the other way around; if we have a proxy group of 4 proxies and we set **Minimum number of proxies** to 3, the whole proxy group will go down if we lose just two proxies. Make sure to find the correct balance between the number of proxies in a group and the minimum proxies required for the group to handle the full load of monitoring.

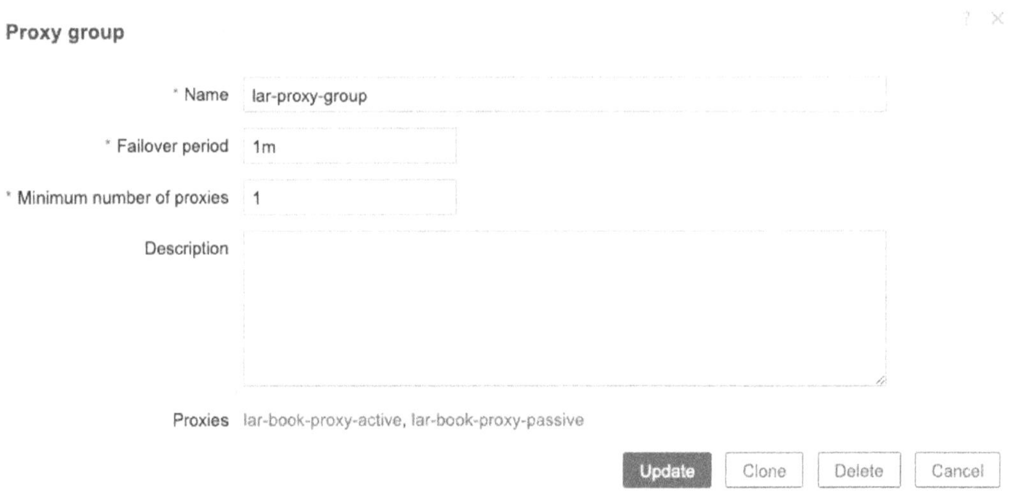

Proxy group   ? ✕

* Name	lar-proxy-group
* Failover period	1m
* Minimum number of proxies	1
Description	

Proxies   lar-book-proxy-active, lar-book-proxy-passive

[Update] [Clone] [Delete] [Cancel]

Figure 8.28 – Our proxy group settings

Furthermore, we can change the **Failover period** value to determine how fast a failover needs to happen.

> **Important note**
>
> Keep in mind that if you're working with certain other checks, you might need to do some additional configuration with things such as floating IPs or virtual IPs. For a check with SNMP traps, for example, you might need to send your traps to all proxies in a group if supported by your device, adding additional load to the network.

Using discovery with Zabbix proxies

In *Chapter 7, Using Discovery for Automatic Creation*, we talked about Zabbix and discovery. It is a very good idea to edit your discovery rules if you followed along with that chapter. Let's see how this would work in this recipe.

Getting ready

You'll need to have finished *Chapter 7, Using Discovery for Automatic Creation*, or have some discovery rules and active agent autoregistration set up.

I'll be using the `lar-book-lnx-agent-auto`, `lar-book-disc-lnx`, and `lar-book-disc-win` hosts in this example. We will also need our Zabbix server.

How to do it...

Let's start with editing our discovery rule and then move on to editing our active agent to autoregister to the proxy.

Discovery rules

Starting with Zabbix discovery rules, let's look at how to make sure we do this from the Zabbix proxy:

1. Log in to the CLI of `lar-book-disc-lnx` and edit the `/etc/zabbix/ zabbix_agent2. conf` file. Edit the following lines to include our Zabbix proxy address:

    ```
    Server=127.0.0.1,10.16.16.152,10.16.16.160,10.16.16.161
    ServerActive=10.16.16.160
    ```

2. Restart your Zabbix agent by executing the following command:

    ```
    systemctl restart zabbix-agent2
    ```

3. Now, make sure to log in to `lar-book-disc-win` and edit the `C:\Program Files\ Zabbix agent\zabbix_agent2` file. Edit the following lines to include our Zabbix proxy address:

    ```
    Server=127.0.0.1,10.16.16.152,10.16.16.160,10.16.16.161
    ServerActive=10.16.16.160
    ```

> **Important note**
>
> On the `ServerActive` lines in our configuration files, make sure to only include the Zabbix proxy we want to send data to. The Zabbix agent will actively try to send data to all our Zabbix proxies or Zabbix servers listed here, so we should only list the one we want to use.

4. Restart your Zabbix agent by executing the following commands in the Windows command line:

    ```
    zabbix_agent2.exe --stop
    zabbix_agent2.exe --start
    ```

    Or, use the Windows **Services** window to restart the agent.

    Next, navigate to **Data collection | Hosts** and delete the discovered hosts:

    ```
    lar-book-disc-lnx
    lar-book-disc-win
    ```

    We do this to prevent duplicate hosts.

5. Now, navigate to **Data collection | Discovery**.

6. Click on **Discover Zabbix Agent hosts** and change the **Discovered by proxy** field, as shown in the following screenshot:

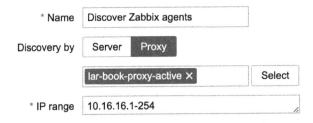

Figure 8.29 – Alerts | Actions, Discovery Actions by proxy lar-book-proxy-active

7. Click on the blue **Update** button, and that's all there is to editing your discovery rule to be monitored by a proxy.

8. You can now check out your newly discovered hosts under **Configuration | Hosts** and see that they are monitored by the proxy:

Figure 8.30 – Data collection | Hosts page with our discovered hosts

Active agent autoregistration

Moving on to active agent autoregistration, let's see how we can do this from our Zabbix proxy:

1. Start by navigating to **Data collection | Hosts** and deleting `lar-book-lnx- agent-auto`.

   To do active agent autoregistration to a proxy, we have to log in to our `lar-book-lnx-agent-auto` host CLI.

2. Edit the Zabbix agent configuration file with the following command:

   ```
   vim /etc/zabbix/zabbix_agent2.conf
   ```

3. Make sure to edit the following line to the Zabbix proxy address instead of the Zabbix server address:

   ```
   ServerActive=10.16.16.160
   ```

4. Restart the Zabbix agent:

   ```
   systemctl restart zabbix-agent2
   ```

We can now see our host autoregister to the Zabbix proxy instead of the Zabbix server:

☐	lar-book-disc-lnx_2	Applications 11	Items 42	Triggers 14	Graphs 8	Discovery 3	Web	10.16.16.156: 10050	lar-book-proxy-active
☐	lar-book-disc-win_2	Applications 11	Items 31	Triggers 12	Graphs 5	Discovery 4	Web	10.16.16.157: 10050	lar-book-proxy-active

Figure 8.31 – Data collection | Hosts page with our two auto registered hosts

How it works...

Discovery with a Zabbix proxy works the same as discovery with a Zabbix server. The only thing that changes is the location of where we are registering to or discovering from.

If you want to learn more about the process of discovery and autoregistration, check out *Chapter 7, Using Discovery for Automatic Creation*, if you haven't already.

Monitoring your Zabbix proxies

A lot of Zabbix users – or even monitoring users in general – forget a very important part of their monitoring. They forget to monitor the monitoring infrastructure. I want to make sure that when you set up Zabbix proxies, you also know how to monitor the health of these proxies.

Let's check out how to do so in this recipe.

Getting ready

For this recipe, we will need our new `lar-book-proxy-active` Zabbix proxy. We will also need our Zabbix server to monitor the Zabbix proxy.

How to do it...

We are going to build some monitoring in our Zabbix frontend, but we'll also check the integrated monitoring options for Zabbix proxies. Let's start by building our own.

Monitoring the proxy with Zabbix

We can monitor our Zabbix proxy with the Zabbix proxy itself to make sure we know exactly what's going on:

1. Let's start by logging in to our `lar-book-proxy-active` Zabbix proxy CLI.

2. Issue the following command to install Zabbix agent 2 for RHEL-based systems:

    ```
    dnf install zabbix-agent2
    ```

    For Ubuntu, issue this command:

    ```
    apt install zabbix-agent2
    ```

3. Edit the Zabbix agent configuration file by issuing the following command:

    ```
    vim /etc/zabbix/zabbix_agent2.conf
    ```

4. Edit the following lines to point toward `localhost`:

    ```
    Server=127.0.0.1
    ServerActive=127.0.0.1
    ```

5. Also, make sure to add the hostname to the Zabbix agent file:

    ```
    Hostname=lar-book-proxy-active
    ```

6. Now, log in to the Zabbix frontend and navigate to **Data collection** | **Hosts**.

7. Click on the blue **Create host** button in the top-right corner and add the following host:

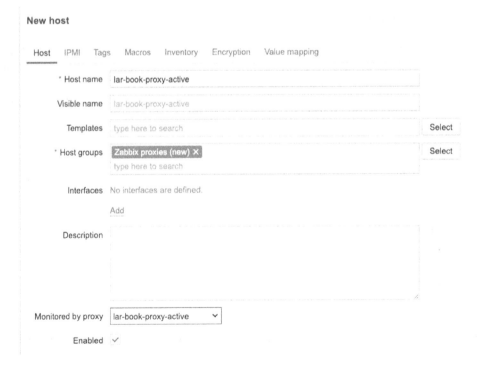

Figure 8.32 – Data collection | Hosts, Create host popup, lar-book-proxy-active

Take extra care at the **Monitored by proxy** field – we want to monitor from the proxy because we are doing Zabbix internal checks, which need to be handled by the Zabbix daemon that received this configuration.

8.   Before clicking the blue **Add** button, make sure to add **Templates**. Add the following templates to the host:

Figure 8.33 – Data collection | Hosts, Create host popup templates for host lar-book-proxy-active

9.   We can now click the blue **Add** button to create the host.

10.   Now, navigate to **Monitoring** | **Latest data** and add the following filters:

Figure 8.34 – Monitoring | Latest data page with filters, host lar-book-proxy-active

11.   We can now see our Zabbix proxy statistics, such as the number of processed values and utilization of certain internal processes:

	Host	Name ▲	Last check	Last value	Change
☐	lar-book-proxy-active	System uptime	16s	00:44:44	
☐	lar-book-proxy-active	Total memory	10s	1.77 GB	
☐	lar-book-proxy-active	Total swap space	18s	1.6 GB	
☐	lar-book-proxy-active	Version of Zabbix agent running			
☐	lar-book-proxy-active	Zabbix agent ping	48s	Up (1)	
☐	lar-book-proxy-active	Zabbix proxy: Configuration cache, % used	3s	2.6318 %	-0.001431...
☐	lar-book-proxy-active	Zabbix proxy: History index cache, % used	57s	0.2283 %	
☐	lar-book-proxy-active	Zabbix proxy: History write cache, % used	58s	0 %	

Figure 8.35 – Monitoring | Latest data page with data from our Zabbix proxy

Monitoring the proxy remotely from our Zabbix server

We can also monitor our Zabbix proxy remotely from our Zabbix server, so, let's see how that works:

1. Let's start by logging in to our `lar-book-proxy-active` host CLI and editing the following file:

    ```
    vim /etc/zabbix/zabbix_agent2.conf
    ```

2. Edit the following lines to match your Zabbix server address (every node in a Zabbix server HA cluster):

    ```
    Server=127.0.0.1,10.16.16.152
    ServerActive=127.0.0.1,10.16.16.152
    ```

3. Also, edit the following file:

    ```
    vim /etc/zabbix/zabbix_proxy.conf
    ```

4. Edit the following line to match your Zabbix server address:

    ```
    StatsAllowedIP=127.0.0.1,10.16.16.152
    ```

5. Now, navigate to the Zabbix frontend and go to **Data collection | Hosts**.

6. Click on the blue **Create host** button in the top-right corner and add the following host:

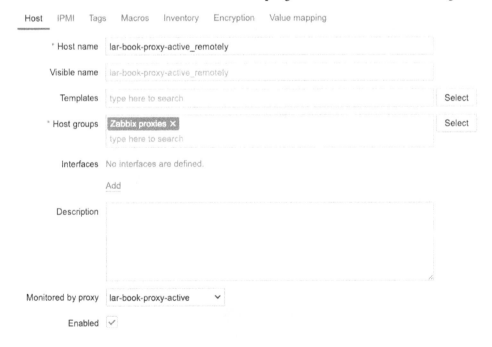

Figure 8.36 – Data collection | Hosts, Create host popup, lar-book-proxy-active_remotely

7. Before clicking the blue **Add** button, make sure to add the right **Templates**. Add the following templates to the host:

* Host name	lar-book-proxy-active_remotely
Visible name	lar-book-proxy-active_remotely
Templates	Linux by Zabbix agent active ×	Remote Zabbix proxy health × Select
	type here to search

Figure 8.37 – Data collection | Hosts, create new host popup templates, lar-book-proxy-active_ remotely

8. We can now click the blue **Add** button to create the host.

    Back at **Data collection | Hosts**, click on your new `lar-book-proxy-active_ remotely` host.

9. Go to **Macros** and add the following two macros:

Macro	Value	
{$ZABBIX.PROXY.ADDRESS}	10.16.16.153	T ˅
{$ZABBIX.PROXY.PORT}	10051	T ˅

Figure 8.38 – Data collection | Hosts, Edit host popup, Macros tab, lar-book-proxy-active_remotely

Now, click on the blue **Update** button, and that's it for this host.

10. If we navigate to **Monitoring | Latest data** and check the items for this host, we can see data coming in:

Name ▲	Last check	Last value
Zabbix stats ?		
Zabbix stats queue	26s	0
Zabbix stats queue over $5	7m 27s	0

Figure 8.39 – Monitoring | Latest data page for host lar-book-proxy-active_remotely

Monitoring the proxy from the Zabbix frontend

1. Let's start this off by navigating to **Administration | Queue**.

2. Use the drop-down menu to go to **Queue overview by proxy**:

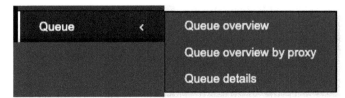

Figure 8.40 – Administration | Queue menu

This will bring us to the page shown in the following screenshot:

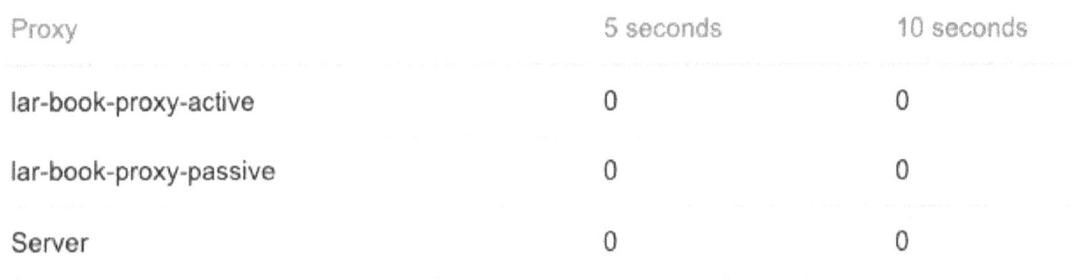

Proxy	5 seconds	10 seconds
lar-book-proxy-active	0	0
lar-book-proxy-passive	0	0
Server	0	0

Figure 8.41 – Administration | Queue overview by proxy page

Let's see how this works in the next section of our recipe.

How it works...

Monitoring your Zabbix proxies is an important task, thus we need to make sure that whenever we add a new Zabbix proxy, we are taking care of it like we would any other host.

Monitoring the proxy with Zabbix

By adding the Zabbix proxy as a host, we can make sure the Linux system that is running our Zabbix proxy is healthy. We also make sure that the Zabbix proxy applications running on this server are in good health.

Besides having the right triggers in these templates, we also get a load of options to troubleshoot issues with the Zabbix proxy.

The Zabbix proxy works just like the Zabbix server when it comes to monitoring. This means that just as with the Zabbix server, we need to keep the proxies in great health by tweaking the Zabbix proxy configuration file to our needs.

Scaling your proxies becomes a lot easier once you figure out what's going on with them. So, this is why we make sure to always monitor them. We monitor them from the proxy itself to make sure that we get the right information with the Zabbix internal checks.

Monitoring the proxy remotely from our Zabbix server

Now, when we monitor with **Remote Zabbix proxy health**, things go a little differently. Instead of doing our checks from the Zabbix proxy itself, we do them remotely from the Zabbix server by defining the Zabbix proxy address and port in the macros. The Zabbix internal check type will still be used for this, executing the checks from the Zabbix server to the Zabbix proxy remotely in this case.

On top of that, it is of course still recommended that we also keep our Zabbix agent running in either passive or active mode.

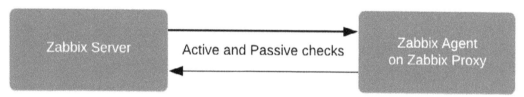

Figure 8.42 – Zabbix agent running on Zabbix proxy monitored by Zabbix server

This way, our Zabbix server is the one requesting and receiving information. Even when the proxy is having issues, the checks will still be done by the Zabbix server.

We can use this template as a way to keep a closer eye on our proxy if we suspect issues with internal checks being performed locally, or we can use this template to bypass certain firewall setups. Both are valid reasons.

Monitoring the proxy from the Zabbix frontend

From the frontend, we can use the **Administration | Queue** page to monitor our proxies. The Zabbix **Queue** page is an important page, but a lot of new users neither know nor fully utilize this page.

When a part of Zabbix starts performing poorly, such as our example Zabbix proxy here, that's when we can see stuff happening in the queue. There are six options on the Zabbix **Queue** page:

- **5 seconds**
- **10 seconds**
- **30 seconds**
- **1 minute**
- **5 minutes**
- **More than 10 minutes**

What the options in the **Queue** mean is that the Zabbix proxy has been waiting on receiving a value that's configured more than expected. I would state that anything up to one minute doesn't necessarily have to be an issue, but this depends on the type of check. The **5 minutes** or **More than 10 minutes** options can mean serious performance issues with your Zabbix proxy, and you would have to troubleshoot this issue. Make sure to keep a good eye on the Zabbix queue when you suspect issues, which are also included as triggers in the templates we added to monitor our Zabbix proxies.

9

Integrating Zabbix with External Services

In this chapter, we are going to set up some of the useful external service integrations that Zabbix has to offer. We can use these external services to notify our Zabbix users of ongoing problems.

We will start by learning how to set up in-company chat applications such as Slack and Microsoft Teams. Then, we will learn how to use a personal chat application such as Telegram before learning how to integrate Atlassian Opsgenie for even more extensive alerting.

Once you've completed these recipes, you will be able to effectively integrate certain services with Zabbix. This is a good starting point for working with external services in general and the easiest way to set up Slack, Teams, Opsgenie, and Telegram.

In this chapter, we will cover the following recipes:

- Setting up Slack alerting with Zabbix
- Setting up Microsoft Teams alerting with Zabbix
- Using Telegram bots with Zabbix
- Integrating Atlassian Opsgenie with Zabbix

Let's get started!

Technical requirements

For this chapter, we are going to need our Zabbix server, preferably how we set it up throughout this book, though any Zabbix 7 server with some alerts on it will do.

We will also need access to a few external services, as follows:

- Slack (free, to an extent):

  `https://slack.com/`

- Microsoft Teams (free, to an extent):

  `https://www.microsoft.com/microsoft-teams`

- Opsgenie (free, to an extent):

  `https://www.atlassian.com/software/opsgenie/`

- Telegram (free):

  `https://telegram.org/`

We will not cover how to set up the services themselves, only how to integrate them with Zabbix. Make sure you have set up the required service by following a guide and that you have some knowledge of the services in general.

Setting up Slack alerting with Zabbix

Slack is a widely used tool for easy text messaging, voice/video chat, and collaboration. In this recipe, we will learn how to use Zabbix Slack integration to send our Zabbix problem information to Slack so that we can gain a good overview of issues.

Getting ready

Make sure you have Slack set up. You can go to `https://slack.com/intl/en-in/` and set it up for free there. We will also need a Zabbix server with some active problems.

How to do it...

1. Once you have set up and opened Slack, you should see the following page:

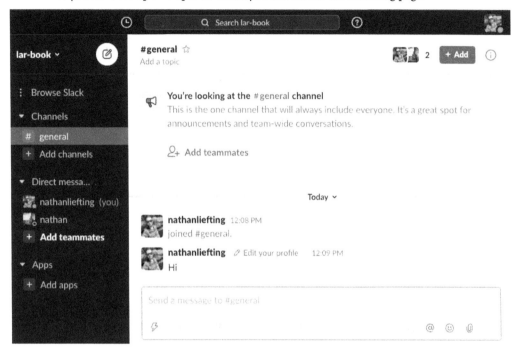

Figure 9.1 – Slack default page

2. Let's create a new channel for our Zabbix notifications by clicking the + **Add channels** button. Then, from the dropdown that appears, click **Create a new channel**:

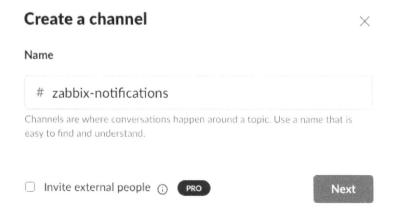

Figure 9.2 – Slack, Create a channel window

3. Click the **Next** button. Then, on the next window, click the **Create** button to make this a public (or private) channel:

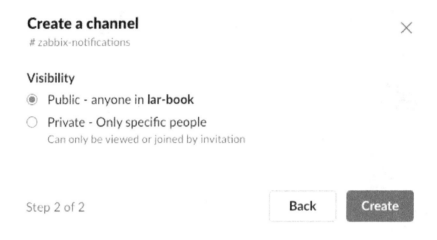

Figure 9.3 – Slack, setting the new channel visibility

4. You'll be presented with one more step to add either specific or all people within your Slack to the new channel. In my test, I'll add everyone, so I'll just click on **Done**.

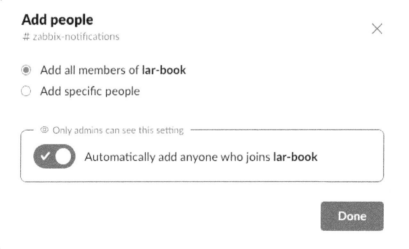

Figure 9.4 – Slack, Add people window

5. Now, navigate to the following link to create a Slack bot for working with Zabbix: `https://api.slack.com/apps`.

6. You will see the **Create New App** option on this page. Click on it, which brings us here:

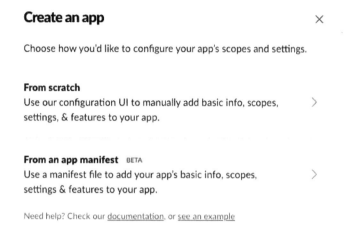

Figure 9.5 – Slack, API Create an app page

7. After clicking the **Create New App** button, we have to click on **From scratch**.

8. Then, we'll see a pop-up window where we can set up our new Slack bot:

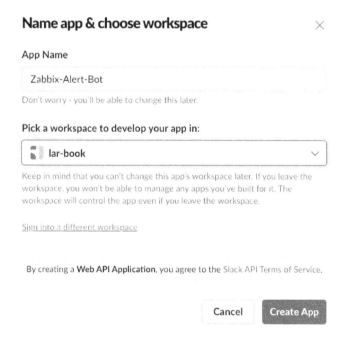

Figure 9.6 – Slack, API Name app & choose workspace window

9. Click on **Create App**. This will take you to the **Basic Information** page. On this page, click on **Bots**, as highlighted in the following screenshot:

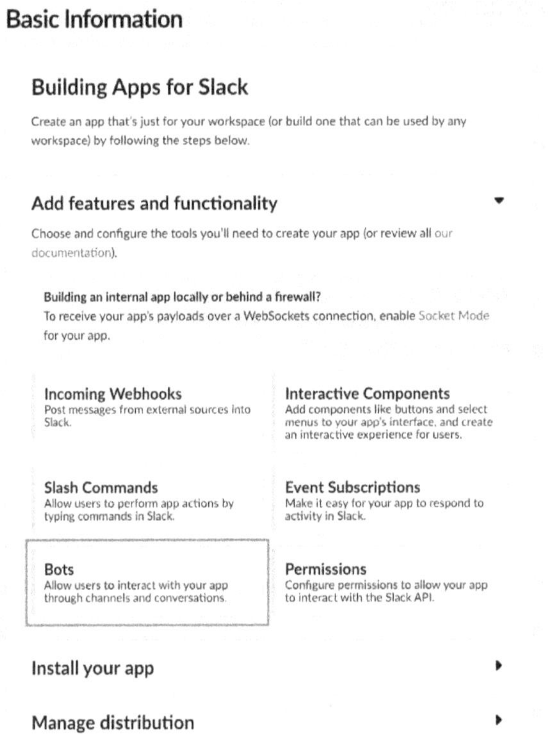

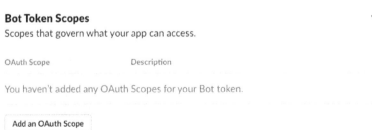

Figure 9.7 – Slack, API Add features and functionality page

10. This will take you to the new app's **App Home** page. On the left-hand side of the page, click the **OAuth & Permissions** option.

11. Scroll down to **Scopes** and click on **Add an OAuth Scope**:

Scopes

A Slack app's capabilities and permissions are governed by the scopes it requests.

Bot Token Scopes

Scopes that govern what your app can access.

OAuth Scope   Description

You haven't added any OAuth Scopes for your Bot token.

Add an OAuth Scope

Figure 9.8 – Slack, API scopes

12. From the drop-down menu, click on **chat:write** to allow our bot to write to a channel:

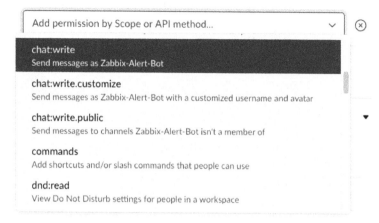

Figure 9.9 – Slack, API Add an OAuth Scope dropdown

13. Do the same for **im:write** and **groups:write**.

14. Scroll back up and click on the **Install to Workspace** button to finish setting up this app:

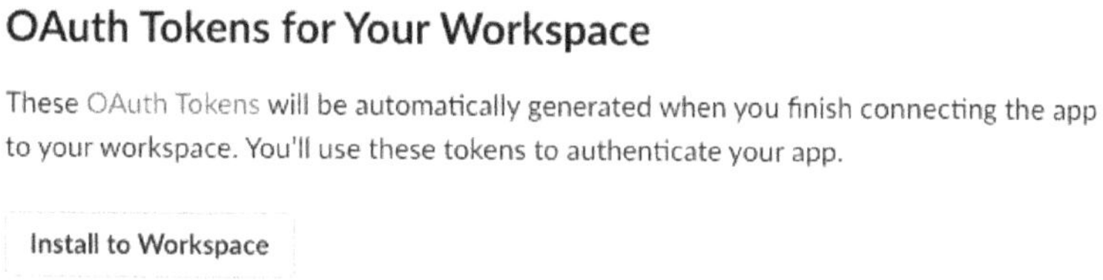

OAuth Tokens for Your Workspace

These OAuth Tokens will be automatically generated when you finish connecting the app to your workspace. You'll use these tokens to authenticate your app.

Install to Workspace

Figure 9.10 – Slack, API Install to Workspace button

15. Next, you will see a pop-up message. Click the green **Allow** button that appears.

16. After clicking **Allow**, you will see your new token. Copy the token by clicking the **Copy** button:

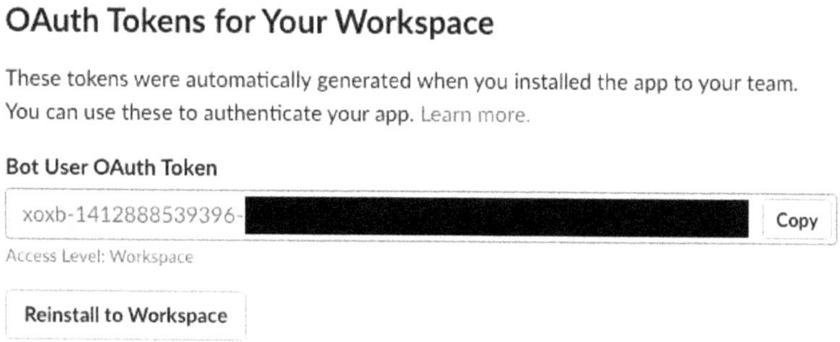

OAuth Tokens for Your Workspace

These tokens were automatically generated when you installed the app to your team. You can use these to authenticate your app. Learn more.

Bot User OAuth Token

Access Level: Workspace

Reinstall to Workspace

Figure 9.11 – Slack, API Our new Bot User OAuth Token

17. Lastly, add your bot to the **# zabbix-notifications** channel by going back to your Slack channel, clicking on the members in the top-right corner, and selecting **Integrations**:

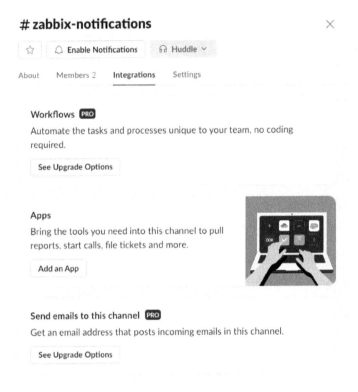

Figure 9.12 – Slack, connect an app to a channel

18. Click on **Add an App**.

19. Simply add **Zabbix-Alert-Bot** by clicking the white **Add** button:

Figure 9.13 – Slack, connect an app with a bot

20. Now, navigate to your Zabbix frontend and go to **Alerts | Media types**.

21. Click on **Slack** to edit the Slack media type. You will see a whole list of preconfigured parameters. We need to paste our OAuth token into the bot_token parameter, like this:

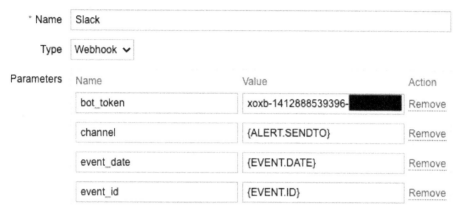

Figure 9.14 – Zabbix, Slack media type Edit page

22. Also, make sure **Media type** is enabled by scrolling all the way to the bottom.

23. On the **Message templates** tab, we already have five message types configured. We can edit these to our liking if we would like to do so.

Figure 9.15 – Zabbix, media type Slack Edit page for message types

24. You can now click on **Update** to save the changes.

25. Now, let's create a new user group for our media types by navigating to **Users | User groups** and clicking the **Create user group** button in the top-right corner. Add the following user group:

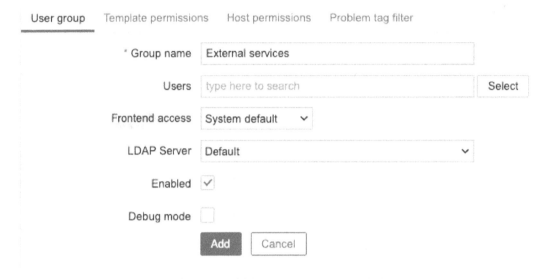

Figure 9.16 – Zabbix, Create user group page for the External services group

26. Click on the **Host permissions** tab and then click on **Select**. Make sure that you select all the groups and subgroups with at least **Read** permissions, like so:

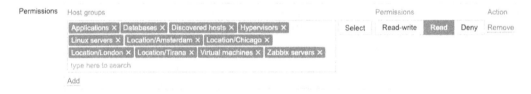

Figure 9.17 – Zabbix, Create user group Host permissions for the External services group

> **Important note**
>
> When applying permissions to the user group, make sure that you only add the host groups you want to receive notifications from. In my lab and even production environments, I add all groups, but sometimes, we want to filter the notifications down. One way to do this is to use different user groups and users so that you only receive notifications from host groups in certain channels, although Actions are also a great way to do this. For a more in-depth look into Zabbix user permissions and triggers, check out *Chapter 2, Getting Things Ready with Zabbix User Management*, and *Chapter 4, Working with Triggers and Alerts*, respectively.

27. Now, click on the blue **Add** button and finish creating this user group.

28. Next, navigate to **Users | User** to create a Slack user. Click on the blue **Create user** button in the top-right corner.

29. Add the following user to your Zabbix server and make sure to give it a secure password:

Figure 9.18 – Zabbix Create user page for the Slack user

Tip

When creating users for things such as alerting or just API access, it is best practice to also add the user to a user group that disables the frontend access. This is why we added the **No access to the frontend** group here.

---

30. Now, click on the **Media** tab and click on the underlined **Add** text. We will add the following media to this user:

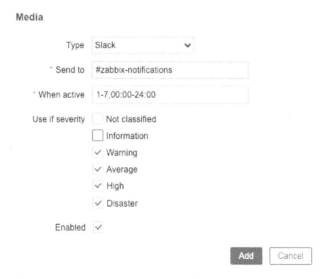

Figure 9.19 – Zabbix Create user media page for the Slack user

31. All users will also need a user role. To add it, go to **Permissions** and add the following:

Figure 9.20 – Zabbix Create user media page for the Slack user, Permissions

> **Important note**
>
> For convenience, we are adding the user in the **Super admin** user role in the example. This overrides the read-only permissions we assigned to the **External Services** user group. For security reasons, you might want to limit the permissions by choosing a **User** or **Admin** role for your user, which will adhere to the host group permissions we assigned earlier.

32. Click on the blue **Add** button at the bottom of the window and then on the blue **Add** button at the bottom of the page.

33. We will also need to add a macro to **Administration | Macros**. Let's add the following macro, which contains your Zabbix URL:

Figure 9.21 – Zabbix Administration | Macros page with Zabbix URL

34. Click on the blue **Update** button.

35. Last but not least, go to **Alerts | Actions**, and on the **Trigger Actions** page, click on the blue **Create action** button.

36. Use `Notify external services` for the name of the action. We won't set up any conditions for this example, but it's recommended to do so in production environments. If you do not set any conditions, all problems will be matched on this action.

37. Go to **Operations** and add the following operations:

Figure 9.22 – Zabbix, Create action operations page for Notify external services

---

Tip

We could also use **Notify all involved** here to send a message to all the users involved in the **Operations** steps.

---

38. Now, click on the blue **Add** button. With that, you're done. You can now view any new problems (once they are generated by Zabbix) in your Slack channel:

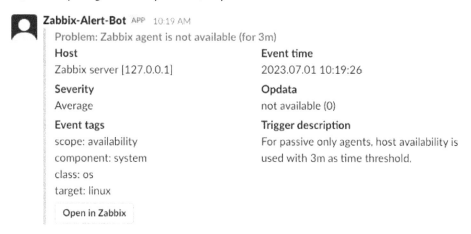

Figure 9.23 – Slack notification sent from our Zabbix server

How it works...

Working with media types might be something completely new to you, or you might have done it in the past. Regardless, starting from Zabbix 5, the process has changed a bit. In the days of Zabbix before Zabbix 5, we had to find the right media types online or make them ourselves.

Now, with Zabbix 7, we get a lot of preconfigured media types that are ready to be used. We just need to do the necessary setup and fill in the right information, just like we just did for Slack. We are then ready to send our problem-related information from Zabbix 7 to Slack every time a problem is created.

In this recipe, we told our Zabbix server to only send problem-related information with a severity of warning or higher to Slack, as shown in the following screenshot:

Figure 9.24 – Zabbix Media page for the Slack user

We can fully customize these severities, but we can also fully customize what severities we send to our Slack setup.

What we configured in this recipe is called a **Zabbix webhook**. A problem gets created in Zabbix and this problem matches our configured criteria, such as its severity. Our action matches the problem, and the media type will then be executed by the **Action** operations and sent to the configured links:

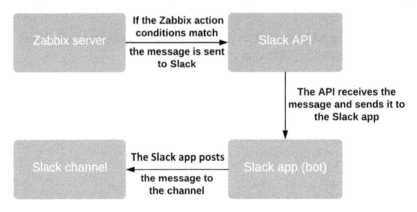

Figure 9.25 – Zabbix Slack integration diagram

Zabbix sends the problem to the Slack API, and then the API processes the problem. Then, the app we configured in Slack posts the problem to our channel.

Now that we've completed this recipe, we can view problems in Slack and keep an eye on our Zabbix alerts from there.

See also

If you want to do more with this integration, check out the Slack API documentation. There's a lot we can do with this API, and we can build a number of awesome apps/bots for our channels: `https://api.slack.com/`.

Setting up Microsoft Teams alerting with Zabbix

The previous edition of this book was written during the COVID-19 pandemic. At that time, a lot of IT companies had been requested to make their employees work from home. Due to this, we noticed a rise in the use of Microsoft Teams and similar applications. Suddenly, a lot of companies started using Microsoft Teams and others to make working from home and collaboration easier. Even after the pandemic ended, this is still the case, which is why we added this recipe.

Let's learn how we can make working with Microsoft Teams even better by integrating our Zabbix alerting into it.

Getting ready

We will need our Zabbix server to be able to create some problems for us. For this, you can use `lar-book-rocky` from our previous chapters or any Zabbix server that you prefer.

We will also need general Microsoft Teams knowledge and, of course, Microsoft Teams itself set up and ready to go.

How to do it...

1. Let's start by opening our Microsoft Teams application on either Windows, macOS, or Linux and creating a new channel. Go to **Teams** and click on the three dots (**…**) next to your team name, as shown in the following screenshot:

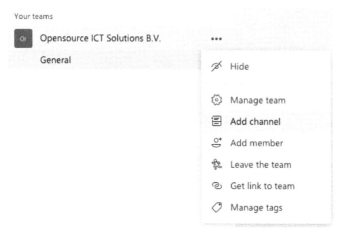

Figure 9.26 – MS Teams, Add channel option

2.   In the **Add channel** window, fill in the following information to create our new channel:

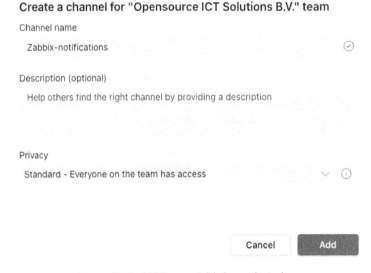

Figure 9.27 – MS Teams, Add channel window

3.   Now, click the purple **Add** button to add the channel. Upon doing this, we will be able to see our new channel in the list.

4.   Click on the three dots (**…**) next to your channel, as shown in the following screenshot:

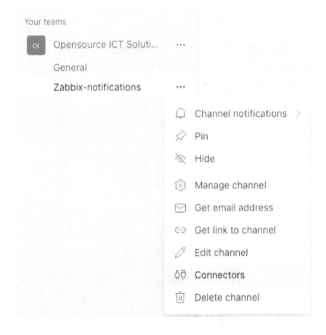

Figure 9.28 – MS Teams, Connectors option

5. We want to select the **Connectors** option from this drop-down menu. This allows us to add our Microsoft Teams connector to this channel.

6. We are using the search field here to find the official **Zabbix Webhook** connector:

Keep your group current with content and updates from other services.

zabbix   Search Results   Sort by: Popularity ∨

MANAGE   Connectors for your team

Configured   Z   **Zabbix Webhook**   Configure
My Accounts   Integrate Zabbix with Microsoft Teams by using this connector

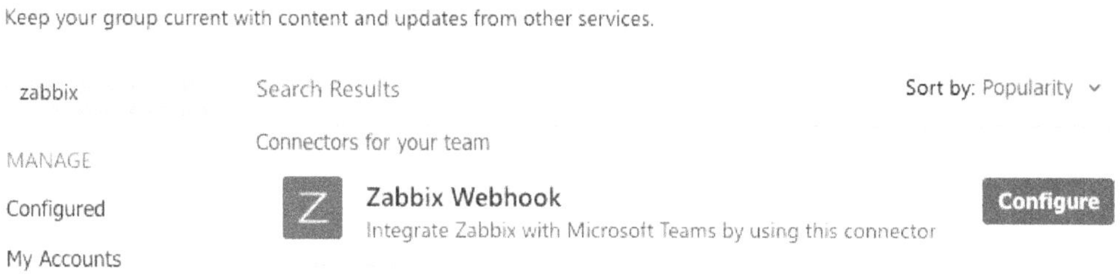

Figure 9.29 – MS Teams, Add connectors window

7. Click on the **Configure** button next to the **Zabbix Webhook** connector to add this connector to our channel. This will open a pop-up window in which you need to copy the webhook URL.

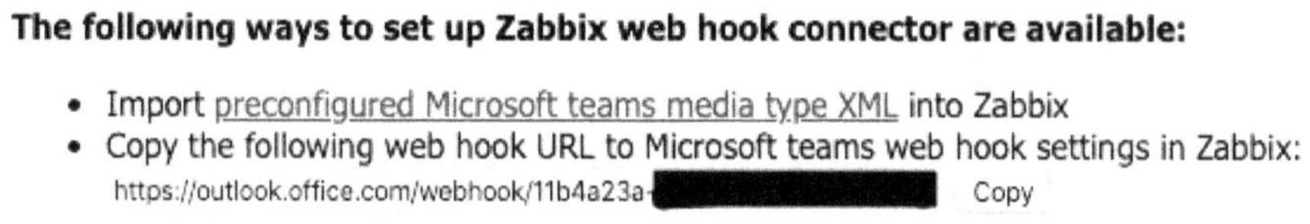

The following ways to set up Zabbix web hook connector are available:

- Import preconfigured Microsoft teams media type XML into Zabbix
- Copy the following web hook URL to Microsoft teams web hook settings in Zabbix:
  https://outlook.office.com/webhook/11b4a23a-   Copy

Figure 9.30 – MS Teams – Webhook URL

8. Now, click the **Save** button. Upon doing this, you can close the pop-up window.

9. Go to the Zabbix frontend and navigate to **Alerts** | **Media types**. Click on the **MS Teams** media type here.

10. Scroll down until you see **teams_endpoint**. Paste the URL you copied previously here, as shown in the following screenshot:

host_ip	{HOST.IP}	Remove
host_name	{HOST.NAME}	Remove
teams_endpoint	https://outlook.office.com/webhook|	Remove
trigger_description	{TRIGGER.DESCRIPTION}	Remove

Figure 9.31 – Zabbix Alerts | Media types, edit MS Teams page

11. Make sure to scroll down and enable the media type before saving.

12. Now, click the blue **Update** button at the bottom of the page.

13. If you didn't follow the previous recipe, then create a new user group for our media types by navigating to **Users | User groups** and clicking the **Create user group** button in the top-right corner. Add the following user group:

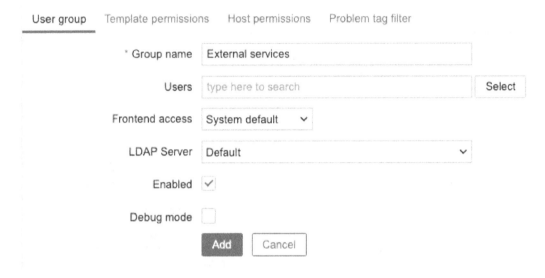

Figure 9.32 – Zabbix, Create user group page, the External services group

14. Click on the **Host permissions** tab and click on **Select**. Make sure that you select all the groups and subgroups with **Read** permissions. The **Permissions** tab will look like this:

Figure 9.33 – Zabbix, Create user group permissions page, the External services group

15. Now, click on the blue **Add** button and finish creating this user group.

16. Navigate to **Users** | **Users** and click on the blue **Create user** button in the top-right corner. Add the following user:

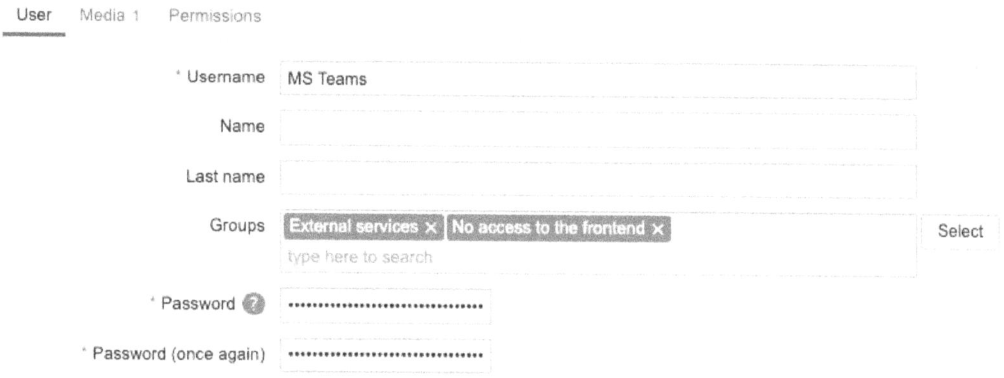

User   Media 1   Permissions

* Username	MS Teams
Name	
Last name	
Groups	External services ✕ No access to the frontend ✕ Select
	type here to search
* Password ?	••••••••••••••••••••••••••••
* Password (once again)	••••••••••••••••••••••••••••

Figure 9.34 – Zabbix Users | Users, Create new user page, MS Teams

---

Tip

When creating users for things such as alerting or just API access, it is best practice to also add the user to a user group that disables the frontend access. This is why we added the **No access to the frontend** group here.

---

17. Next, go to the **Media** tab of the **Create user** page. Click the underlined **Add** text here to create the following media:

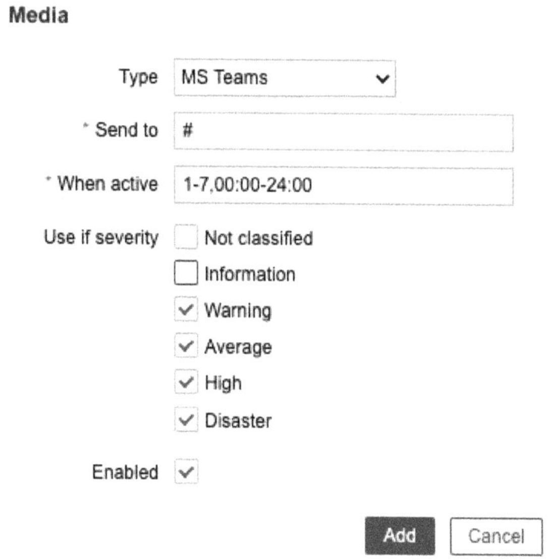

Media

Type	MS Teams ⌄
* Send to	#
* When active	1-7,00:00-24:00
Use if severity	☐ Not classified
	☐ Information
	✔ Warning
	✔ Average
	✔ High
	✔ Disaster
Enabled	✔

Add   Cancel

Figure 9.35 – Zabbix Users | Users, Create new user media page, MS Teams

18. All users will also need a **User** role. To add it, go to **Permissions** and add the following:

Figure 9.36 – Zabbix – Create user media page, MS Teams user, Permissions

> **Important note**
>
> For convenience, we are adding the user in the **Super admin** user role in the example. This overrides the read-only permissions we assigned to the **External services** user group. To limit the permissions, choose a **User** or **Admin** role for your user, which will adhere to the host group permissions we assigned earlier.

19. Once you've filled in this information, click the blue **Add** button at the bottom of the window and then the blue **Add** button at the bottom of the page.

20. If you didn't follow the previous recipe, then you will also need to add a macro to **Administration | Macros** and add the following macro. This macro will contain your Zabbix URL:

Add

Figure 9.37 – Zabbix Administration | Macros page, Zabbix URL for use with MS Teams

21. Click on the blue **Update** button. You will also need to go to **Alerts | Actions** if you didn't follow the previous recipe and, on the **Trigger Actions** page, click on the blue **Create action** button.

22. Use `Notify external services` for the name of the action. We won't set up any conditions for this example, but it's recommended to do so in production environments. If you do not set any conditions, all problems will be matched on this action.

23. Go to **Operations** and add the following operations:

Action   Operations

* Default operation step duration   1h

Operations   Steps   Details   Start in   Duration   Action
1   **Send message to user groups: External services via all media**   Immediately   Default   Edit Remove
Add

Recovery operations   Details   Action
Notify all involved   Edit Remove
Add

Update operations   Details   Action
Notify all involved   Edit Remove
Add

Pause operations for symptom problems   ✓

Pause operations for suppressed problems   ✓

Notify about canceled escalations   ✓

* At least one operation must exist.

Add   Cancel

Figure 9.38 – Zabbix, Create action Operations page, Notify external services for use with MS Teams

24. Now, click on the blue **Add** button and you'll be done. You can now view new problems as they occur in your MS Teams channel:

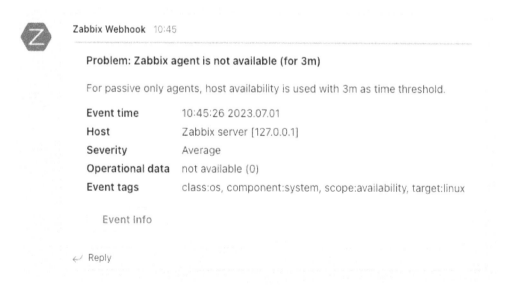

Zabbix Webhook   10:45

Problem: Zabbix agent is not available (for 3m)

For passive only agents, host availability is used with 3m as time threshold.

Event time	10:45:26 2023.07.01
Host	Zabbix server [127.0.0.1]
Severity	Average
Operational data	not available (0)
Event tags	class:os, component:system, scope:availability, target:linux

Event Info

↩ Reply

Figure 9.39 – Zabbix problem in an MS Teams channel

How it works...

Microsoft Teams works in about the same way as our Slack setup. A problem is created in the Zabbix server and, if that problem matches our configured conditions in Zabbix, we send that problem to our Microsoft Teams connector.

For instance, we configured Zabbix so that it only sends problems with a severity of warning or higher to Microsoft Teams, as shown here:

Media	Type	Send to	When active	Use if severity	Status	Action
	MS Teams	#	1-7,00:00-24:00	N I **W A H D**	Disabled	Edit Remove
	Add					

Figure 9.40 – Zabbix Media page for MS Teams users

Our Microsoft Teams connector catches our problem and, since this connector is configured directly on our channel, it posts a notification to the channel:

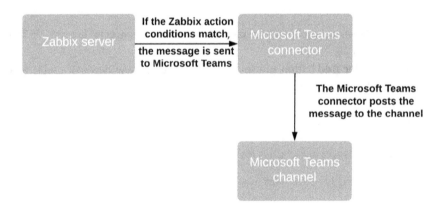

Figure 9.41 – Zabbix Microsoft Teams integration diagram

Now, we can see our Teams notifications in our channel and keep up to date with all our Zabbix issues directly via Microsoft Teams.

See also

For more information about the Zabbix webhook connector, check out this page: `https://appsource.microsoft.com/en-us/product/office/WA200001837?tab=Overview`.

Using Telegram bots with Zabbix

If you love automation in chat applications, you might have heard of or used Telegram Messenger. Telegram has an extensive API and amazing bot features.

In this recipe, we are going to use a Telegram bot to create a Telegram group for Zabbix alerts. Let's get started.

Getting ready

Make sure you have your Zabbix server ready. You can use `lar-book-rocky` or any Zabbix server capable of sending some alerts.

It would be useful if you have some knowledge of Telegram, but I'll be describing how to set it up step by step, even for those of you who have never used Telegram bots. Just make sure that you have the Telegram app on your computer and that your account is set up.

How to do it...

1. First, let's create a new channel in Telegram. Click on the create icon next to the search box and select **New Group**:

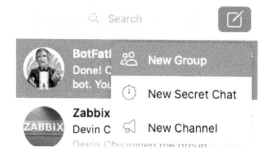

Figure 9.42 – Telegram New Group button

2. Add another user – someone in your team who needs to get notifications as well. Fill in a group name and add a picture if you want:

Figure 9.43 – Telegram New Group page

3. Now, click on the **Create** button in the top-right corner. This will take you to the **New Group** page.

4. Working with Telegram bots is made easy with the @BotFather user on Telegram. We can start creating our bot by searching for botfather and clicking the specified contact:

Figure 9.44 – Telegram, BotFather user

5. Let's start by issuing the /start command in the chat. This will provide you with a list of commands you can use:

Figure 9.45 – Telegram, BotFather user help list

6. Now, let's immediately create a new bot by typing /newbot into the **BotFather** chat. Press *Enter* to send your message. This will give us the following result:

Figure 9.46 – Telegram, BotFather /newbot command

7. Type in the new name of the bot; we will call it zabbix-notfication-bot. Press *Enter* to send your name:

Figure 9.47 – Telegram, BotFather bot username

8. You will then be asked what username you want to give the bot. I will use lar_ zbx_ notfication_bot:

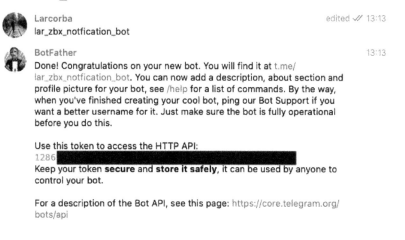

Figure 9.48 – Telegram, BotFather bot token

> **Important note**
> Your bot username must be unique, so you can't use `lar_zbx_ notification_bot` here. Pick a unique bot name that suits you and then use that name throughout this recipe.

9. Make sure that you save the **HTTP API** key somewhere safe.

10. Let's go back to our **Zabbix notifications** group and add our bot. Click on the group in your list of chats and click on the group's name.

11. Now, click on **Add** to add the bot, as follows:

Figure 9.49 – Telegram, Add user to group button

12. Next, you will need to search for your bot using its username, as shown in the following screenshot:

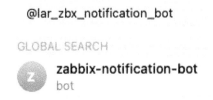

Figure 9.50 – Telegram, Add user to group page

13. Click on the bot and click on the **Add** button. With that, your bot has been added to the channel.

14. Let's navigate to the Zabbix frontend and go to **Alerts | Media types**. Click on the media type titled **Telegram**.

15. Here, you must add the **HTTP API** key you generated earlier to the **Token** field of our media type:

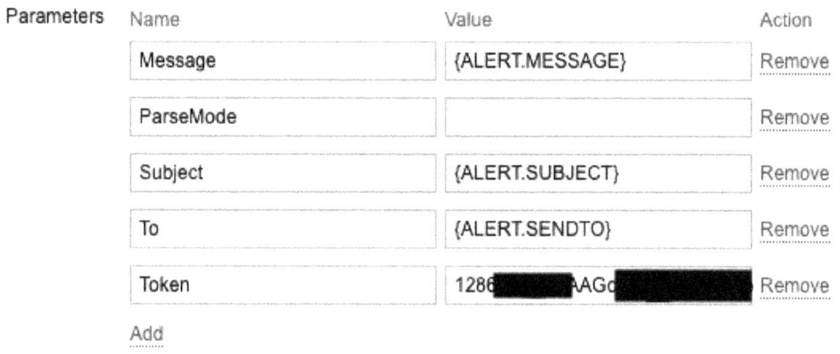

Parameters	Name	Value	Action
	Message	{ALERT.MESSAGE}	Remove
	ParseMode		Remove
	Subject	{ALERT.SUBJECT}	Remove
	To	{ALERT.SENDTO}	Remove
	Token	1286▉▉▉▉AAGd▉▉▉▉	Remove
	Add		

Figure 9.51 – Zabbix Alerts | Media types, Edit Telegram Media type page

16. Also, scroll down and make sure this media type is enabled.

17. Click on the blue **Update** button at the bottom of the page to finish editing the **Telegram** media type.

18. Now, let's go back to Telegram and add another bot to our group. Go to our new group and click on the group's name. Click on **Add** to add the **IDBot** user:

Figure 9.52 – The Add user page for a Telegram group

19. Click on the user and click on **Add**. Then, navigate back to the Zabbix frontend.

20. If you haven't followed any of the preceding recipes, create a new user group for our media types by navigating to **Users | User groups** and clicking the **Create user group** button in the top-right corner. Add the following user group:

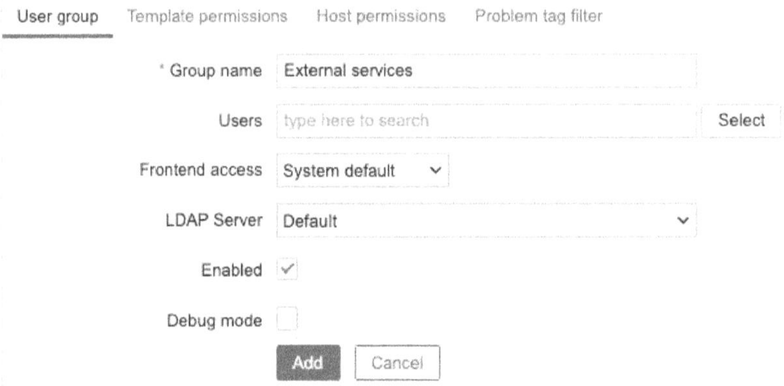

Figure 9.53 – Zabbix Create user group page and the External services group for use with Telegram

21. Click on the **Host permissions** tab and click on **Select**. Make sure that you select all the groups and subgroups with **Read** permissions, as follows:

Figure 9.54 – Zabbix Create user group Permissions page, External services for use with Telegram

22. Now, click on the blue **Add** button and finish creating this user group.

23. At this point, you must create a new user in Zabbix. However, to create this user, you are going to need our new group ID. Go back to Telegram and issue /getgroupid@myidbot in the group chat. You will receive a value that you will need to copy:

Larcorba
/getgroupid@myidbot

IDBot
Larcorba
/getgroupid@myidbot
Your group ID is: –976

Figure 9.55 – Telegram user group ID

24. Let's navigate to **Users | Users** and click the blue **Create user** button. Add the following user:

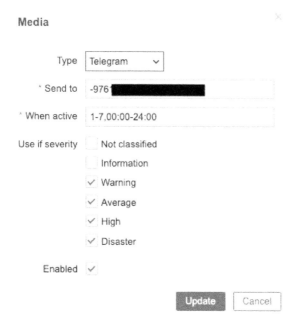

User   Media 1   Permissions

* Username	Telegram
Name	
Last name	
Groups	External services ✕ No access to the frontend ✕ Select
	type here to search
* Password ❓	••••••••••••••••••••••••••
* Password (once again)	••••••••••••••••••••••••••

Figure 9.56 – Zabbix Create user page, Telegram user

Tip

When creating users for things such as alerting or just API access, it is best practice to also add the user to a user group that disables the frontend access. This is why we added the **No access to the frontend** group here.

---

25. Now, select the **Media** tab and click on the underlined **Add** text. Add the following media:

Media   ✕

Type	Telegram ⌄
* Send to	-9761 ██████████
* When active	1-7,00:00-24:00
Use if severity	☐ Not classified
	☐ Information
	✓ Warning
	✓ Average
	✓ High
	✓ Disaster
Enabled	✓

Update   Cancel

Figure 9.57 – Zabbix Create user media page, Telegram user

26. All users will also need a **User** role. To add it, go to **Permissions** and add the following:

Figure 9.58 – Zabbix Create user media page, Telegram user, Permissions

> **Important note**
>
> For convenience, we are adding the user in the **Super admin** user role in the example. This overrides the read-only permissions we assigned to the **External services** user group. To limit the permissions, choose a **User** or **Admin** role for your user, which will adhere to the host group permissions we assigned earlier.

27. Make sure that you add the group ID to the **Send to** field with the – text and click on the blue **Add** button.

28. If you haven't followed the previous recipes, you will also need to go to **Alerts | Actions**. Then, on the **Trigger Actions** page, click on the blue **Create action** button.

29. Use `Notify external services` for the name of the action. We won't set up any conditions for this example, but it's recommended to do so in production environments. If you do not set any conditions, all problems will be matched on this action.

30. Go to **Operations** and add the following operations:

Figure 9.59 – Zabbix – Create action operations page, Notify external services for use with Telegram

31. Now, click on the blue **Add** button. With that, you're done. You can now view new problems in your Telegram group:

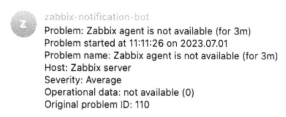

Figure 9.60 – Zabbix notifications in Telegram chat

How it works...

Slack apps, Microsoft connectors, and Telegram bots all work kind of the same in the end. There's just another backend (API) provided by the respective companies, but the Zabbix webhook remains.

Now that we've added our Zabbix Telegram integration, we are receiving notifications in our Telegram group via the Zabbix webhook:

Figure 9.61 – Zabbix Users | Users, Edit user media page

However, we will only receive these notifications if they match our configured settings. For instance, we've added our media type so that it only sends problems with a severity of warning or higher to our Telegram bot:

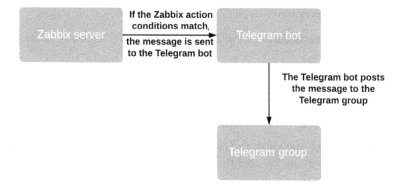

Figure 9.62 – Zabbix Telegram integration diagram

Our Zabbix server is now sending our problems that match the **Action** conditions to our Telegram bot. The bot catches these problems successfully. Because our bot is in our Telegram group, the problems are posted in our Telegram group.

There's more...

There's a very cool Zabbix community group on Telegram. Now that you have Telegram, do not forget to join using the following invitation link: `https://t.me/ZabbixTech`.

See also

Make sure that you check out all the awesome features Telegram bots have to offer. There's a lot of information available directly from Telegram, and you can build amazing integrations by using them: `https://core.telegram.org/bots`.

Integrating Atlassian Opsgenie with Zabbix

Atlassian Opsgenie is so much more than just another integration service for receiving notifications. Opsgenie offers us a call system, an SMS system, iOS and Android apps, two-way acknowledgments, and even an on-call schedule.

I think Opsgenie (and PagerDuty) are the best tools for replacing old-school call and SMS systems and fully integrating them with Zabbix. So, let's get started with Opsgenie and see how we can get this amazing tool set up.

Getting ready

Ensure that your Zabbix server is ready. I'll be using the `lar-book-rocky` server, but any Zabbix server ready to send problems should work.

You are also going to need an Atlassian Opsgenie account with Opsgenie ready to go. This integration is slightly different from the previous examples, as we'll also be utilizing some scripts.

I won't show you how to create accounts, but we'll start this recipe with Opsgenie ready to go.

How to do it...

1. Let's start by logging in to our Atlassian Opsgenie setup and going to **Settings** on the home page. From the left sidebar, click on **Notifications**.

2. Make sure that you add your email and phone number here by using the + **Add email** and + **Add phone number** buttons. We need these in order to receive notifications:

Figure 9.63 – Opsgenie profile, Contact methods

3. Your settings will be automatically saved once you've added them, which means we can navigate away from **Settings** to **Teams** using the top bar.

4. From the **Teams** tab, click on the blue **Add team** button in the top-right corner. Then, add the following information:

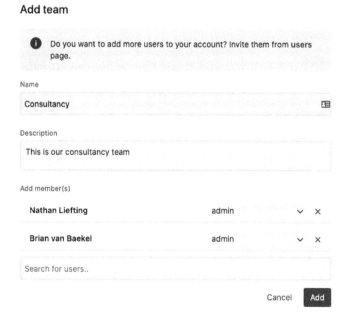

Figure 9.64 – Opsgenie, Add team window

5. I've set up our `Consultancy` team with two users that are part of this team. Click on the blue **Add** button at the bottom of the window to add the new team.

6. This will take you to the new **Consultancy** team page. Click on **Integrations** and click on the blue **Add integration** button in the top-right corner.

7. When we use the search field to search for the Zabbix integration, we can see it immediately, as shown in the following screenshot:

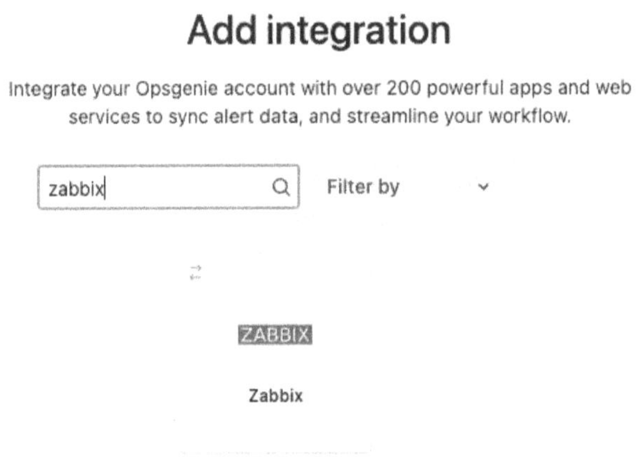

Add integration

Integrate your Opsgenie account with over 200 powerful apps and web services to sync alert data, and streamline your workflow.

> zabbix   🔍   Filter by   ⌄

ZABBIX

Zabbix

Figure 9.65 – Opsgenie, Add integration page

8. Click on the Zabbix integration. This will take you to the next page, where a **Name** and **API Key** value will be generated:

Settings

Name:

> Consultancy_Zabbix

API Key: ⊘

> a0ce55e2-███████████████████████   🗐   ♻

Enabled: ⊘

☑

Send Via OEC: ⊘

☑

Figure 9.66 – Opsgenie, Add Zabbix integration page

9. Copy the **API Key** information and scroll to the bottom of the page. Here, click on the blue **Save integration** button.

10. Now, navigate to your Zabbix server CLI and execute the following code.

For RHEL-based systems, use the following:

```
wget https://github.com/opsgenie/oec-scripts/releases/download/
Zabbix-1.1.6_oec-1.1.3/opsgenie-zabbix-1.1.6.x86_64.rpm
```

For Ubuntu systems, use the following:

```
wget https://github.com/opsgenie/oec-scripts/releases/download/
Zabbix-1.1.6_oec-1.1.3/opsgenie-zabbix_1.1.6_amd64.deb
```

11. We can now install the downloaded Zabbix Opsgenie plugin by issuing the following command(s):

For RHEL-based systems, use the following:

```
rpm -i opsgenie-zabbix-1.1.6.x86_64.rpm
```

For Ubuntu systems, use the following:

```
dpkg -i opsgenie-zabbix_1.1.6_amd64.deb
```

12. Once you've installed the plugin, go to the Zabbix frontend. If you haven't followed any of the previous recipes, create a new user group for our media types by navigating to **Users | User groups** and clicking the **Create user group** button in the top-right corner. Add the following user group:

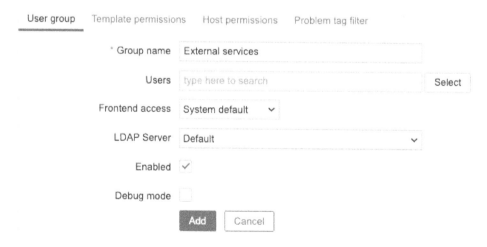

Figure 9.67 – Zabbix Create user group page, the External services group for use with Opsgenie

13. Click on the **Permissions** tab and click on **Select**. Make sure that you select all the groups and subgroups with **Read** permissions, as follows:

Figure 9.68 – Zabbix Create user group permissions page, the
External services group for use with Opsgenie

14. Now, click on the blue **Add** button and finish creating this user group.

15. Let's navigate to **Users | Users** and click the blue **Create user** button. Add the following user:

Figure 9.69 – Zabbix Create user page, Opsgenie user

> **Tip**
> When creating users for things such as alerting or just API access, it is best practice to also add the user to a user group that disables the frontend access. This is why we added the **No access to the frontend** group here.

16. All users will also need a **User** role. To add it, go to **Permissions** and add the following:

Figure 9.70 – Zabbix Create user media page, Opsgenie user, Permissions

> **Important note**
>
> For convenience, we are adding the user in the **Super admin** user role in the example. This overrides the read-only permissions we assigned to the **External services** user group. To limit the permissions, choose a **User** or **Admin** role for your user, which will adhere to the host group permissions we assigned earlier.

17. For use in our action, we need to set up a script; to do so, navigate to **Alerts | Scripts** and click on **Create script** in the top-right corner.

18. We'll name the script `Opsgenie connector`. Set **Scope** to **Action operation**, **Type** to **Script**, and **Execute on** to **Zabbix server**.

19. Paste the contents of the /home/opsgenie/oec/opsgenie-zabbix/actionCommand. txt file on the CLI into **Commands**. It should look like the following screenshot:

Figure 9.71 – Zabbix Opsgenie connector script

20. Don't forget to press the **Add** button at the bottom of the page to save this new script.

21. Next, navigate to **Alerts | Actions**. On the **Trigger actions** page, click on the blue **Create action** button in the top-right corner.

22. In the **Name** field, type `Opsgenie action` and add the following items to **Conditions**:

Action   Operations

* Name   Opsgenie action

Conditions   Label   Name   Action

A   Trigger severity is greater than or equals *Warning*   Remove

Add

Enabled ☑

* At least one operation must exist.

Figure 9.72 – Zabbix Create new action page, Opsgenie action

23. Now, click on the **Operations** tab.

24. Click the underlined **Add** text option next to **Operations** to add your first operation. For the **Operation** dropdown, select **Opsgenie connector**.

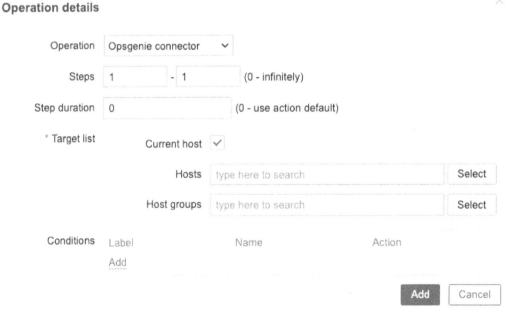

Figure 9.73 – Zabbix Create action operations window, Opsgenie connector

25. Repeat *step 24* for **Recovery operations** and **Update operations**. It should look like this:

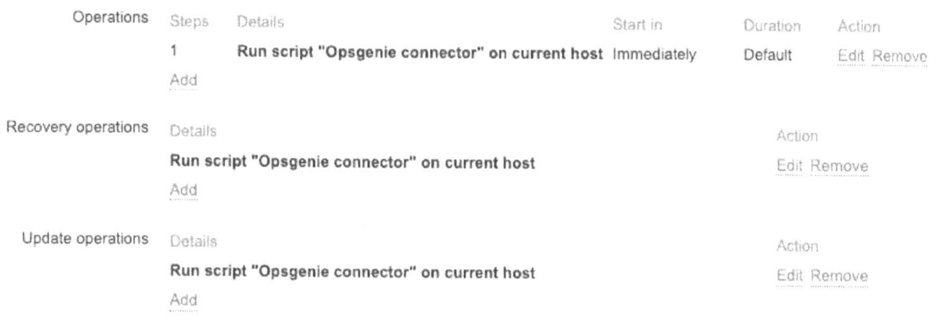

Operations	Steps	Details	Start in	Duration	Action
	1	Run script "Opsgenie connector" on current host	Immediately	Default	Edit Remove
	Add				
Recovery operations	Details			Action	
	Run script "Opsgenie connector" on current host			Edit Remove	
	Add				
Update operations	Details			Action	
	Run script "Opsgenie connector" on current host			Edit Remove	
	Add				

Figure 9.74 – Zabbix Create action operations page, Opsgenie action

26. Click on the blue **Add** button at the bottom of the page to finish setting up the action.

27. Now, you must configure the Opsgenie Zabbix integration. Edit the `config` file with the following command:

```
vim /home/opsgenie/oec/conf/config.json
```

28. Make sure that you edit the `apiKey`, `command_url`, `user`, and `password` lines, as shown in the following screenshot:

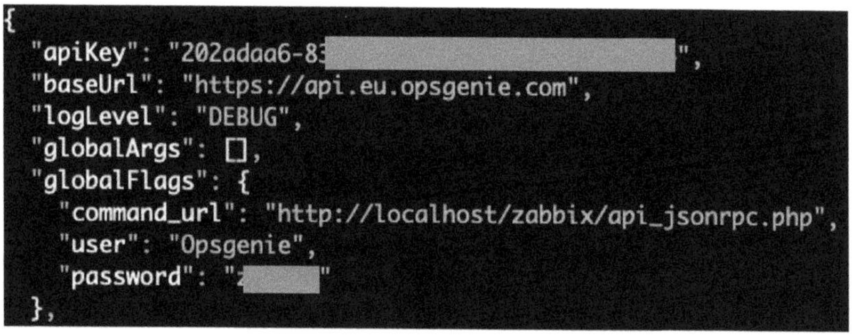

```
{
  "apiKey": "202adaa6-83                    ",
  "baseUrl": "https://api.eu.opsgenie.com",
  "logLevel": "DEBUG",
  "globalArgs": [],
  "globalFlags": {
    "command_url": "http://localhost/zabbix/api_jsonrpc.php",
    "user": "Opsgenie",
    "password": "z        "
  },
```

Figure 9.75 – Opsgenie config.json file

> **Important note**
>
> You will need to edit `baseUrl` if you are not located in the United States. I am in Europe, so I changed it to `https://api.eu.opsgenie.com`.

29. That's it! You can now see your alerts coming in and acknowledge them from Opsgenie:

☐ #2   [Zabbix] Zabbix agent is not available   Nathan Liefting   ACK'ED

   x1   👥 Consultancy   Close   UnAck   •••

                                                      Jul 1, 2023 12:05 PM (GMT+02:00)

Figure 9.76 – Opsgenie alert from Zabbix

How it works...

When an alert is created in Zabbix, it is sent to Opsgenie via the Zabbix integration. This integration utilizes the Opsgenie API to catch our Zabbix information and send back a reply if required. This way, we have two-way communication between the two applications:

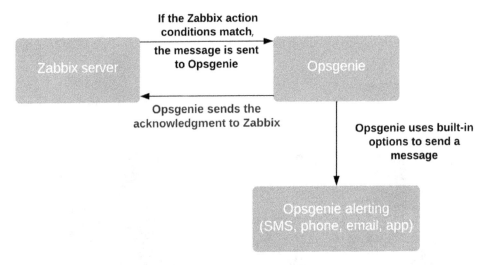

Figure 9.77 – Opsgenie setup diagram

Opsgenie is an amazing tool that can take several tasks away from your Zabbix server. I've used it in the past to migrate away from another monitoring tool to Zabbix. Opsgenie makes it easy to receive alerts from our products and centralize notifications:

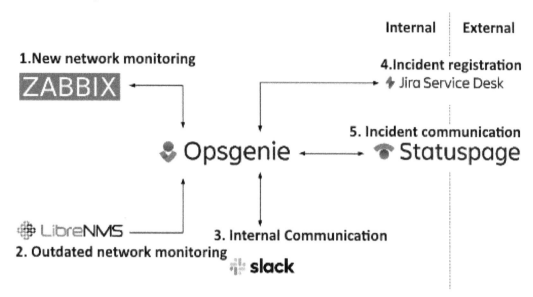

Figure 9.78 – Opsgenie setup example, inspired by Wadie

Another great feature of Atlassian Opsgenie is the integration it offers with other Atlassian products. We can build a setup like the one shown in the preceding diagram to integrate all the products used in our company.

There's more...

Opsgenie doesn't just allow us to send notifications. It also enables us to define entire on-call schedules within the product, making it super easy for us to create schedules within the program, where Zabbix might be a bit too static. Check out the following link for more information about how to do that:

```
https://support.atlassian.com/opsgenie/docs/manage-on-call-schedules-and-rotations/
```

Furthermore, there are also competitive products that integrate just as well as Opsgenie. PagerDuty is one of these, with native Zabbix integration. It has a similar feature set and it all boils down to preference, other software we might already have, and... price!

Extending Zabbix Functionality with Custom Scripts and the Zabbix API

Zabbix offers a lot of functionality out of the box. But where Zabbix really shines is customization, not only through the default frontend but especially with scripts and the Zabbix API.

In this chapter, I will go over the basics of using the Zabbix API. We will then see how a Python script can utilize the API to build something cool, such as a jumphost. After that, we'll use some scripts written by *Brian van Baekel* to enable and disable hosts with limited permissions from a Zabbix map.

After following these recipes, you'll be more than ready to tackle the Zabbix API and you'll know how to use scripts to extend Zabbix functionality. This chapter will expand your possibilities with Zabbix to almost endless proportions and you'll be ready to become a professional Zabbix user yourself.

In this chapter, we will cover the following recipes:

- Setting up and managing API tokens
- Using the Zabbix API for extending functionality
- Building a jumphost using the Zabbix API and Python
- Enabling and disabling a host from Zabbix maps

Technical requirements

We are going to need a Zabbix server as well as some new Linux hosts. We will also need to have general knowledge of scripting and programming. We are going to use Python to extend some functionality of Zabbix, which we'll provide scripts for.

The code required for the chapter can be found at the following link:

```
https://github.com/PacktPublishing/Zabbix-7-IT-Infrastructure-
Monitoring-Cookbook/tree/main/chapter10
```

Make sure to keep all of this ready and you'll be sure to nail these recipes.

Setting up and managing API tokens

Let's start off our chapter by doing some configuration for working with APIs in Zabbix. If you've worked with the Zabbix API before, you might know it can be quite a hassle to use API calls to authenticate and get an API token for using it in your scripts. This is no longer the case, as we can generate API tokens using the Zabbix frontend.

Getting ready

For this recipe, all we'll need is the Zabbix setup running. We'll be using the frontend to generate the API token. From here, we can use the API token in any of our integrations further on in this chapter.

How to do it...

1. First, let's log in to the Zabbix frontend as a Super admin user.
2. Navigate to **Users | User groups** and click the blue **Create user group** button in the top-right corner.
3. Here, we'll create a new user group. Fill in the **Group name** field as API Users.
4. Switch to the **Host Permissions** tab and give your API user group permission to every host by clicking the **Select** button and selecting every host group.

Host groups

☑ Name

✓ Applications

✓ Databases

✓ Discovered hosts

✓ Hypervisors

✓ Linux servers

✓ Location/Amsterdam

✓ Location/Chicago

✓ Location/London

✓ Location/Tirana

✓ Virtual machines

✓ Zabbix servers

[Select] [Cancel]

Figure 10.1 – Zabbix Users | User groups, creating user group host permissions, API Users

5. Move on to the **Template permissions** tab and do the same thing here.

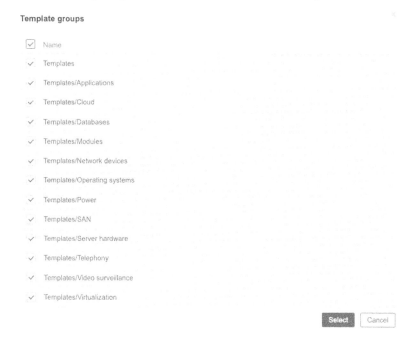

Template groups

☑ Name

✓ Templates

✓ Templates/Applications

✓ Templates/Cloud

✓ Templates/Databases

✓ Templates/Modules

✓ Templates/Network devices

✓ Templates/Operating systems

✓ Templates/Power

✓ Templates/SAN

✓ Templates/Server hardware

✓ Templates/Telephony

✓ Templates/Video surveillance

✓ Templates/Virtualization

[Select] [Cancel]

Figure 10.2 – Zabbix Users | User groups, creating user group template permissions, API Users

6. Click the blue **Select** button at the bottom of this popup and click on **Read-write** followed by the small dotted **Add** button. It should now look like this:

Figure 10.3 – Zabbix Users | User groups, user group permissions page, API Users

> **Tip**
> Instead of creating the API user as Super admin, we can also limit the permissions by limiting the host and template group access on the **API Users** user group. This could be preferred in production environments, as you might want to limit API access. Use whatever fits your preference.

7. Click the blue **Add** button at the bottom of the page to add this new user group.

8. Now, let's go to **Users | Users** and click the blue **Create user** button in the top-right corner.

9. Here we will create a new user called the API user. Create the user as follows:

Figure 10.4 – Zabbix Users | Users, user creation page, API user

10. Before adding the user, switch to the **Permissions** tab and add **Super admin role**.

User   Media   Permissions

Role	Super admin role ✕		Select
User type	Super admin		
Permissions	Group	Type	Permissions
	All groups	Hosts	Read-write
	All groups	Templates	Read-write

Permissions can be assigned for user groups only.

Figure 10.5 – Zabbix Users | Users, user permissions page, API user

11. We can now add the user by clicking the blue **Add** button at the bottom of the page.

12. Next up, we need to create some API tokens for this user. Navigate to **Users | API tokens**.

13. Let's click the blue **Create API token** button in the top-right corner and fill in the **User** field as API and the **Name** field as API book key. Set **Expires at** to somewhere far in the future or disable expiration entirely – whatever you think might be secure. It will look like this:

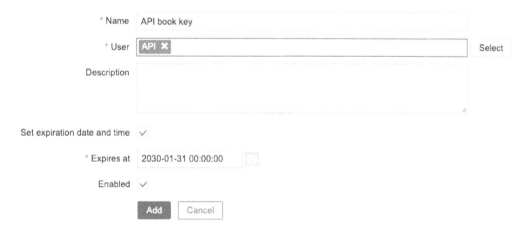

Figure 10.6 – Administration | General | API token, API token creation page

14. Click the blue **Add** button at the bottom of the page to generate the new API token. This will bring us to the next page:

Figure 10.7 – Zabbix API user API token generated page

15. Make sure to save the **Auth token** value to a secure location, such as a password vault. It will be important later in the labs.

16. We can click the **Close** button at the bottom of this page now. This will bring us back to the **API tokens** page where we can manage all of our created API tokens.

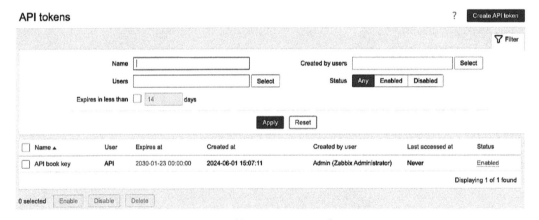

Figure 10.8 – Zabbix API user API tokens page

How it works...

Because Zabbix now comes with built-in API token management, it has become a lot easier to work with the Zabbix API. Using a dedicated API user, we can manage all of our tokens in a single location or we can set up private API tokens under our own user account.

In this case, we created a new API Users group. This is important because our API tokens are still part of a user account meaning they will respect that user its permissions. If we create an API user under any other user type than Super admin, we can restrict our API access using the **API Users** group.

Make sure that you apply the role to the user and the permissions to the user group as you see fit in your environment. Also, make sure to set a reasonable expiration date for your API tokens so we can regenerate them from time to time.

There's not much else to say about setting up and managing your API tokens, but let's see how we can apply what we have learned in this recipe in the next recipes.

Using the Zabbix API for extending functionality

An API is your gateway to getting started with extending the functionality of any piece of software. Luckily, Zabbix offers a solid working API that we can use to extend our functionality with ease. Zabbix has also released `zabbix-utils` for Python, making building scripts a lot easier for everyone. It's an amazing addition, but as it's not allowed in every environment and we want to keep the dependencies to a minimum, we won't utilize it in our test here. Nevertheless, check out the library here:

`https://github.com/zabbix/python-zabbix-utils`

In this recipe, we'll explore the use of the Zabbix API to do some tasks, creating a good basis to start working with the Zabbix API in your actual production environments.

Getting ready

We are going to need a Zabbix server with some hosts. I'll be using our `lar-book-centos` host from the previous chapters, but feel free to use any Zabbix server. I will also use another Linux host to do the API calls from, but this can be done from any Linux host.

We will need to install Python 3 on the Linux host, though, as we'll be using this to create our API calls.

Also, make sure you have an API user with an API token. It is recommended to use the one we created in the first recipe.

How to do it...

1. First, on our Linux CLI, let's move to a new directory:

    ```
    cd /home/zabbix/
    ```

2. Install Python 3 on the host with the following command.

    For RHEL-based systems, use this command:

    ```
    dnf install python3
    ```

For Ubuntu systems, use this command:

```
apt install python3
```

3. Python `pip` should've been installed with this package by default as well. If not, issue the following command.

For RHEL-based systems, use this command:

```
dnf install python3-pip
```

For Ubuntu systems, use this command:

```
apt install python3-pip
```

> **Important note**
>
> It is possible to have an older version of Python(3) shipped with your Linux distribution. If you run into any errors with the scripts later in the chapter, make sure to check for any error messages indicating that your Python version might not support certain functionality.

4. Now. let's install our dependencies using Python `pip`. We'll need these dependencies as they'll be used in the script:

```
pip3 install requests
```

5. Download the start of our script from the Packt GitHub repo of this book by issuing the following command:

```
wget https://raw.githubusercontent.com/PacktPublishing/Zabbix-7-
IT-Infrastructure-Monitoring-Cookbook/main/chapter10/api_test.py
```

6. If you can't use wget from your host, you can download the script at the following URL: `https://github.com/PacktPublishing/Zabbix-7-IT-Infrastructure-Monitoring-Cookbook/tree/main/chapter10/api_test.py`.

7. Next up, we are going to edit our newly downloaded script by executing the following command:

```
vim api_test.py
```

8. First, let's change the IP address, 10.16.16.152, in the `url` variable to the IP or DNS of your Zabbix server. Then, make sure to edit the `api_token` variable by replacing PUT_YOUR_TOKEN_HERE with the API token we generated in the first recipe of this chapter:

```
url = "http://10.16.16.152/api_jsonrpc.php"
api_token = "c01ce8726bfdbce02664ec8750f99da
1bbbcb3cb295d924932e2f2808846273"
```

9. We will also add some lines of code to our script to retrieve our host ID, hostname, and the interfaces of all our Zabbix hosts. Make sure to add your new code between the comments shown in the following screenshot:

#Add new code below here

#Add new code above here

Figure 10.9 – Comments showing where to put code

10. Now, add the following lines of code:

```
#Function to retrieve the hosts and interfaces
def get_hosts(api_token, url):
    payload = {
    "jsonrpc": "2.0",
    "method": "host.get",
    "params": {
        "output": [
            "hostid",
            "host"
        ],
        "selectInterfaces": [
            "interfaceid",
            "ip",
            "main"
        ]
    },
    "id": 2,
    "auth": api_token
    }
    resp = requests.post(url=url, json=payload )
    out = resp.json()
    return out['result']
```

11. Then, we'll also add lines to write the requested information to a file so we can see what happens after execution:

```
#Write the results to a file
def generate_host_file(hosts,host_file):
    hostname = None
    f = open(host_file, "w")
```

```
#Write the host entries retrieved from Zabbix
for host in hosts:
    hostname = host['host']
    for interface in host["interfaces"]:
        if interface["main"] == "1":
            f.write(hostname + " " + interface["ip"] + "\n")
f.close()
return
```

12. You should be able to execute this now by executing the following:

    python3 api_test.py

13. This should run but it won't give you any output. If this doesn't work, make sure to retrace your steps.

14. Let's check out the file to see what happened by executing the following:

    cat /home/zabbix/results

15. The output of the preceding command should look something like this:

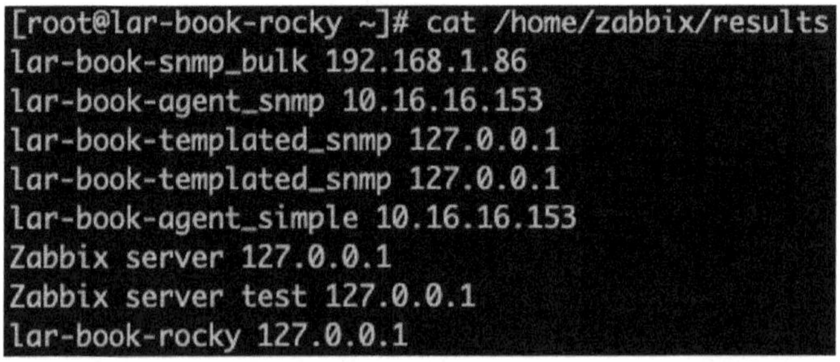

```
[root@lar-book-rocky ~]# cat /home/zabbix/results
lar-book-snmp_bulk 192.168.1.86
lar-book-agent_snmp 10.16.16.153
lar-book-templated_snmp 127.0.0.1
lar-book-templated_snmp 127.0.0.1
lar-book-agent_simple 10.16.16.153
Zabbix server 127.0.0.1
Zabbix server test 127.0.0.1
lar-book-rocky 127.0.0.1
```

Figure 10.10 – The cat command with our results showing in the file

We've now written a short script in Python to use the Zabbix API.

How it works...

Coding with the Zabbix API can be done with Python, but it's definitely not our only option. We can use a wide variety of coding languages, including Perl, Go, C#, and Java.

In our example, though, we've used Python, so let's see what we do here. If we look at the script, we have two main functions:

- get_hosts
- generate_host_file

First, we filled in our `api_token` and `url` variables, which are used to authenticate against the Zabbix API. We then used these to call on the `get_hosts` function to retrieve information from the Zabbix API:

```
payload = {
"jsonrpc": "2.0",
"method": "host.get",
"params": {
    "output": [
        "hostid",
        "host"
    ],
    "selectInterfaces": [
        "interfaceid",
        "ip",
        "main"
    ]
},
"id": 2,
"auth": api_token
}
```

Figure 10.11 – Python function Zabbix API payload

Looking at the code, we used a JSON payload to request information such as `host` for the hostname, `hostid` for the host ID, and `ip` for the interface's IP address.

Now, if we look at our last function, `generate_host_file`, we can see that we write the host with an interface IP to the `/home/results` file. This way, we have a solid script for writing host information to a file.

If you're not familiar with Python or coding in general, working with the Zabbix API might be a big step to take. Let's take a look at how the API actually works:

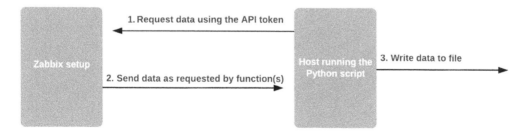

Figure 10.12 – Python script Zabbix API functionality diagram

In *step 1*, we make an API call, using our target URL and API token as specified in our variables for authentication. Next, in *step 2*, we receive the data as requested in our Python function from Zabbix to further use in our Python script.

Step 3 is our data processing step. We can do anything we want with the data received from the Zabbix API, but in our case, we format the data and write it to a file. That's how we use the Zabbix API for extending functionality. This is the step where our file gets filled with hostnames and IP information.

See also

If you are interested in learning more about the Zabbix API and its available functionality, check out the Zabbix documentation at `https://www.zabbix.com/documentation/current/en/manual/api`.

Building a jumphost using the Zabbix API and Python

A lot of organizations have a jumphost (sometimes referred to as a bastion host) to access servers, switches, and their other equipment from a host. A jumphost generally has all the firewall rules needed to access everything important. Now, if we keep our monitoring up to date, we should have every single host in there as well.

My friend, ex-colleague, and fellow Zabbix geek, *Yadvir Singh*, had the amazing idea to create a Python script to export all Zabbix hosts with their IPs to the `/etc/hosts` file on another Linux host. Let's see how we can build a jumphost just like his.

Getting ready

We are going to need a new host for this recipe with Linux installed and ready. We'll call this host `lar-book-jump`. We will also need our Zabbix server, for which I'll use `lar-book-centos`.

Also, it is important to navigate to *Yadvir's* GitHub account, drop him a follow, and star his repository if you think this is a cool script: `https://github.com/cheatas/zabbix_scripts`.

> **Important note**
> Setting up this script will override your `/etc/hosts` file every time the script is executed. Only use this script when you understand what it's doing, make sure you use an empty host for this lab, and check the default `/etc/hosts` settings.

How to do it...

1. If you haven't already created an API user with an API token, make sure to check out the first recipe of this chapter first.

2. Install Python 3 on the host CLI with the following command.

    For RHEL-based systems, use this command:

    ```
    dnf install python3
    ```

    For Ubuntu systems, use this command:

    ```
    apt-get install python3
    ```

3. Python `pip` should've been installed with this package by default as well. If not, issue the following command

    For RHEL-based systems, use this command:

    ```
    dnf install python3-pip
    ```

    For Ubuntu systems, use this command:

    ```
    apt-get install python3-pip
    ```

4. Now, let's install our dependencies using Python `pip`. We'll need these dependencies as they'll be used in the script:

    ```
    pip3 install requests
    ```

5. First things first, log in to our new Linux host, `lar-book-jump`, and download Yadvir's script to your Linux host with the following command:

    ```
    wget https://raw.githubusercontent.com/cheatas/zabbix_scripts/
    main/host_pull_zabbix.py
    ```

6. If you can't use wget from your host, you can download the script at the following URL: `https://github.com/cheatas/zabbix_scripts/blob/main/host_pull_zabbix.py`.

    As a backup, we also provide this script in the Packt repository. You may download this version at `https://github.com/PacktPublishing/Zabbix-7-IT-Infrastructure-Monitoring-Cookbook/tree/main/chapter10/host_pull_zabbix.py`.

7. Now, let's edit the script by executing the following command:

    ```
    vim host_pull_zabbix.py
    ```

8. First, let's edit the `zabbix_url` variable by replacing `https://myzabbix.com/api_jsonrpc.php` with the IP address or DNS name of our Zabbix frontend:

    ```
    zabbix_url = "http://10.16.16.152/api_jsonrpc.php"
    ```

9. We do not need to fill out our username and password as that was only required on older Zabbix versions. Instead, we will need an API token, as generated in the first recipe in this chapter. Fill in the `api_token` variable in the script as follows:

    ```
    api_token = "c01ce8726bfdbce02664ec8750f99da1bbbcb3cb295
    d924932e2f2808846273"
    ```

You can find this variable at the bottom of the file.

10. We also need to uncomment the following lines:

```
zabbix_hosts = get_hosts(api_token,zabbix_url)
generate_host_file(zabbix_hosts,"/etc/hosts")
```

11. The end of the script should now look like this:

```
#If you do not have a API token yet, use the following line to aquire it.
#Once printed, copy the token and paste it in the variable below.

#print(get_api_token(zabbix_url))

api_token = "c01ce8726bfdbce02664ec8750f99da1bbbcb3cb295d924932e2f2808846273"

#once the API token has been set, comment the print line again and uncomment the follwoing lines.

zabbix_hosts = get_hosts(api_token,zabbix_url)
generate_host_file(zabbix_hosts,"/etc/hosts")
```

Figure 10.13 – End of the script after receiving the API token and with commenting removed

12. Last but not least, make sure to comment and uncomment the right lines for your Linux distro. It will look like the following figures.

For Ubuntu, it will look like this:

```
    #For Debian based hosts
    f.write('''127.0.0.1 localhost\n\n''')

    #For RHEL based hosts
#    f.write('''127.0.0.1    localhost localhost.localdomain localhost4 localhost4.localdomain4
#::1          localhost localhost.localdomain localhost6 localhost6.localdomain6\n\n''')

# The following lines are desirable for IPv6 capable hosts
    f.write('''::1 localhost ip6-localhost ip6-loopback
fe00::0 ip6-localnet
ff00::0 ip6-mcastprefix
ff02::1 ip6-allnodes
ff02::2 ip6-allrouters\n\n\n''')
```

Figure 10.14 – Print to file for Ubuntu systems

For RHEL-based systems, it will look like this:

```
#For Debian based hosts
#f.write('''127.0.0.1 localhost\n\n''')

#For RHEL based hosts
    f.write('''127.0.0.1     localhost localhost.localdomain localhost4 localhost4.localdomain4
::1             localhost localhost.localdomain localhost6 localhost6.localdomain6\n\n''')

# The following lines are desirable for IPv6 capable hosts
    f.write('''::1 localhost ip6-localhost ip6-loopback
fe00::0 ip6-localnet
ff00::0 ip6-mcastprefix
ff02::1 ip6-allnodes
ff02::2 ip6-allrouters\n\n\n''')
```

Figure 10.15 – Print to file for RHEL-based systems

13. That's all there is to do, so we can now execute the script again and start using it. Let's execute the script as follows:

    ```
    python3 host_pull_zabbix.py
    ```

14. Test whether it worked by looking at the host file with the following command:

    ```
    cat /etc/hosts
    ```

    This should give us an output like that shown in the following screenshot:

```
[root@lar-book-jump ~]# cat /etc/hosts
127.0.0.1     localhost localhost.localdomain localhost4 localhost4.localdomain4
::1             localhost localhost.localdomain localhost6 localhost6.localdomain6

::1 localhost ip6-localhost ip6-loopback
fe00::0 ip6-localnet
ff00::0 ip6-mcastprefix
ff02::1 ip6-allnodes
ff02::2 ip6-allrouters


lar-book-centos 127.0.0.1
lar-book-agent_passive 10.16.16.153
lar-book-agent_passive 10.16.16.153
lar-book-agent 10.16.16.153
```

Figure 10.16 – /etc/hosts filled with our script information

15. We can now try to SSH directly to the name of a host, instead of having to use the IP, by issuing the following command:

```
ssh lar-book-agent_passive
```

16. We can also use it to find hosts from the file with the following command:

```
cat /etc/hosts | grep agent
```

17. Let's do one more thing. We want this script to be as up to date as possible. So, let's add a cronjob. Issue the following command to add a cronjob:

```
crontab -e
```

18. Then add the following line, making sure to fill in the right script location for your setup:

```
*/15 * * * * $(which python3) /home/host_pull_zabbix.py >> ~/
cron.log 2>&1
```

That's it – we will now have an up-to-date /etc/hosts file all the time with our new Python script and Zabbix.

How it works...

If your organization uses Zabbix as the main monitoring system, you now have the skills and knowledge to create an organized, reliably up-to-date, and easy-to-use jumphost.

Jumphosts are super useful when set up correctly, but it's important to keep them clean so that they are easy to update.

By using this script, we only add Python 3 and a simple script as a requirement to the server, but the end result is a jumphost that knows about all hosts in the environment.

If you've followed along with the previous *Using the Zabbix API for extending functionality* recipe, then you might notice that it works in roughly the same way. We can see in the following diagram how we utilize the script:

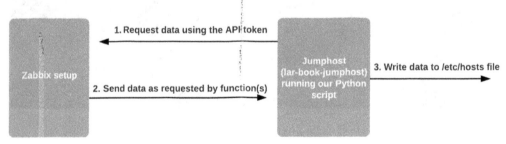

Figure 10.17 – Jumphost using script functionality diagram

After editing, our script will start at *step 1* of the diagram to request data with an API call, where we use the API token to authenticate. We receive this data in *step 2*. In the script, we add our default values and then write all the hostnames and IP addresses to the /etc/hosts file.

Now, because a Linux host uses the /etc/hosts file for hostname-to-IP translation, we can use the real names of servers in Zabbix to SSH to the hosts. This makes it easier for us to use the jumphost, as we can use the same name as the hostname we know from the Zabbix frontend.

See also

Yadvir will keep updating the script after writing this recipe (we've been using version 1.0 so far). Make sure to follow his GitHub account and star his repository to get the updates. Also, if you have some cool ideas for additions, you can always open a pull request.

The Zabbix community is all about sharing cool ideas and useful scripts like this one. As *Yadvir* has shown, we can get very valuable stuff from each other. Be like *Yadvir* – use the Zabbix community GitHub and support other Zabbix users by adding to their GitHub repositories. You can find the Zabbix community GitHub using the link here:

```
https://github.com/zabbix/community-templates
```

Enabling and disabling a host from Zabbix maps

We've noticed that it is not possible to enable and disable hosts as a Zabbix user. For some companies, this may be a requirement, so we've created an extension for it. In this recipe, I will show you just how to work with this Python script and execute it from a map.

Getting ready

For this recipe, all we are going to need is our Zabbix server, some knowledge of Python, and some knowledge of the Zabbix API.

How to do it...

1. First, let's log in to our Zabbix server CLI and create a new directory:

   ```
   mkdir /etc/zabbix/frontendscripts
   ```

2. Change to the new directory:

   ```
   cd /etc/zabbix/frontendscripts
   ```

3. Now, download the public script from the *Opensource ICT Solutions* GitHub:

   ```
   wget https://github.com/OpensourceICTSolutions/zabbix-toggle-
   hosts-from-frontend/archive/v2.0.tar.gz
   ```

4. If you can't use wget from your host, you can check out the script here: https://github.com/OpensourceICTSolutions/zabbix-toggle-hosts-from-frontend/releases/tag/v2.0.

5. Unzip the file with the following command:

```
tar -xvzf v2.0.tar.gz
```

6. Remove the tar file using the following command:

```
rm v2.0.tar.gz
```

7. Move the script over from the newly created folder with the following command:

```
mv zabbix-toggle-hosts-from-frontend-2.0/enable_disable-host.py
./
```

8. We are going to need Python to use this script, so let's install it as follows.

For RHEL-based systems, use this command:

```
dnf install python3 python3-pip
```

For Ubuntu systems, use this command:

```
apt-get install python3 python3-pip
```

9. We will also need the requests module from pip. Install it as follows:

```
pip3 install requests
```

10. Now, let's edit the script with the following command:

```
vim enable_disable-host.py
```

11. In this file, we will change the url and token variables. Change the url variable to match your own Zabbix frontend IP or DNS name. Then, replace PUT_YOUR_TOKEN_HERE with your Zabbix API token. I will fill in the following, but be sure to enter your own information:

```
url = 'http://10.16.16.152/api_jsonrpc.php?'
token = " c01ce8726bfdbce02664ec8750f99da1bbbcb3cb295d
924932e2f2808846273 "
```

12. Now, we can move on to our Zabbix frontend to add a frontend script. Navigate to **Alerts | Scripts**, then click the blue **Create script** button at the top right.

13. Add the following script:

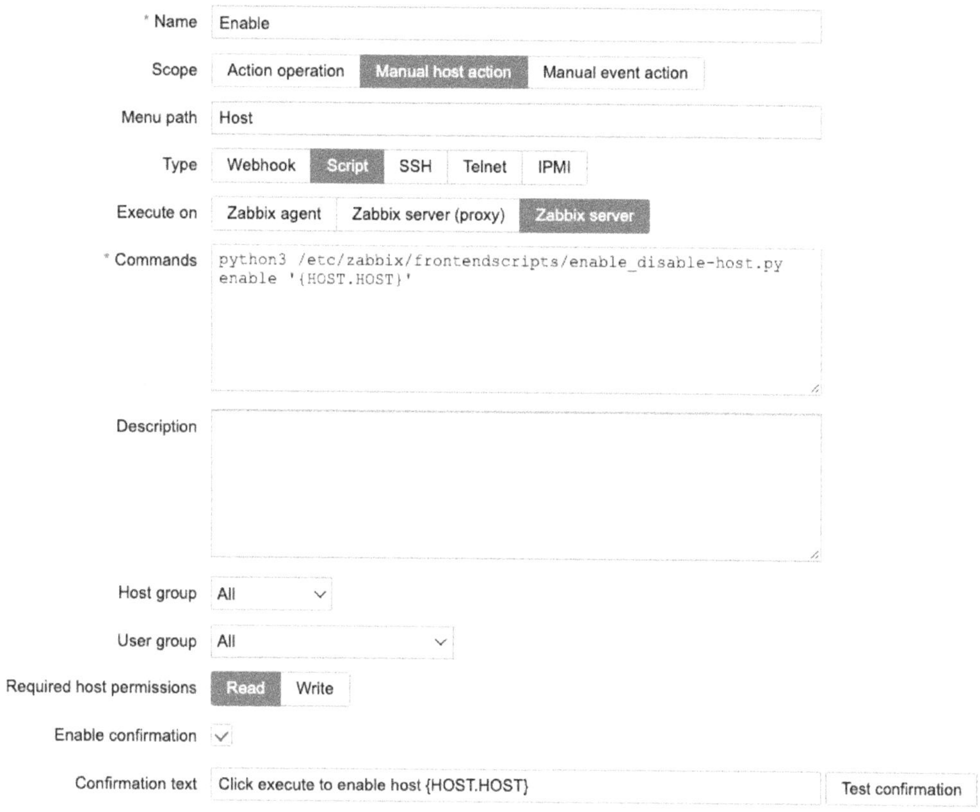

* Name	Enable
Scope	Action operation **Manual host action** Manual event action
Menu path	Host
Type	Webhook **Script** SSH Telnet IPMI
Execute on	Zabbix agent Zabbix server (proxy) **Zabbix server**
* Commands	`python3 /etc/zabbix/frontendscripts/enable_disable-host.py enable '{HOST.HOST}'`
Description	
Host group	All
User group	All
Required host permissions	**Read** Write
Enable confirmation	✓
Confirmation text	Click execute to enable host {HOST.HOST} Test confirmation

Figure 10.18 – Zabbix Alerts | Scripts, the Create script page, Enable

14. Click on the blue **Add** button and then, on the next page, click the blue **Create script** button in the top-right corner again.

15. Now, add the second and final script as follows:

Figure 10.19 – Zabbix Alerts | Scripts, the Create script page, Disable

16. Now, navigate to **Monitoring | Maps**, and you should see a map called **Local network** here, as it's included with Zabbix by default. Click this map (or any other map with hosts in it).

17. Now, if you click on a host on the map, you will see a drop-down menu like this:

Figure 10.20 – Zabbix Monitoring | Maps, the Local network map drop-down menu

18. If we click on **Disable** here, we will get a pop-up message as follows:

Figure 10.21 – Zabbix script confirmation window

19. Click on the blue **Execute** button and this host will be disabled. Navigate to **Monitoring | Hosts** to confirm whether this worked. You should see that the host is set to **Disabled**.

20. Back at **Monitoring | Maps**, you can enable the host again with the same drop-down menu. This time, select **Enable**.

How it works...

The script we just used was built in Python utilizing the Zabbix API. With this script, we can now enable and disable hosts from the Zabbix frontend as a Zabbix user.

This works because the **Monitoring | Maps** option is available even to Zabbix users. This script uses the API user for execution, though. Since our Zabbix API user has more user permissions, it can execute the script that gets host information from a Zabbix database and creates a maintenance period using the information. As we can see in the following diagram, our script follows roughly the same steps as the other Zabbix API utilities:

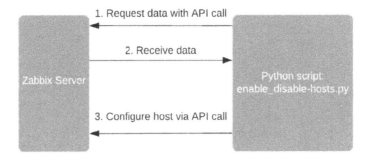

Figure 10.22 – Python script maintenance.py execution diagram

Because the Zabbix API is very flexible, we can pull data and write data to do almost anything we could do from the frontend.

We can now use this cool function from anywhere in the Zabbix frontend where we see a dotted line with the hostname, even from **Monitoring | Hosts**.

There's more...

In Zabbix 7, it is possible to provide input from the frontend onto your scripts. If you'd like to test this, we can edit the script slightly:

1. Go to **Alerts | Scripts** and edit the **Host/Enable** and **Host/Disable** scripts.
2. In the **Commands** field, replace {HOST.HOST} with {MANUALINPUT} to make this script accept dynamic input.

3. Then open up **Advanced configuration** and fill in the **Input prompt**, **Default input string**, and **Input validation rule** fields. It should now look like the following figure:

```
* Commands   python3 /etc/zabbix/frontendscripts/enable_disable-host.py
             enable '{MANUALINPUT}'
```

Advanced configuration

Enable user input ✓

* Input prompt   Which host would you like to enable?   Test user input

Input type   [String] Dropdown

Default input string   hostname_here

* Input validation rule   ([A-Za-z0-9-_.]+)

Enable confirmation

Confirmation text   Test confirmation

Figure 10.23 – Manual script input

4. Now, when we execute the script, it will ask us for input. Fill in a hostname here and the host that we wrote in the field will be enabled or disabled.

Manual input   ✕

Which host would you like to enable?

Zabbix server

Cancel   **Execute**

Figure 10.24 – Manual script input execution

As you can see, with this new script input method, it is now possible to make the whole execution process a lot more flexible. We can execute scripts like this in a non-static way, allowing us to provide user input data before execution. Neat!

See also

Brian van Baekel created this script for a customer at *Opensource ICT Solutions* and then open sourced it. Because Zabbix has a very cool community that continues to extend the possibilities of Zabbix even further, we too upload some of our scripts. Sharing is caring, so check out the other open source scripts at `https://github.com/OpensourceICTSolutions`.

11

Maintaining Your Zabbix Setup

Like any good piece of software, Zabbix needs to be maintained in order to keep working over the years. A lot of users have been running their setups since the days of Zabbix 2.0. It's perfectly viable to do this if you bring the right knowledge of Zabbix to the equation.

In this chapter, we are going to see how to do some of the most important parts of Zabbix maintenance to make sure you can keep your setup available and running smoothly. We are going to cover creating maintenance periods, how to make backups, how to upgrade Zabbix and various Zabbix components, and how to do some performance maintenance.

We'll cover these in the following recipes:

- Setting Zabbix maintenance periods
- Backing up your Zabbix setup
- Upgrading the Zabbix backend from older PHP versions to PHP 8.2 or higher
- Upgrading a Zabbix database from older MariaDB versions to MariaDB 10.11
- Upgrading your Zabbix setup
- Maintaining Zabbix performance over time

Technical requirements

We are going to need several important servers for these recipes. First of all, we are going to need a running Zabbix 7 server for which to set up maintenance periods and do performance tuning.

For the upgrade part, we will need one of the following servers:

- A Rocky Linux 8 server running Zabbix server 6, a PHP version before 8.2, and a MariaDB version before 11.4
- An Ubuntu 22.04 server running Zabbix server 6 a PHP version before 8.3, and a MariaDB version before 11.4

I will call the upgrade server `lar-book-zbx6`, which you can run with a distribution of your choice.

If you do not have any prior experience with Zabbix, this chapter may prove a good challenge, as we are going to go into the more advanced Zabbix processes in depth.

Setting Zabbix maintenance periods

When we are working on our Zabbix server or on other hosts, it's super useful to set up maintenance periods in the Zabbix frontend. With maintenance periods we can make sure that our Zabbix users don't get alerts going off because of our maintenance. One improvement you'll find in Zabbix 7 is the inclusion of near-instant maintenance periods. As we no longer have to wait long for the config cache reload, Zabbix also altered the timer process to instantly enable a new maintenance period.

Let's see how we can schedule maintenance periods in this recipe.

Getting ready

All we are going to need in this recipe is our Zabbix server, for which I'll use `lar-book-rocky`. The server will need at least some hosts and host groups to create maintenance periods for. Furthermore, we'll need to know how to navigate the Zabbix frontend.

How to do it...

1. Let's get started with this recipe by logging in to our frontend and navigating to **Data collection | Maintenance**.

2. We are going to click on the blue **Create maintenance period** button in the top-right corner.

3. This will show us a popup, where we can set up our maintenance period. Let's start by defining the maintenance period parameters. Fill in the following:

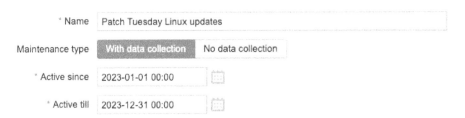

Figure 11.1 – Zabbix Data collection | Maintenance, create maintenance window, Patch Tuesday

4. Now, at the **Periods** part, we'll create a new maintenance period. We need to click on the underlined **Add** text.

5. This will bring us to another pop-up window where we can set the maintenance period. We need to fill in the following information:

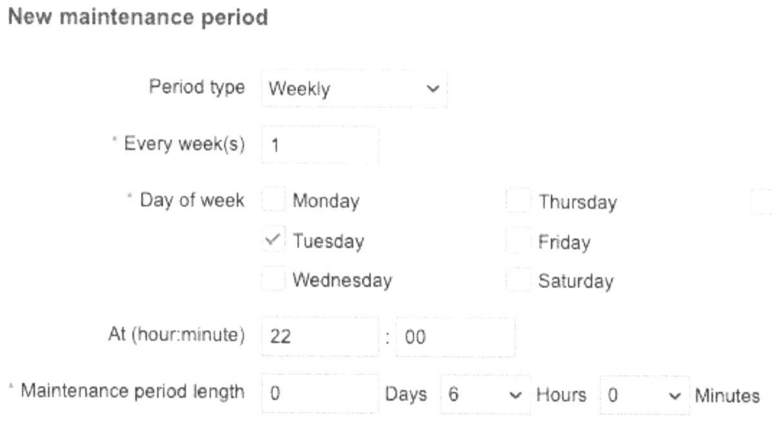

New maintenance period

Period type	Weekly ⌄
* Every week(s)	1
* Day of week	☐ Monday ☐ Thursday ☐ Sunday
	✓ Tuesday ☐ Friday
	☐ Wednesday ☐ Saturday
At (hour:minute)	22 : 00
* Maintenance period length	0 Days 6 ⌄ Hours 0 ⌄ Minutes

Figure 11.2 – Zabbix Data collection | Maintenance, create maintenance period window, Patch Tuesday

6. Now click the blue **Add** button to continue. You should now see that our maintenance period is filled in:

* Periods	Period type	Schedule	Period	Action
	Weekly	At 22:00 Tuesday of every week	6h	Edit Remove
	Add			

Figure 11.3 – Zabbix Configuration | Maintenance, create maintenance period page, Patch Tuesday

7. Now, next to the **Host groups** field, click on the **Select** button and select the **Linux servers** host group. Our page should look like this:

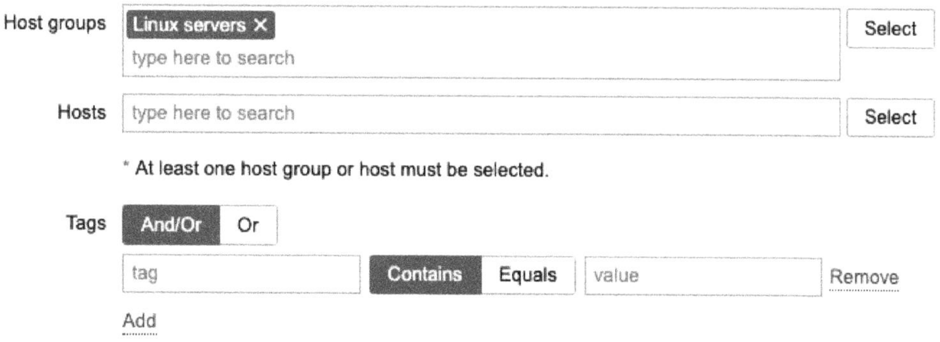

Host groups	Linux servers ✕ type here to search	Select
Hosts	type here to search	Select

* At least one host group or host must be selected.

Tags	And/Or Or
	tag Contains Equals value Remove
	Add

Figure 11.4 – Zabbix Configuration | Maintenance, add hosts to maintenance page, Patch Tuesday

8. You can still add a **Description** if you'd like.

9. Next, click on the blue **Add** button at the bottom of the page to finish creating the maintenance period. This will bring us back to our **Maintenance periods** page, where we should see that our maintenance window has been created.

How it works...

When configuring actions in Zabbix, we tell Zabbix to do a certain defined operation when a trigger is fired. Maintenance periods (with data collection) work by suppressing these Zabbix operations for the time period defined in the maintenance period. We do this to make sure that no Zabbix users are notified of any problems going on as maintenance is being done on a host. Of course, it's a good idea to only use this during the time that we are actually working on the hosts in question. This only works if the **Pause operations for suppressed problems** checkbox is ticked on the action, though.

In the case of this recipe, we've created a recurring maintenance period (with data collection) for the entire year of 2023. Let's say the organization we're working for has a lot of Linux hosts that need to be patched weekly. We set up the maintenance period to recur weekly every Tuesday between 22:00 and 04:00.

Now keep in mind that after December 31, 2023, Zabbix will stop this maintenance period as it won't be active any longer. We have two time/date values to bear in mind when setting up scheduled maintenance. The **Active since**/**Active till** time/date value of the maintenance period and the **Periods** time/date value of the maintenance period. This allows us to create more flexible periods, along with recurring ones as we just did.

Also, note that this maintenance period is **With data collection**. We can also create a maintenance period with the option of **No data collection**. When we use the **With data collection** option, we will keep collecting data but won't send any alerts to Zabbix Users. If you want to stop collecting the data, simply select the **No data collection** option. Keep in mind, no data collection also means your triggers won't fire anymore, nor will they resolve. The `nodata` trigger function is also affected by maintenance and it won't fire in both situations.

Backing up your Zabbix setup

Before working on any Zabbix setup, it is vital to make a backup of everything important. In this recipe, we will go through some of the most important steps you should always take before doing maintenance on your Zabbix setup.

Getting ready

We are going to need our Zabbix server, for which I'll use `lar-book-rocky`. Make sure to get the CLI to the server ready, as this whole recipe will use the Linux CLI.

How to do it...

1. Let's start by logging in to our Zabbix server via the Linux CLI and create some new directories that we are going to use for our Zabbix backups. Preferably, this directory would be on another partition:

```
mkdir /opt/zbx-backup/
mkdir /opt/zbx-backup/database/
mkdir /opt/zbx-backup/zbx-config/
mkdir /opt/zbx-backup/nginx/
mkdir /opt/zbx-backup/lib/
mkdir /opt/zbx-backup/shared/
mkdir /opt/zbx-backup/shared/zabbix/
mkdir /opt/zbx-backup/shared/doc/
```

2. It's important to back up all of our Zabbix configuration data, which is located at /etc/zabbix/. We can manually copy the data from our current folder to our new backup folder by issuing the following command:

```
cp -r /etc/zabbix/ /opt/zbx-backup/zbx-config/
```

3. Now, let's do the same for our nginx configuration:

```
cp -r /etc/nginx/ /opt/zbx-backup/nginx/
```

> **Important note**
>
> Please note that if you are using Apache, your web configuration location might be different. Adjust your command accordingly. For Red Hat-based systems it's usually /etc/httpd and for Debian-based systems, /etc/apache2.

4. It's also important to keep our Zabbix PHP files and binaries backed up. We can do that using the following commands:

```
cp -r /usr/share/zabbix/ /opt/zbx-backup/shared/zabbix/
cp -r /usr/share/doc/zabbix-* /opt/zbx-backup/shared/doc/
```

5. Lastly, let's make sure to also back up the Zabbix files in /usr/lib:

```
cp -r /usr/lib/zabbix/ /opt/zbx-backup/lib/
```

6. We could also create a cronjob to automatically compress and back up these files for us every day at 00:00. Simply issue the following command:

```
crontab -e
```

7. And add the following information:

```
0 0 * * * tar -zcvf /opt/zbx-backup/zbx-config/zabbix.tar.gz /
etc/zabbix/ >/dev/null 2>&1
0 0 * * * tar -zcvf /opt/zbx-backup/web-config/zabbix-web.tar.gz
/etc/nginx/ >/dev/null 2>&1
0 0 * * * tar -zcvf /opt/zbx-backup/shared/zabbix/zabbix_usr_
share.tar.gz /usr/share/zabbix/ >/dev/null 2>&1
0 0 * * * tar -zcvf /opt/zbx-backup/shared/doc/zabbix_usr_share_
doc.tar.gz /usr/share/doc/ >/dev/null 2>&1
0 0 * * * tar -zcvf /opt/zbx-backup/lib /zabbix_usr_lib.tar.gz /
usr/lib/zabbix/ >/dev/null 2>&1
```

8. These are all of the most important files we need to back up from our Zabbix stack. Let's move on to our database. We could now additionally use a rotation tool such as `logrotate` to manage our files.

9. Backing up our database is quite easy. We can simply use the built-in tools provided by MySQL and PostgreSQL. Issue the following command for your respective database (make sure to fill in the right username, database name, and password):

For MySQL databases:

```
mysqldump --add-drop-table --add-locks --extended-insert
--single-transaction --quick -u zabbixuser -p zabbixdb > /opt/
zbx-backup/database/backup_zabbixDB_<DATE>.sql
```

For PostgreSQL databases:

```
pg_dump zabbixdb > /opt/zbx-backup/database/backup_
zabbixDB_<DATE>.bak
```

10. Make sure to add the right location, as the database dump will be quite large if the database itself is large. Preferably, dump to another disk/partition or even better, another machine. As such, /opt/ might not be the best location.

11. We can also do this with a cronjob by issuing the following command:

```
crontab -e
```

12. Then for MySQL, add the following line where -u is the username, -p is the password, and the database name is zabbix. This is the command for MySQL:

```
55 22 * * 0 mysqldump -u'zabbixuser' -p'password' zabbixdb > /
opt/zbx-backup/database/backup_zabbixDB.sql
```

13. If you want to back up a PostgreSQL database with a cronjob, we will need to create a file in our user's home directory:

```
vim ~/.pgpass
```

14. We add the following to this file, where `zabbixuser` is the username and `zabbixdb` is the database name:

```
#hostname:port:database:username:password
localhost:5432:zabbixdb:zabbixuser:password
```

15. Then we can add a cronjob for PostgreSQL as follows:

```
55 22 * * 0 pg_dump --no-password -U zabbixuser zabbixdb > /opt/
zbx-backup/database/backup_zabbixDB_date.bak
```

16. We can also add a cronjob to only keep a certain number of days' worth of backups. Issue the following command:

```
crontab -e
```

17. Then add the following line, where `+60` is the number of days you want to keep backups for:

```
55 22 * * 0 find /opt/zbx-backup/database/ -mtime +60 -type f
-delete
```

18. That concludes our demonstration of backing up our Zabbix components the easy way.

> **Important note**
>
> For MySQL databases, there are also tools such as ExtraBackup, and for Postgres we could use PGBarman. It's never a bad idea to look into tools such as these to create backups for your system, but the built-in examples provided here can prove to be just as useful.

How it works...

A Zabbix setup consists of several components. We have the Zabbix frontend, Zabbix server, and Zabbix database. These components in this setup require different pieces of software to run on, as shown in the following diagram:

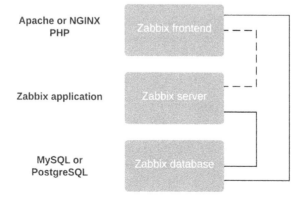

Figure 11.5 – Zabbix key components setup diagram

Looking at the preceding diagram, we can see that our Zabbix frontend runs on a web engine such as NGINX or Apache. We also need PHP to run our Zabbix web pages. This means that we have to back up two components:

- The web engine: NGINX, Apache, or another

- PHP

The Zabbix server is the application designed by Zabbix, so we only need to back up one thing here: the **Zabbix server config files**.

Then last, but definitely not least, we need to make a backup of our database. The most common databases used are MySQL and PostgreSQL, so we only need to do one thing for this: **create a dump of the Zabbix database**.

There's more...

Backing up your Zabbix setup like this is one thing, but of course, it's not everything. Make sure you take the correct backups of your Linux system using snapshots and other technologies.

When you follow standard backup implementations, you should be prepared for any unforeseen circumstances with your Zabbix setup.

Upgrading the Zabbix backend from older PHP versions to PHP 8.2 or higher

RHEL7, Ubuntu 20.04, and Debian 9 (Stretch) are no longer supported by Zabbix, thus our upgrade recipe no longer includes any information about the upgrade path from PHP versions before 7.2 to version 8.2 or higher. Newer Linux versions already ship with PHP8.0 or higher, which means that when we are upgrading a Zabbix setup from Zabbix version 6 to Zabbix 7, we can upgrade immediately.

The PHP requirement for Zabbix 7 is different than it was for Zabbix 6, meaning that if we are running PHP 7.2, we actually have a mandatory upgrade to do before we can run the latest Zabbix 7 release. I also like to work in a *future-proofing* kind of way, so in this recipe, we will go over how to upgrade PHP 7.2 to 8.2 which is the latest supported version on RHEL8-based systems at the time of writing.

Getting ready

For this recipe, we will need our server installed with a RHEL8-based system, which will be running Zabbix server 6 with PHP version 7.2.

Another possibility is that you have a server running a Debian-based distribution such as Ubuntu 20.04, Debian 11, or a newer version of those Linux distributions. These include PHP version 7.2 or higher by default.

I will refer to both possible servers as `lar-book-zbx6` throughout this recipe.

Lastly, make sure to take backups of your system and read the release notes for the new version you're installing.

How to do it...

This recipe is split into two different sections, one for RHEL8-based systems and another for Ubuntu systems. We will start by going through the steps for RHEL8.

RHEL8-based systems

If you are already running PHP version 7.2 on a RHEL8-based system, the upgrade process is a bit simpler. Let's check out how we can upgrade our `lar-book-zbx6` server in this scenario:

1. First, always verify what PHP version we are running with the following command:

   ```
   php-fpm --version
   ```

2. If the version is older than 8.2, we can continue with the next step. We'll execute the following:

   ```
   dnf module list php
   ```

3. This will show us something like the following screenshot:

```
[root@lar-book-zbx6 ~]# dnf module list php
Last metadata expiration check: 0:06:00 ago on Sun 16 Jul 2023 11:08:44 AM CEST.
Rocky Linux 8 - AppStream
Name          Stream          Profiles                      Summary
php           7.2 [d][e]      common [d], devel, minimal    PHP scripting language
php           7.3             common [d], devel, minimal    PHP scripting language
php           7.4             common [d], devel, minimal    PHP scripting language
php           8.0             common [d], devel, minimal    PHP scripting language

Hint: [d]efault, [e]nabled, [x]disabled, [i]nstalled
```

Figure 11.6 – RHEL8 DNF module list for PHP

4. Unfortunately, on RHEL8, the latest stable PHP 8.3 version is not included in the DNF modules from AppStream. This means we will have to find an alternative route for RHEL8-based systems. If you want to install PHP 8.3 or higher, continue to *step 9*.

5. Since PHP 8.2 is included in the AppStream list, reset your already available PHP modules:

   ```
   dnf module reset php
   ```

6. Make sure to answer with Y. Then we will enable the latest PHP version with the following command:

   ```
   dnf module enable php:8.2
   ```

7. Answer with Y again to enable PHP 8.2 and then we can upgrade our PHP version by using the following command:

    ```
    dnf update
    ```

8. Answer Y again and your PHP version will now be running the latest PHP 8.2 version.

9. If we cannot use the `dnf module enable` method to reach the version you want to install, we are going to have to rely on different means of getting PHP, the most popular route being the REMI repositories.

10. Make sure your system is up to date with the following command:

    ```
    dnf update
    ```

11. REMI depends on the EPEL repository, so we will have to add that first:

    ```
    dnf install epel-release
    ```

12. After installing `epel-release`, make sure to exclude Zabbix from it. This ensures that Zabbix is only downloaded and updated from the official Zabbix repositories:

    ```
    [epel]
    excludepkgs=zabbix*
    ```

13. .Then we install the REMI repository with the following command:

    ```
    sudo dnf -y install http://rpms.remirepo.net/enterprise/remi-
    release-8.rpm
    ```

14. Reset the PHP modules and enable the REMI PHP 8.3 version:

    ```
    dnf module reset php -y
    dnf module install php:remi-8.3
    ```

15. Enter Y or Yes everywhere during the installation procedure.

16. Then, verify whether the upgrade was successful:

    ```
    php-fpm -v
    ```

17. Make sure to restart NGINX (or Apache) and `php-fpm`:

    ```
    systemctl restart nginx php-fpm
    ```

These steps have been tested on a Rocky Linux RHEL system, but they should work with any RHEL8-based system, be it in Stream or when it's a full rebuild as with Alma Linux.

Consider upgrading to RHEL9-based systems for further support for newer versions of the PHP packages.

Ubuntu systems

Let's upgrade to the latest version of PHP available on our Ubuntu system:

1. First, start by adding the PPA repository to our host with the following command:

    ```
    apt install software-properties-common
    add-apt-repository ppa:ondrej/php
    ```

2. Now update the repositories with the following command:

    ```
    apt update
    ```

3. On some installations, the key for the repository might not be available, in which case we might see an error reading Key is not available. We can fix this with the following command, where PUB_KEY_HERE is the key shown in the error:

    ```
    apt-key adv --keyserver keyserver.ubuntu.com --recv- keys PUB_
    KEY_HERE
    ```

4. Now we can install PHP version 8.3 with the following command:

    ```
    apt install -y php8.3
    apt upgrade -y php
    apt autoremove
    ```

5. That's it, the version of PHP should now be the one we want. Check the version of PHP with the following command:

    ```
    php --version
    ```

How it works...

Because Zabbix 7 requires us to install PHP version 8.0 or higher, we need to upgrade the PHP version if we are still using PHP 7.2 for our Zabbix 6 install. It's a different requirement than for Zabbix 6, making the upgrade process fairly long in some cases. If you are still running RHEL7, Ubuntu 20.04, or Debian 9 (Stretch), then you will need to upgrade your Linux system first as well. Zabbix has dropped support for these older Linux versions in favor of installation simplicity in terms of package management, security, and support.

Now, it is still possible to run Zabbix on older Linux versions by building from sources, but it is not recommended.

In this recipe, we did an upgrade from PHP 7.2 to PHP 8.2 or 8.3, which are some of the latest supported stable versions at the time of writing. Doing this upgrade will not break our current Zabbix server installation. As mentioned, this is a mandatory upgrade as PHP versions below 8.0 are too old to run Zabbix 7. Even if the upgrade was optional, it is always good to consider running the latest stable release of software to make sure that we are ready for the future.

Now that we have upgraded PHP, we are ready to move on to upgrading the Zabbix database engine.

Upgrading a Zabbix database from older MariaDB versions to MariaDB 11.4

For our Zabbix 7 installation, we are going to need MariaDB 10.5 or a newer supported version, so, it is a good idea to keep your database version up to date. MariaDB regularly makes improvements to how it handles certain aspects of performance.

This recipe details how to upgrade MariaDB to the latest stable LTS version, which is MariaDB 11.4 at the time of writing.

Getting ready

For this recipe, we will need our server which we called `lar-book-zbx6`. At this point, the server is running a RHEL8-based distribution.

Another option is to have a server running a Debian-based distribution such as Ubuntu 22.04, Debian 12, or a newer version of those Linux distributions. We will be upgrading the MariaDB instance on this server to version 11.4.

If you've followed the *Upgrading the Zabbix backend from older PHP versions to PHP 8.2 or higher* recipe, your server will now be running PHP version 8.2 or higher. If not, it's a good idea to follow that recipe first.

Also, make sure to take backups of your system and read the release notes for the new version you're installing. We covered this in the *Backing up your Zabbix setup* recipe.

How to do it...

1. First things first, let's log in to our Linux host CLI to check out our versions. Issue the following commands:

   For Zabbix server:

   ```
   zabbix_server --version
   ```

   For PHP:

   ```
   php-fpm --version
   ```

   For MariaDB:

   ```
   mysql --version
   ```

2. After verifying our versions match the versions mentioned in the *Getting ready* section of this recipe, let's move on to upgrade our version of MariaDB.

RHEL-based systems

1. On our RHEL-based server, the first thing we'll do after checking the versions is to stop our Zabbix environment:

   ```
   systemctl stop mariadb nginx zabbix-server
   ```

2. Now set up a repository file for MariaDB with the following command:

   ```
   vim /etc/yum.repos.d/mariadb.repo
   ```

3. We will add the following code to this new file. Make sure to add the correct architecture after baseurl if using anything other than amd64:

   ```
   [mariadb-main]
   name = MariaDB Server
   baseurl = https://dlm.mariadb.com/repo/mariadb-server/11.4/yum/
   rhel/8/x86_64
   gpgkey = file:///etc/pki/rpm-gpg/MariaDB-Server-GPG-KEY
   gpgcheck = 1
   enabled = 1
   module_hotfixes = 1


   [mariadb-maxscale]
   # To use the latest stable release of MaxScale, use "latest" as
   the version
   # To use the latest beta (or stable if no current beta) release
   of MaxScale, use "beta" as the version
   name = MariaDB MaxScale
   baseurl = https://dlm.mariadb.com/repo/maxscale/latest/yum/
   rhel/8/x86_64
   gpgkey = file:///etc/pki/rpm-gpg/MariaDB-MaxScale-GPG-KEY
   gpgcheck = 1
   enabled = 1


   [mariadb-tools]
   name = MariaDB Tools
   baseurl = https://downloads.mariadb.com/Tools/rhel/8/x86_64
   gpgkey = file:///etc/pki/rpm-gpg/MariaDB-Enterprise-GPG-KEY
   gpgcheck = 1
   enabled = 1
   [mariadb-main]
   name = MariaDB Server
   baseurl = https://dlm.mariadb.com/repo/mariadb-server/11.4/yum/
   ```

```
rhel/8/x86_64
gpgkey = file:///etc/pki/rpm-gpg/MariaDB-Server-GPG-KEY
gpgcheck = 1
enabled = 1
module_hotfixes = 1


[mariadb-maxscale]
# To use the latest stable release of MaxScale, use "latest" as
the version
# To use the latest beta (or stable if no current beta) release
of MaxScale, use "beta" as the version
name = MariaDB MaxScale
baseurl = https://dlm.mariadb.com/repo/maxscale/latest/yum/
rhel/8/x86_64
gpgkey = file:///etc/pki/rpm-gpg/MariaDB-MaxScale-GPG-KEY
gpgcheck = 1
enabled = 1


[mariadb-tools]
name = MariaDB Tools
baseurl = https://downloads.mariadb.com/Tools/rhel/8/x86_64
gpgkey = file:///etc/pki/rpm-gpg/MariaDB-Enterprise-GPG-KEY
gpgcheck = 1
enabled = 1
```

4. Alternatively and probably the best method is to use the MariaDB setup script:

```
curl -LsS https://r.mariadb.com/downloads/mariadb_repo_setup |
sudo bash -s -- --mariadb-server-version="mariadb-11.4"
```

5. Now upgrade your MariaDB server with the following command:

```
dnf upgrade MariaDB*
```

6. Restart the MariaDB service with the following command:

```
systemctl start mariadb zabbix-server nginx
```

7. That's it, MariaDB should now be upgraded to the intended version. Check the version again with the following command to make sure:

```
mariadb --version
```

Ubuntu systems

1. On our Ubuntu server, the first thing we'll do after checking the versions is to stop our Zabbix server environment:

```
systemctl stop mariadb zabbix-server nginx
```

2. Check for the MariaDB repository file at /etc/apt/sources.list.d/mariadb.list. To check whether it is on version 11.4, edit it with the following command:

```
vim /etc/apt/sources.list.d/mariadb.list
```

3. The file should look like the following code block. If it doesn't look right, edit it to match. Make sure to add the correct architecture on the deb lines if using anything other than amd64:

```
# MariaDB Server
# To use a different major version of the server, or to pin to a
specific minor version, change URI below.
deb [arch=amd64,arm64] https://dlm.mariadb.com/repo/mariadb-
server/11.4/repo/ubuntu jammy main

deb [arch=amd64,arm64] https://dlm.mariadb.com/repo/mariadb-
server/11.4/repo/ubuntu jammy main/debug


# MariaDB MaxScale
# To use the latest stable release of MaxScale, use "latest" as
the version
# To use the latest beta (or stable if no current beta) release
of MaxScale, use "beta" as the version
deb [arch=amd64,arm64] https://dlm.mariadb.com/repo/maxscale/
latest/apt jammy main


# MariaDB Tools
deb [arch=amd64] http://downloads.mariadb.com/Tools/ubuntu jammy
main
```

4. Alternatively, we can use the MariaDB repository setup script to update to the right repository. Execute the following command:

```
curl -LsS https://r.mariadb.com/downloads/mariadb_repo_setup |
sudo bash -s -- --mariadb-server-version="mariadb-11.4"
```

5. We need to remove our old MariaDB packages with the following command:

```
apt remove mariadb-server mariadb-client
```

6. Now upgrade the MariaDB server version with the following command:

```
apt install mariadb-server mariadb-client
```

7. Restart MariaDB with the following command:

```
systemctl restart mariadb
```

8. Then issue the upgrade command:

```
mariadb-upgrade
```

9. Now start Zabbix back up:

```
systemctl restart zabbix-server nginx
```

10. That's it, MariaDB should now be upgraded to the correct version. Check the version again with the following command:

```
mariadb --version
```

How it works...

Now, while it might not always be a requirement, it is a smart idea to upgrade your database version regularly. New versions of your database engine might include improvements to stability and performance, both of which could improve your Zabbix server greatly.

Do keep the release notes and bug reports on your radar though. MariaDB 11.4 is, at the time of writing, the newest LTS version on the market. You might want to stay behind one or two releases as these are still supported and have been running in production for a while already. After all, nobody likes unforeseen issues such as bugs.

For Zabbix 7, we do need to install at least MariaDB 10.5 or a newer supported version though, so keep that in mind.

There's more...

If you really cannot upgrade to MariaDB version 10.5 or, if you are running another database, the supported version for that one, then there's a new Zabbix feature. Zabbix 7 allows us to run unsupported database versions. When we edit the Zabbix server configuration files at /etc/zabbix/zabbix_server.conf, we can add the following parameter:

```
AllowUnsupportedDBVersions=1
```

This will allow you to run an older or newer version of your database that is not officially tested and supported by Zabbix yet, but keep in mind that it is not recommended to do so. Check out the current Zabbix LTS installation requirements here:

```
https://www.zabbix.com/documentation/current/en/manual/installation/
requirements
```

Upgrading your Zabbix setup

As we've seen throughout the book already, Zabbix 7 offers a great deal of cool new features. Zabbix 7.0 is a **Long-Term Support** (**LTS**) release, so just like 5.0 and 6.0, you will receive long-term support for it. Let's see how we can upgrade a Zabbix server from version 6.0 to version 7.0.

Getting ready

For this recipe, we will need our server called `lar-book-zbx6`. At this point, your server will be running either a RHEL8-based Linux distribution or a Debian-based distribution like Ubuntu 22.04, Debian 12, or newer versions of those distributions.

If you followed the *Upgrading the Zabbix backend from older PHP versions to PHP 8.2 or higher* recipe, your server will now be running PHP version 8.2 or higher. If not, it's a good option to follow that recipe first.

If you followed the *Upgrading a Zabbix database from older MariaDB versions to MariaDB 11.4* recipe, it will now be running MariaDB version 11.4. If not, it's wise to follow that recipe first.

Also, make sure to take backups of your system and read the release notes for the new version you're installing. We covered this in the *Backing up your Zabbix setup* recipe.

How to do it...

First things first, let's log in to our Linux host CLI to check out our software versions:

1. Issue the following commands to check the respective software versions:

    For Zabbix server:

    ```
    zabbix_server --version
    ```

    For PHP:

    ```
    php-fpm --version
    ```

    For MariaDB:

    ```
    mariadb --version
    ```

2. After verifying our versions match the versions mentioned in the *Getting ready* section of this recipe, let's move on to upgrade our Zabbix server.

RHEL-based systems

First, we will start with upgrading the Zabbix server on a RHEL-based system:

1. Let's stop our Zabbix server components with the following command:

    ```
    systemctl stop zabbix-server zabbix-agent2
    ```

2. On our server, let's issue the following command to add the new Zabbix 7.0 repository:

    ```
    rpm -Uvh https://repo.zabbix.com/zabbix/7.0/rhel/8/ x86_64/
    zabbix-release-7.0-1.el8noarch.rpm
    ```

3. Run the following command to clean the repositories:

    ```
    dnf clean all
    ```

4. Now upgrade the Zabbix setup with the following command:

    ```
    dnf upgrade zabbix-server-mysql zabbix-web-mysql zabbix-agent2
    ```

5. Additionally, install the Zabbix NGINX configuration:

    ```
    dnf install zabbix-nginx-conf
    ```

6. Start the Zabbix components with the following command:

    ```
    systemctl restart zabbix-server zabbix-agent2
    ```

7. When we check if the server is running, it should say Active (running) when we issue the following command:

    ```
    systemctl status zabbix-server
    ```

8. If not, we check the logs with the following command, so we can see what is happening:

    ```
    tail -f /var/log/zabbix/zabbix_server.log
    ```

9. Check the log file for any notable errors and if you find any, fix them before continuing.

10. If we start the server again, this error should be gone and the Zabbix server should keep running:

    ```
    systemctl restart zabbix-server
    ```

11. Now restart the Zabbix components with the following command:

    ```
    systemctl restart nginx php-fpm zabbix-server mariadb
    ```

12. Now everything should be working as expected and we should see the new Zabbix 7 frontend, as shown in the following screenshot:

Figure 11.7 – Zabbix 7 frontend after the upgrade on RHEL

Ubuntu systems

1. First, let's stop our Zabbix server components with the following command:

```
systemctl stop zabbix-server zabbix-agent2
```

2. Now add the new repository for Zabbix 7 on Ubuntu with the following commands:

```
wget https://repo.zabbix.com/zabbix/7.0/ubuntu/pool/main/z/
zabbix-release/zabbix-release_7.0-1+ubuntu22.04_all.deb
dpkg -i zabbix-release_7.0-1+ubuntu22.04_all.deb
```

> **Important note**
> Always check zabbix.com/download to get the right repository for your systems. In the example, I used the Ubuntu repository. Switch this out for the right repository for your system.

3. Update the repository information with the following command:

```
apt update
```

4. Now upgrade the Zabbix server components with the following command:

```
apt install --only-upgrade zabbix-server-mysql zabbix-frontend-
php zabbix-agent2
```

5. Make sure to not overwrite your Zabbix server configuration. If you do overwrite your configuration file, you can restore it from the backup taken in the *Backup up your Zabbix setup* recipe.

6. Then install the new Zabbix NGINX configuration with the following command:

```
apt install zabbix-nginx-conf
```

7. Restart the Zabbix server components with the following command and you should be done:

```
systemctl restart zabbix-server nginx zabbix-agent2
```

8. We check the logs with the following command so we can see what is happening:

```
tail -f /var/log/zabbix/zabbix_server.log
```

9. Check the log file for any notable errors and if you find any, fix them before continuing.

10. If we start the server again, this error should be gone and the Zabbix server should keep running:

```
systemctl restart zabbix-server
```

11. This should conclude the upgrade process, and if we go to the frontend, we should see the new Zabbix 7 frontend:

Figure 11.8 – Zabbix 7 frontend after the Ubuntu upgrade

How it works...

Upgrading Zabbix can be an easy task when we are running the latest version of Linux. When we are running older versions of software though, we might run into some issues.

The recipe we've just followed shows us the upgrade process for a Zabbix 6 instance resulting in a setup running Zabbix 7, along with the most common issues we might run into.

> **Important note**
> While upgrading, make sure to keep an eye on your `zabbix_server.log` file, as this file will tell you if something has gone wrong during the upgrade process.

We made sure to upgrade PHP to a version higher than 8.0 as this was the requirement for Zabbix 7, making the upgrade process from Zabbix 6 a bit more complicated if we ran an older PHP version. For the database, Zabbix kept the same requirements between Zabbix 6 and Zabbix 7, requiring MariaDB 10.5 or a newer supported version for your Zabbix setup.

Now that you've upgraded all the components, you should be ready to work with Zabbix 7 and your setup will be future-proof for a while – of course, until Zabbix 8 comes out, when we might see some new requirements come along.

See also

Make sure to check out the Zabbix documentation for the versions you are upgrading from and to. Zabbix always includes detailed descriptions of the requirements and processes to make it as easy as possible for you to upgrade. Check out the right documentation for your version:

```
https://www.zabbix.com/documentation/current/en/manual/installation/
upgrade
```

Maintaining Zabbix performance over time

It's important to make sure that your Zabbix setup keeps performing well over time. There are several key components that are important to keep your Zabbix setup performing optimally. Let's see how to work on some of these components and keep your Zabbix setup running smoothly.

Getting ready

All we are going to need for this recipe is a Zabbix 7 server.

How to do it...

We will go through three of the main problems people face whilst maintaining Zabbix server performance. First things first, let's look at the Zabbix processes and how to edit them.

Zabbix processes

A regular problem people face is a Zabbix process being too busy. Let's log in to our Zabbix frontend and check out how this problem might look.

First, let's start by logging in to our Zabbix server frontend and check out some messages:

1. When we navigate to **Monitoring | Dashboard** and then select the default dashboard **Global view**, we might see something like this:

Figure 11.9 – Zabbix problem from our Zabbix server, discoverer processes 75% busy

2. Then we navigate to **Monitoring | Hosts** and click on **Latest data** for the Zabbix server host (which, in my case, is called `lar-book-rocky`). This will take us to the latest data for our host.

3. For the filters, type `discovery` in the **Name** field, then click on **Graph** for the discovery worker item. This will show you the following graph:

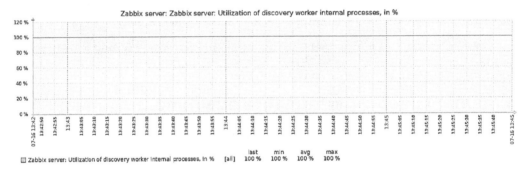

Figure 11.10 – Zabbix server discoverer graph, Utilization of discoverer data collector in %

This graph is at 100% almost all the time, which explains why we see the problem shown in *Figure 11.9* on our dashboard.

4. Let's log in to the Linux CLI of our Zabbix server to edit this process.

5. Edit the following file on your Zabbix server:

```
vim /etc/zabbix/zabbix_server.conf
```

6. Now, if we want to give our Zabbix server's `discoverer` process more room, we need to edit the correct parameter. Scroll down until you see the following:

```
### Option: StartDiscoverers
#          Number of pre-forked instances of discoverers.
#
# Mandatory: no
# Range: 0-250
# Default:
# StartDiscoverers=1
```

Figure 11.11 – Zabbix server configuration file, StartDiscoverers default

7. Now add a new line under this and add the following line:

```
StartDiscoverers=2
```

8. If your file now looks like the following screenshot, you can save and exit the file:

```
### Option: StartDiscoverers
#          Number of pre-forked instances of discoverers.
#
# Mandatory: no
# Range: 0-250
# Default:
# StartDiscoverers=1
StartDiscoverers=2
```

Figure 11.12 – Zabbix server configuration file, StartDiscoverers 2

9. For the changes to take effect, we will need to restart the Zabbix server with the following command:

```
systemctl restart zabbix-server
```

10. Now if we go back to our Zabbix frontend, we should still be at our graph where we can see the following:

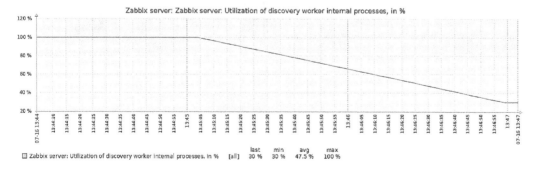

Figure 11.13 – Zabbix server discoverer graph

The utilization of our discoverer process has gone down, which means our utilization problem won't show up anymore. That's how we edit Zabbix server processes.

Zabbix housekeeper

Another very common problem people face is the Zabbix housekeeper process being too busy. Let's log in to our Zabbix frontend and check out the problem:

1. When we navigate to **Monitoring | Dashboard** and then select the default dashboard **Global view**, we might see something like this:

	Time ▾	Severity	Recovery time	Status	Info	Host	Problem
☐	09:42:06 •	Average		PROBLEM		Zabbix server	Utilization of housekeeper processes over 75%

Figure 11.14 – A problem with Zabbix housekeeper

2. Similar to editing any Zabbix process, we can also edit the Zabbix housekeeper process. Let's log in to the Linux CLI of our Zabbix server to edit our process.

3. Let's edit the following file on our Zabbix server:

    ```
    vim /etc/zabbix/zabbix_server.conf
    ```

4. Now, if we want to edit this process, we need to edit the correct parameters. Scroll down until you see the following:

```
### Option: HousekeepingFrequency
#       How often Zabbix will perform housekeeping procedure (in hours).
#       Housekeeping is removing outdated information from the database.
#       To prevent Housekeeper from being overloaded, no more than 4 times HousekeepingFrequency
#       hours of outdated information are deleted in one housekeeping cycle, for each item.
#       To lower load on server startup housekeeping is postponed for 30 minutes after server start.
#       With HousekeepingFrequency=0 the housekeeper can be only executed using the runtime control option.
#       In this case the period of outdated information deleted in one housekeeping cycle is 4 times the
#       period since the last housekeeping cycle, but not less than 4 hours and not greater than 4 days.
#
# Mandatory: no
# Range: 0-24
# Default:
# HousekeepingFrequency=1
```

Figure 11.15 – Zabbix configuration file, HousekeepingFrequency 1

5. This is our first housekeeper parameter. Let's edit this parameter by adding the following line under this block:

    ```
    HousekeepingFrequency=2
    ```

> **Important note**
>
> Making the interval longer is not going to solve your issue; at most, you are delaying the inevitable. It is only recommended to change this setting until the next maintenance window, and it should be avoided as much as possible.

6. Now scroll down until you see the following:

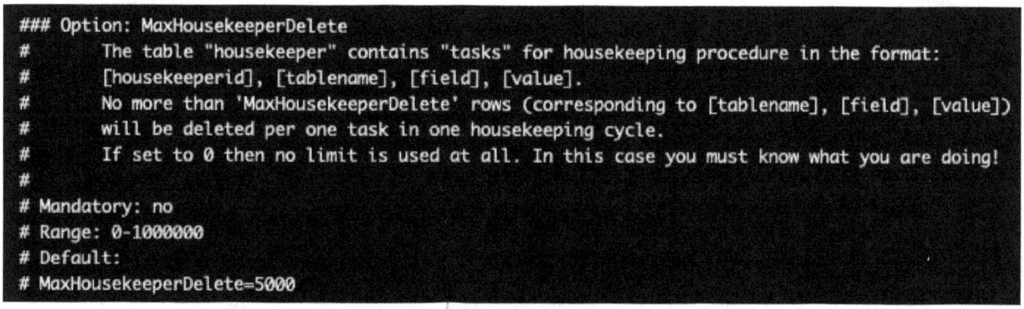

```
### Option: MaxHousekeeperDelete
#        The table "housekeeper" contains "tasks" for housekeeping procedure in the format:
#        [housekeeperid], [tablename], [field], [value].
#        No more than 'MaxHousekeeperDelete' rows (corresponding to [tablename], [field], [value])
#        will be deleted per one task in one housekeeping cycle.
#        If set to 0 then no limit is used at all. In this case you must know what you are doing!
#
# Mandatory: no
# Range: 0-1000000
# Default:
# MaxHousekeeperDelete=5000
```

Figure 11.16 – Zabbix configuration file, HousekeepingDelete 5000

7. The preceding screenshot shows our second housekeeper parameter. Let's edit this parameter by adding the following line under this code block:

    ```
    MaxHousekeeperDelete=20000
    ```

8. For the changes to take effect, we need to restart the Zabbix server with the following command:

    ```
    systemctl restart zabbix-server
    ```

Tuning a MySQL database

1. Let's see how we can tune a MySQL database with ease. First off, let's go to the following link in our browser: https://github.com/major/MySQLTuner-perl.

2. This link brings us to an open source GitHub project started by *Major Hayden*. Be sure to follow the repository and do all you can to help out. Let's download the script from the GitHub repository or simply use the following command:

    ```
    wget https://raw.githubusercontent.com/major/MySQLTuner-perl/
    master/mysqltuner.pl
    ```

3. Now we can execute this script with the following command:

    ```
    perl mysqltuner.pl
    ```

4. This will bring us to a prompt for our MySQL database credentials. Fill them out and continue:

```
[root@lar-book-centos ~]# perl mysqltuner.pl
 >>  MySQLTuner 1.7.19 - Major Hayden <major@mhtx.net>
 >>  Bug reports, feature requests, and downloads at http://mysqltuner.pl/
 >>  Run with '--help' for additional options and output filtering

[--] Skipped version check for MySQLTuner script
Please enter your MySQL administrative login: root
Please enter your MySQL administrative password: [OK] Currently running supported MySQL version 10.3.17-MariaDB
```

Figure 11.17 – MySQL tuner script execution

5. Now, the script will output a lot of information that you will need to read carefully, but the most important part is at the end – everything after `Variables to adjust`:

```
Variables to adjust:
  *** MySQL's maximum memory usage is dangerously high ***
  *** Add RAM before increasing MySQL buffer variables ***
    query_cache_size (=0)
    query_cache_type (=0)
    query_cache_limit (> 1M, or use smaller result sets)
    join_buffer_size (> 256.0K, or always use indexes with JOINs)
    performance_schema = ON enable PFS
    innodb_buffer_pool_size (>= 2.9G) if possible.
    innodb_log_file_size should be (=16M) if possible, so InnoDB total log files size equals to 25% of buffer pool size.
```

Figure 11.18 – MySQL tuner script output

> **Important note**
> *DO NOT* simply copy over the output from this script. The script is simply giving us an indicator of what might be tuned in our MySQL settings. Always look up the settings suggested and read about the best practices for those settings.

6. We can edit these variables in the MySQL `my.cnf` file. In my case, I edit it with the following command:

```
vim /etc/my.cnf.d/server.cnf
```

7. Now, you simply edit or add the variables that are suggested in the script and then restart your MySQL server:

```
systemctl restart mariadb
```

How it works...

We've just done three of the main performance tweaks we can do for a Zabbix server, but there's a lot more to do. Let's take a look at what we've just edited, consider why we've edited it, and find out whether it's really that simple.

Zabbix processes

Zabbix processes are a big part of your Zabbix server setup and must be edited with care. In this recipe, we've only just edited the discoverer process on a small installation. This problem was easy as the server had more than enough resources to account for another process running.

Now if we look at the following diagram, we can see the situation as it was before we added a new discoverer process:

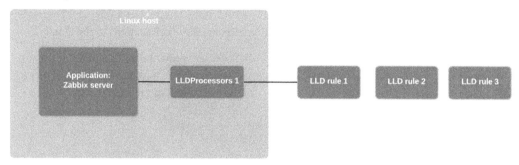

Figure 11.19 – Zabbix server single-process setup diagram

We can see our **Linux host** running our **Zabbix server** application and we can see our **LLDProcessors 1** process discovering **LLD rule 1**. **LLD rule 2** and **LLD rule 3** are queueing up as one LLDProcessor subprocess can only handle one rule at a time.

As we've seen that this is apparently too heavy for our system, we have added another LLDProcessor:

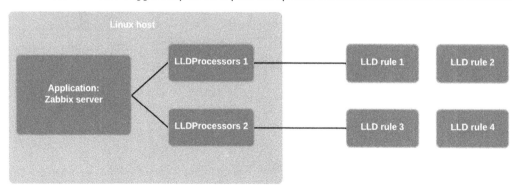

Figure 11.20 – Zabbix server multiple-process setup diagram

Our new setup will balance the load to a certain extent. It's only possible for a discovery rule to be handled by a single discoverer process. This means that if we have multiple discovery rules, we can add discoverers like this to make sure there are enough resources available per discovery rule. It works the same for the other processes – more processes mean better distributions of tasks.

However, there are several things to be careful of here. First of all, not all issues can be solved by simply throwing more resources at them. Some Zabbix setups are configured poorly, where there's something in the configuration making our processes unnecessarily busy. If we deal with the poor configuration aspect, we can take away the high load, thus we need fewer processes.

The second thing I'd like to stress is that we can keep adding processes to our Zabbix server configuration – within limits. Before we reach those limits though, you are definitely going to reach the roof of what our Linux host hardware is capable of. Make sure you have enough RAM and CPU power to actually run all these processes or use Zabbix proxies for offloading. Also keep in mind that adding more processes might require additional database tuning, for example allowing more connections to the database.

Last but not least, keep in mind that changing the Zabbix server configuration requires a restart of the `zabbix-server` process. On large installations, this can take a long time. The Zabbix server might have to do a lot of database writes (for example, of the trend data) to get the `zabbix-server` process to shut down.

Zabbix housekeeper

Now for Zabbix housekeeper, which is a very important process for Zabbix administrators who haven't set up MySQL partitioning or PostgreSQL TimescaleDB partitioning yet. The Zabbix housekeeper process connects to our database and then drops information line by line that has *expired*. You might think, how do you mean expired? Well, we can set limits in the Zabbix server for how long an item should be kept in the database.

If we look at **Administration | Housekeeping**, part of what we will see is shown in the following screenshot:

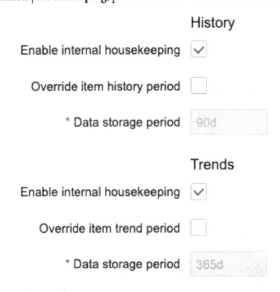

Figure 11.21 – Zabbix server history and trends housekeeping setup

These are our global **History** and **Trends** housekeeping parameters. This defines how long an item's data should be kept in our database. If we look at an item on a template or host, we can also see these parameters:

Figure 11.22 – Zabbix item history and trends housekeeping parameters

These settings override the global settings so you can tweak the housekeeper further. That's how the housekeeper keeps your database in check.

But now, let's look at the tweaks we made in our Zabbix server configuration file, the first of which is HousekeepingFrequency. Housekeeping frequency is how often the housekeeper process is started. We've lowered this from every hour to every two hours. Now you might think that's worse, but it doesn't have to be. A lot of the time, we see that housekeeping is not done after one hour and then it just keeps going on and on.

We also changed the MaxHousekeeperDelete parameter, which is something completely different. This determines how many database rows our Zabbix housekeeper is allowed to delete in each run. The default settings determined that every hour, we can delete 5,000 database rows. With our new settings, we can now delete 20,000 database rows every two hours. Each row will basically just be a single metric we are allowed to delete.

How does this change anything at all? Well, it might not. It completely depends on your setup. Tweaking the Zabbix housekeeper is different for every setup, and you will have to determine your optimum settings for yourself. Try to balance what you see in your graphs with the two settings we've discussed here to see how well you can optimize it.

However, at one point, your Zabbix setup might grow big enough and Zabbix housekeeping won't be able to keep up. This is when you'll need to look at MySQL partitioning or PostgreSQL TimescaleDB. There's no predefined point where the Zabbix housekeeper won't be able to keep up, so it is smarter to just start with MySQL partitioning or PostgreSQL TimescaleDB right from the start. After all, any setup might grow larger than expected, right? More on this subject is explained in *Chapter 12* of this book.

Tuning a MySQL database

Now for tuning your MySQL database with the mysqltuner.pl script. This script does a lot in the background, but we can summarize it as follows: it looks at what the current utilization of your MySQL database is, and then outputs what it thinks the correct tuning variables would be.

Do not take the script output as a given, as with Zabbix housekeeping, there is no way to give you a definitive setup for your database. Databases are simply more complicated than just doing some tweaks and being done with it.

The script will definitely help you tweak your MySQL database to an extent, especially for smaller setups. But make sure to extend your knowledge by reading blogs, guides, and books about databases regularly.

There's more...

We went over how to tune a MySQL database, but we didn't go over how to tune a PostgreSQL instance. There's a wide variety of options out there to do this, so for more on that I recommend checking out the PostgreSQL wiki at `https://wiki.postgresql.org/wiki/Performance_Optimization`. There are different varieties and different preferences at play here. Make sure to check them all out well and pick the one that works the best for you.

There's also a new addition in Zabbix 7.0, which is specifically for the following three pollers:

- Agent poller
- HTTP agent poller
- SNMP poller (for walk[OID] and get[OID] items)

These processes now execute checks asynchronously. What that means is that they can execute multiple (item) checks at the same time. In older versions of Zabbix, these pollers could only execute a single check at the time.

It's still possible to add multiple of these processes with, for example, **StartAgentPollers**, but it now functions differently. This will execute a maximum of 1,000 checks per poller, which is configurable with the **MaxConcurrentChecksPerPoller** parameter.

12
Advanced Zabbix Database Management

Whether you've been using Zabbix for a while or you are looking toward setting up your first production instance, database management is important right from the start. A lot of the time, people set up their Zabbix database and don't know yet that it will be a big database. The Zabbix housekeeper just can't keep up when your database grows beyond a certain size, and that's when we need to look for different options.

In this chapter, we'll look into keeping our Zabbix database from using up 100% disk space when the Zabbix housekeeper is not keeping up. For MySQL users, we'll look into using database partitioning to keep our database in check. For PostgreSQL users, we'll look toward the TimescaleDB support. Last but not least, we'll also check out how to secure our connection between the Zabbix server and the database.

We'll do all this in the following recipes:

- Setting up MySQL partitioning for your Zabbix database
- Using the PostgreSQL TimescaleDB functionality
- Securing your Zabbix MySQL database

Without further ado, let's get started on these recipes and learn all about managing our database.

Technical requirements

We are going to need some new servers for these recipes. One Linux server needs to run Zabbix server 7 with MySQL (MariaDB) set up; we'll call this host `lar-book-mysql- mgmt`. We will also need a Linux server running Zabbix server 7 with PostgreSQL, which we'll call `lar-book-postgresql-mgmt`.

We'll also need two servers for creating a secure Zabbix database setup. One server will be running the MySQL (MariaDB) database; let's call this server `lar-book-secure-db`. Then, connecting externally to a Zabbix database, we'll have our Zabbix server, which we'll call `lar-book-secure-zbx`.

The code files can also be accessed in the GitHub repository here:

`https://github.com/PacktPublishing/Zabbix-7-IT-Infrastructure-Monitoring-Cookbook/tree/main/chapter12`

Setting up MySQL partitioning for your Zabbix database

When working with a MySQL database, the biggest issue we face is how MySQL stores its data by default. There is no real order to the data that we can use if we want to drop large chunks of data. MySQL partitioning solves this issue; let's see how we can configure it to use for our Zabbix database.

> **Important note**
>
> Here at Opensource ICT Solutions, we have fixed the script to work with MySQL 8. The script should work for *any* MySQL setup once more. Check out the link for more information: `https://github.com/OpensourceICTSolutions/zabbix-mysql-partitioning-perl`.

Getting ready

For this recipe, we are going to need a running Zabbix server with a MySQL database. I'll be using MariaDB in my example, but any MySQL flavor should be about the same. The Linux host I'll be using is called `lar-book-mysql-mgmt`, which already meets the requirements.

If you are running these steps in a production environment, make sure to create your database backups first as things can always go wrong.

How to do it...

1. First things first, let's log in to our Linux CLI to execute our commands.
2. It's a good idea to use TMUX because partitioning can take several days for big databases. TMUX will keep the sessions open in the background, even if we lose the SSH connection. If TMUX is not installed, install it first before proceeding.

   The RHEL-based command is as follows:

   ```
   dnf install tmux
   ```

   The Ubuntu command is as follows:

   ```
   apt install tmux
   ```

3. Open a new `tmux` session by issuing the following command:

    ```
    tmux
    ```

> **Important note**
>
> It's not required to run partitioning in a `tmux` window, but it's definitely smart. Partitioning a big database can take a long time. You could move your database to another machine with ample resources (CPU, memory, and disk speed) to partition, or if that's not a possibility stop the Zabbix server process for the duration of the partitioning process.

4. Now, let's log in to the MySQL application as the root user with the following command:

    ```
    mysql -u root -p
    ```

5. Now, move to use the Zabbix database with the following command:

    ```
    USE zabbix;
    ```

6. We are going to need to partition some tables here, but to do this, we need to know the UNIX timestamp on our tables:

    ```
    SELECT FROM_UNIXTIME(MIN(clock)) FROM history;
    ```

    You will receive an output like this:

    ```
    MariaDB [zabbix]> SELECT FROM_UNIXTIME(MIN(clock)) FROM history;
    +---------------------------+
    | FROM_UNIXTIME(MIN(clock)) |
    +---------------------------+
    | 2023-06-11 13:14:26       |
    +---------------------------+
    1 row in set (0.047 sec)
    ```

    Figure 12.1 – MySQL returning a timestamp on the table history

7. This timestamp should be about the same for every single table we are going to partition. Verify this by running the same query for the remaining history tables:

    ```
    SELECT FROM_UNIXTIME(MIN(clock)) FROM 'history';
    SELECT FROM_UNIXTIME(MIN(clock)) FROM 'history_uint';
    SELECT FROM_UNIXTIME(MIN(clock)) FROM 'history_str';
    SELECT FROM_UNIXTIME(MIN(clock)) FROM 'history_text';
    SELECT FROM_UNIXTIME(MIN(clock)) FROM 'history_log';
    SELECT FROM_UNIXTIME(MIN(clock)) FROM 'history_bin';
    ```

8. A table might return a different value or even no value at all. We need to take this into account when creating our partitions. A table showing NULL has no data, but an earlier date means we need an earlier partition:

```
MariaDB [zabbix]> SELECT FROM_UNIXTIME(MIN(clock)) FROM `history_log`;
+---------------------------+
| FROM_UNIXTIME(MIN(clock)) |
+---------------------------+
| NULL                      |
+---------------------------+
1 row in set (0.000 sec)
```

Figure 12.2 – MySQL returning a timestamp on the history_log table

9. Let's start with the history table. We are going to partition this table by day, and we are going to do this up until the date it is today; for me, it is 18-06-2023. Let's prepare the following MySQL query (for example, in a notepad):

```
ALTER TABLE history PARTITION BY RANGE ( clock)
(PARTITION p2023_06_11 VALUES LESS THAN (UNIX_
TIMESTAMP("2023-06-12 00:00:00")) ENGINE = InnoDB,
PARTITION p2023_06_12 VALUES LESS THAN (UNIX_
TIMESTAMP("2023-06-13 00:00:00")) ENGINE = InnoDB,
PARTITION p2023_06_13 VALUES LESS THAN (UNIX_
TIMESTAMP("2023-06-14 00:00:00")) ENGINE = InnoDB,
PARTITION p2023_06_14 VALUES LESS THAN (UNIX_
TIMESTAMP("2023-06-15 00:00:00")) ENGINE = InnoDB,
PARTITION p2023_06_15 VALUES LESS THAN (UNIX_
TIMESTAMP("2023-06-16 00:00:00")) ENGINE = InnoDB,
PARTITION p2023_06_16 VALUES LESS THAN (UNIX_
TIMESTAMP("2023-06-17 00:00:00")) ENGINE = InnoDB,
PARTITION p2023_06_17 VALUES LESS THAN (UNIX_
TIMESTAMP("2023-06-18 00:00:00")) ENGINE = InnoDB,
PARTITION p2023_06_18 VALUES LESS THAN (UNIX_
TIMESTAMP("2023-06-19 00:00:00")) ENGINE = InnoDB);
```

> **Tip**
>
> If we only have 7 days of history data, creating this list by hand is not that hard. If we want to do it on a big existing database, it can be a big list to edit by hand. It's easy to create a big list using software such as Excel or by creating a small script.

10. Make sure that the oldest partition here matches the timestamp we collected in *step 9*. In my case, the oldest data was from June 11, 2023, so this is my oldest partition. Also, make sure that your newest partition matches the date you are partitioning on.

11. Copy and paste the prepared MySQL query from *step 9* and press *Enter*. This might take a while, as your table might be quite large. After you're done, you will see the following:

```
MariaDB [zabbix]> ALTER TABLE history PARTITION BY RANGE ( clock)
    -> (PARTITION p2023_06_11 VALUES LESS THAN (UNIX_TIMESTAMP("2023-06-12 00:00:00")) ENGINE = InnoDB,
    -> PARTITION p2023_06_12 VALUES LESS THAN (UNIX_TIMESTAMP("2023-06-13 00:00:00")) ENGINE = InnoDB,
    -> PARTITION p2023_06_13 VALUES LESS THAN (UNIX_TIMESTAMP("2023-06-14 00:00:00")) ENGINE = InnoDB,
    -> PARTITION p2023_06_14 VALUES LESS THAN (UNIX_TIMESTAMP("2023-06-15 00:00:00")) ENGINE = InnoDB,
    -> PARTITION p2023_06_15 VALUES LESS THAN (UNIX_TIMESTAMP("2023-06-16 00:00:00")) ENGINE = InnoDB,
    -> PARTITION p2023_06_16 VALUES LESS THAN (UNIX_TIMESTAMP("2023-06-17 00:00:00")) ENGINE = InnoDB,
    -> PARTITION p2023_06_17 VALUES LESS THAN (UNIX_TIMESTAMP("2023-06-18 00:00:00")) ENGINE = InnoDB,
    -> PARTITION p2023_06_18 VALUES LESS THAN (UNIX_TIMESTAMP("2023-06-19 00:00:00")) ENGINE = InnoDB);
Stage: 1 of 2 'copy to tmp table'  0.229% of stage done
Query OK, 435758 rows affected (0.624 sec)
Records: 435758  Duplicates: 0  Warnings: 0
```

Figure 12.3 – MySQL returning a successful query result for the history table

12. Do the same partitioning for the remaining history tables; make sure to use the other UNIX timestamps for the earliest partition:

    • `history_uint`

    • `history_str`

    • `history_text`

    • `history_log`

    • `history_bin`

13. Once you've partitioned all the history tables, let's partition the `trends` tables. We have two of these called `trends` and `trends_uint`.

14. We are going to check the timestamps again with the following:

    ```
    SELECT FROM_UNIXTIME(MIN(clock)) FROM trends;
    SELECT FROM_UNIXTIME(MIN(clock)) FROM trends_uint;
    ```

15. For these tables, it's important to focus on what the earliest month is. For my tables, this is month 06 of the year 2023.

16. Now, let's prepare and execute the partitioning for this table. Let's do two extra partitions starting from the earliest date seen in the timestamp in *step 14*:

    ```
    ALTER TABLE trends PARTITION BY RANGE ( clock)
    (PARTITION p2023_06 VALUES LESS THAN (UNIX_TIMESTAMP("2023-07-01
    00:00:00")) ENGINE = InnoDB,
    PARTITION p2023_07 VALUES LESS THAN (UNIX_TIMESTAMP("2023-08-01
    00:00:00")) ENGINE = InnoDB,
    PARTITION p2023_08 VALUES LESS THAN (UNIX_TIMESTAMP("2023-09-01
    00:00:00")) ENGINE = InnoDB);
    ```

17. Again, we partition from the earliest collected UNIX timestamp, up until the current month. But there's no harm in creating some new partitions for future data:

```
MariaDB [zabbix]> ALTER TABLE trends PARTITION BY RANGE ( clock)
    -> (PARTITION p2023_06 VALUES LESS THAN (UNIX_TIMESTAMP("2023-07-01 00:00:00")) ENGINE = InnoDB,
    -> PARTITION p2023_07 VALUES LESS THAN (UNIX_TIMESTAMP("2023-08-01 00:00:00")) ENGINE = InnoDB,
    -> PARTITION p2023_08 VALUES LESS THAN (UNIX_TIMESTAMP("2023-09-01 00:00:00")) ENGINE = InnoDB);
Query OK, 163660 rows affected (0.235 sec)
Records: 163660  Duplicates: 0  Warnings: 0
```

Figure 12.4 – MySQL returning a successful query result for the trends table

18. Do the same thing for the `trends_uint` table.

19. That concludes the actual partitioning of the database. Let's make sure our partitions remain managed. On your Zabbix database Linux host, download the partitioning script with the following command:

```
wget https://raw.githubusercontent.com/OpensourceICTSolutions/
zabbix-mysql-partitioning-perl/main/mysql_zbx_part.pl
```

20. If you can't use wget, simply download the script from the following link: https://github.com/OpensourceICTSolutions/zabbix-mysql-partitioning-perl/blob/main/mysql_zbx_part.pl.

Alternatively, you can download the partitioning script using the Packt GitHub here:

https://github.com/PacktPublishing/Zabbix-7-IT-Infrastructure-Monitoring-Cookbook/tree/main/chapter12/mysql_zbx_part.pl

21. Now, create the directory and move the script to the /usr/lib/zabbix/ folder with the following command:

```
mkdir /usr/lib/zabbix/
mv mysql_zbx_part.pl /usr/lib/zabbix/
```

22. We are going to customize some details in the script. Edit the script with the following:

```
vim /usr/lib/zabbix/mysql_zbx_part.pl
```

We need to edit some text in the following part:

```
my $db_schema = 'zabbix';
my $dsn = 'DBI:mysql:'.$db_schema.':mysql_socket=/var/lib/mysql/mysql.sock';
my $db_user_name = 'zabbix';
my $db_password = 'password';
my $tables = {  'history' => { 'period' => 'day', 'keep_history' => '60'},
                'history_log' => { 'period' => 'day', 'keep_history' => '60'},
                'history_str' => { 'period' => 'day', 'keep_history' => '60'},
                'history_text' => { 'period' => 'day', 'keep_history' => '60'},
                'history_uint' => { 'period' => 'day', 'keep_history' => '60'},
# Comment the history_bin line below if you're running Zabbix versions older than 7.0
                'history_bin' => { 'period' => 'day', 'keep_history' => '60'},
                'trends' => { 'period' => 'month', 'keep_history' => '12'},
                'trends_uint' => { 'period' => 'month', 'keep_history' => '12'},
```

Figure 12.5 – MySQL Zabbix partitioning script user parameters

23. Edit $db_schema to match your Zabbix database name.

24. Edit $db_user_name to match your Zabbix database username.

25. Edit $db_password to match your Zabbix database password.

26. Now, at the $tables variable, we are going to add some of the most important details. This is where we'll add how many days of history data we want to keep and how many months of trends data. Add your values; the default settings keep 30 days of history data and 12 months of trends data.

27. Also, make sure to edit the my $curr_tz = Etc/UTC; line to match your own time zone. I will use Europe/Amsterdam, for example.

> **Tip**
>
> If you are using a version of Zabbix before 2.2 or a MySQL version before 5.6 or if you are running MySQL 8, then there are some extra lines of configuration that need to be commented and uncommented in the script. If this applies to you, read the comments in the mysql_zbx_part.pl script file and edit it. Additionally, check out the GitHub repo mentioned in the introduction of this recipe.

28. Before executing the script, we are going to need to install some Perl dependencies. On RHEL-based systems, we need additional repositories.

The RHEL8-based commands are as follows:

```
dnf config-manager --set-enabled powertools
```

For RHEL9-based systems, use this command:

```
dnf config-manager --enable crb
```

29. Install the dependencies with the following commands.

The RHEL-based commands are as follows:

```
dnf update
dnf install perl-Sys-Syslog
dnf install perl-DateTime
dnf install perl-DBD-mysql
dnf install perl-DBI
```

The Ubuntu commands are as follows:

```
apt install liblogger-syslog-perl
apt install libdatetime-perl
```

30. Make the script executable with the following command:

```
chmod +x /usr/lib/zabbix/mysql_zbx_part.pl
```

31. Then, this is the moment where we should be ready to execute the script to see whether it is working. Let's execute it:

```
/usr/lib/zabbix/mysql_zbx_part.pl
```

32. Once your script has finished running, let's see whether it was successful with the following command:

```
journalctl -t mysql_zbx_part
```

33. You should see an output like this:

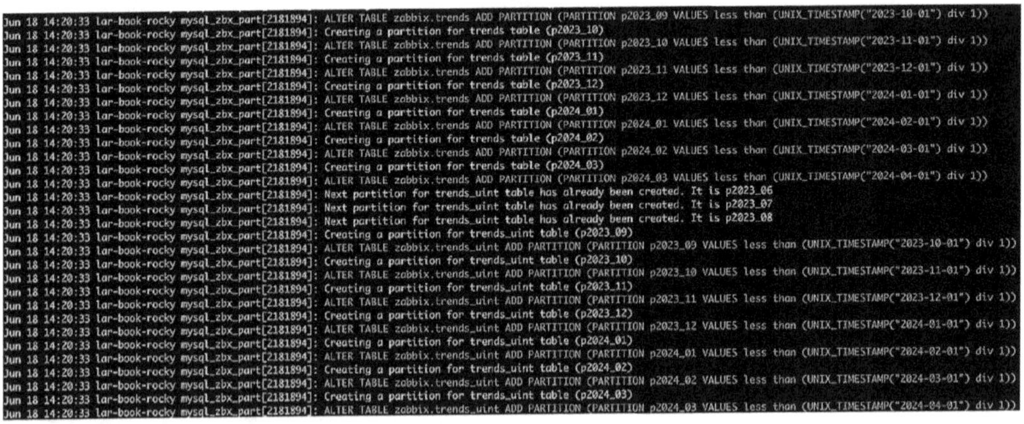

Figure 12.6 – MySQL Zabbix partitioning script results

34. Now, execute the following command:

```
crontab -e
```

35. To automate the execution of the script, add the following line to the file:

```
55 22 * * * /usr/lib/zabbix/mysql_zbx_part.pl
```

36. The last thing we are going to need to do is to go to the Zabbix frontend. Navigate to **Administration | Housekeeping**.

37. As the script will take over database history and trend deletion, the housekeeping for the **History** and **Trends** tables must be disabled. It will look like the following:

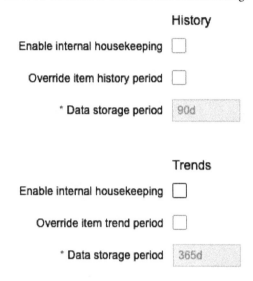

Figure 12.7 – Zabbix Administration | General | Housekeeping disabled for History and Trends

That concludes our Zabbix database partitioning setup.

How it works...

Database partitioning seems like a daring task at first, but once you break it down into chunks, it is not that hard to do. It is simply the process of breaking down our most important Zabbix database tables into time-based partitions. Once these partitions are set up, we simply need to manage these tables with a script and we're ready.

Look at the following figure, and let's say today is **19-06-2023**. We have a lot of partitions managed by the script. All of our **history** data today is going to be written to the partition for this day and all of our **trends** data is going to be written into the partition for this month:

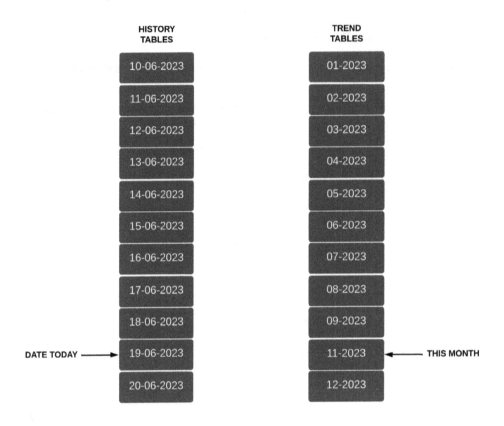

Figure 12.8 – Zabbix partitioning illustration

The actual script does only two things. It creates new partitions and it deletes old partitions.

For deleting partitions, once a partition reaches an age older than specified in the $tables variable, it drops the entire partition.

For creating partitions, every time the script is run, it creates 10 partitions in the future starting from today, except, of course, when a partition already exists.

This is better than using the housekeeper for one clear reason. It's simply faster! The Zabbix housekeeper goes through our database data line by line to check the UNIX timestamp and then it deletes that line when it reaches data older than specified. This takes time and resources. Dropping a partition, though, is almost instant.

One downside of partitioning a Zabbix database, though, is that we can no longer use the frontend item history and trend configuration. This means we can't specify different history and trends for different items; it's all global now.

See also

When I first started using Zabbix, I did not have a book like this one. Instead, I relied heavily on the resources available online and my own skillset. There are loads of great guides for partitioning and other stuff available on the internet. If something isn't mentioned in this book, make sure to Google it and see if there's something available online. You might also want to check out some amazing books written by our Zabbix peers and, of course, if you've figured out something by yourself, sharing is caring!

Using the PostgreSQL TimescaleDB functionality

TimescaleDB is an open source relational PostgreSQL database extension for time-based series data. Using PostgreSQL TimescaleDB is a solid way to work around using the Zabbix housekeeper to manage your PostgreSQL database. In this recipe, we will go over the installation of PostgreSQL TimescaleDB on a new server and how to set it up with Zabbix.

Getting ready

We will need an empty Linux server. I'll be using my server called `lar-book- postgresql-mgmt`.

How to do it...

We have a bit of a different process for RHEL-based and Ubuntu systems, which is why we have split this *How to do it...* section in two. We will start with Ubuntu systems.

Ubuntu installation

1. Let's log in to our Linux CLI and add the PostgreSQL repo with the following commands:

    ```
    apt install gnupg postgresql-common apt-transport-https
    lsb-release wget
    /usr/share/postgresql-common/pgdg/apt.postgresql.org.sh
    ```

2. Now, add the TimescaleDB repository:

    ```
    echo "deb https://packagecloud.io/timescale/timescaledb/ubuntu/
    $(lsb_release -c -s) main" | sudo tee /etc/apt/sources.list.d/
    timescaledb.list
    wget --quiet -O - https://packagecloud.io/timescale/timescaledb/
    gpgkey | sudo apt-key add -
    apt update
    ```

3. Now, install TimescaleDB with the installation command:

```
apt install timescaledb-2-postgresql-15
```

4. Start and enable PostgreSQL 12:

```
systemctl enable postgresql
systemctl start postgresql
```

5. Now, continue with the *TimescaleDB configuration* section of this recipe.

RHEL-based installation

1. Let's start by logging in to our Linux CLI. We will need PostgreSQL version 11 or higher. Let's install version 12; first, disable AppStream:

```
dnf -qy module disable postgresql
```

2. Add the correct repository:

```
dnf install https://download.postgresql.org/pub/repos/yum/
reporpms/EL-8-x86_64/pgdg-redhat-repo-latest.noarch.rpm
```

3. Then, install PostgreSQL:

```
dnf install postgresql15 postgresql15-server
```

4. Make sure to initialize the database:

```
/usr/pgsql-15/bin/postgresql-15-setup initdb
```

5. Now, add the repo information to the file and save it:

```
tee /etc/yum.repos.d/timescale_timescaledb.repo <<EOL
[timescale_timescaledb]
name=timescale_timescaledb
baseurl=https://packagecloud.io/timescale/timescaledb/el/$(rpm
-E %{rhel})/\$basearch
repo_gpgcheck=1
gpgcheck=0
enabled=1
gpgkey=https://packagecloud.io/timescale/timescaledb/gpgkey
sslverify=1
sslcacert=/etc/pki/tls/certs/ca-bundle.crt
metadata_expire=300
EOL
```

6. Install TimescaleDB with the installation command:

```
dnf install timescaledb-2-postgresql-15
```

7. Now, continue with the *TimescaleDB configuration* section of this recipe.

TimescaleDB configuration

In this section, we'll go over how to set up TimescaleDB after finishing the installation process. There's a lot more to configure, so let's check it out:

1. Let's start by running the following command:

```
timescaledb-tune
```

2. Sometimes this does not work, and you want to specify the PostgreSQL location like this:

```
timescaledb-tune --pg-config=/usr/pgsql-15/bin/pg_config
```

3. Go through the steps and answer the questions with yes or no accordingly. For a first-time setup, yes for everything is good.

4. Now, restart PostgreSQL:

```
systemctl restart postgresql-15
```

5. If you haven't already, download and install Zabbix with the following.

   The RHEL-based commands are as follows:

```
rpm -Uvh https://repo.zabbix.com/zabbix/7.0/rhel/9/ x86_64/
zabbix-release-7.0-1.el8.noarch.rpm
dnf clean all
dnf install zabbix-server-pgsql zabbix-web-pgsql zabbix-apache-
conf zabbix-agent2
```

   The Ubuntu commands are as follows:

```
wget https://repo.zabbix.com/zabbix/7.0/ubuntu/pool/main/z/
zabbix-release/zabbix-release_7.0-1+ubuntu22.04_all.deb
dpkg -i zabbix-release_7.0-1+ubuntu22.04_all.deb
apt update
apt install zabbix-server-pgsql zabbix-frontend-phpphp-pgsql
zabbix-apache-conf zabbix-agent2
```

6. Create the initial database with the following:

```
sudo -u postgres createuser --pwprompt zabbix
sudo -u postgres createdb -O zabbix zabbix
```

7. Import the database schema for PostgreSQL:

```
zcat /usr/share/doc/zabbix-server-pgsql*/create.sql.gz | sudo -u
zabbix psql zabbix
```

8. Add the database password to the Zabbix configuration file by editing it:

```
vim /etc/zabbix/zabbix_server.conf
```

9. Add the following lines, where password is your password as set in *step* 6 and DBHost is empty:

```
DBHost=
DBPassword=password
```

10. Now, enable the TimescaleDB extension with the following command:

```
echo "CREATE EXTENSION IF NOT EXISTS timescaledb CASCADE;" |
sudo -u postgres psql zabbix
```

11. Unpack the timescale.sql script located in your Zabbix share folder:

```
gunzip /usr/share/doc/zabbix-sql-scripts/postgresql/timescaledb.
sql.gz
```

12. Now, let's run timescale.sql:

```
cat /usr/share/doc/zabbix-sql-scripts/postgresql/timescaledb.
sql| sudo -u zabbix psql zabbix
```

13. We need to do one more thing before moving to the frontend. We need to edit the pg_hba.
    conf file to allow our Zabbix frontend to connect. Edit the following file:

```
vim /var/lib/pgsql/15/data/pg_hba.conf
```

14. Make sure the following lines match in your file; they need to end with md5:

```
# "local" is for Unix domain socket connections only
local all all
scram-sha256
# IPv4 local connections:
host all  all  127.0.0.1/32
md5
# IPv6 local connections:
host all  all  ::1/128
scram-sha256
```

15. Now, start Zabbix and finish the frontend setup using the following commands:

On RHEL-based systems:

```
systemctl restart zabbix-server zabbix-agent2 httpd php-fpm
systemctl enable zabbix-server zabbix-agent2 httpd php-fpm
```

On Ubuntu systems:

```
systemctl restart zabbix-server zabbix-agent2 apache2 php-fpm
systemctl enable zabbix-server zabbix-agent2 apache2 php-fpm
```

16. Once we navigate to the frontend and we've logged in to our setup, navigate to **Administration | Housekeeping**.

17. We can now edit the following parameters to match our preferences, and TimescaleDB will take care of maintaining the data retention period:

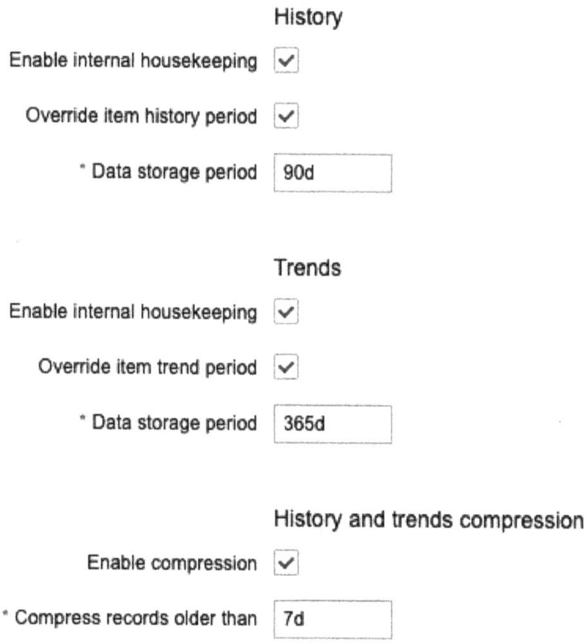

History

Enable internal housekeeping	✔
Override item history period	✔
* Data storage period	90d

Trends

Enable internal housekeeping	✔
Override item trend period	✔
* Data storage period	365d

History and trends compression

Enable compression	✔
* Compress records older than	7d

Figure 12.9 – Zabbix Administration | Housekeeping, TimescaleDB-specific options

How it works...

Using the TimescaleDB functionality with your Zabbix setup is a solid integration with your PostgreSQL database. The extension is supported by Zabbix, and you can expect it to only get better in the near future.

Now, how TimescaleDB works is by dividing up your PostgreSQL hypertable into time-based chunks. If we look at the following figure, we can see how that looks:

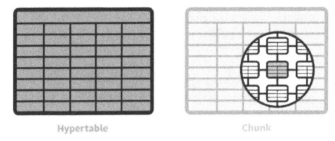

Figure 12.10 – TimescaleDB hypertable chunks diagram

These time-based chunks are a lot faster to drop from the database than using the Zabbix housekeeper. The Zabbix housekeeper goes through our database data line by line to check the UNIX timestamp, and then it drops the line when it reaches data that is older than specified. This takes time and resources. Dropping a chunk, though, is almost instantaneous.

Another great thing about using TimescaleDB with a Zabbix database is that we can still use the frontend item history and trend configuration. On top of that, TimescaleDB can compress our data, to keep databases smaller.

The downside is that we can't specify different history and trends for different items; it's all global now.

See also

This recipe details the installation of PostgreSQL TimescaleDB. As this process is constantly changing, you might need to include some new information from the official TimescaleDB documentation. Check out their documentation here:

```
https://docs.timescale.com/latest/getting-started/installation/rhel-
centos/installation-yum
```

Securing your Zabbix MySQL database

Another great added feature for the Zabbix server is the ability to encrypt data between the database and Zabbix components. This is particularly useful when you are running a split database and the Zabbix server over the network. A **Man-in-the-Middle (MITM)** attack or other attacks can be executed on the network to gain access to your monitoring data.

In this recipe, we'll set up MySQL encryption between Zabbix components and the database to add another layer of security.

Getting ready

We are going to need a Zabbix setup that uses an external database. I'll be using the Linux `lar-book-secure-db` and `lar-book-secure-zbx` hosts.

The new server called `lar-book-secure-zbx` will be used to connect externally to the `lar-book-secure-db` database server. The database servers won't run our Zabbix server; this process will run on `lar-book-secure-zbx`.

Make sure that MariaDB is already installed on the `lar-book-secure-db` host and that you are running a recent supported version that is able to use encryption. If you don't know how to upgrade your database, check out the recipe named *Upgrading Zabbix database from older MariaDB versions to MariaDB 10.5* in *Chapter 11*, *Maintaining Your Zabbix Setup*, or check the documentation online.

How to do it...

1. Make sure your host files on both hosts from the *Getting ready* section contain the hostname and IP for your Linux hosts and edit the file with the following:

   ```
   vim /etc/hosts
   ```

2. Then, fill in the file with your hostnames and IPs. It will look like this:

   ```
   10.16.16.170 lar-book-secure-db
   10.16.16.171 lar-book-secure-zbx
   ```

3. On the `lar-book-secure-db` MySQL server, if you haven't already, create the Zabbix database by logging in to MySQL:

   ```
   mysql -u root -p
   ```

4. Then, issue the following command to create the database:

   ```
   create database zabbix character set utf8mb4 collate
   utf8mb4_ bin;
   ```

5. Also, make sure to create a user that will be able to access the database securely. Make sure the IP matches the IP from the Zabbix server (and one for the Zabbix frontend if they are separate):

   ```
   create user 'zabbix'@'10.16.16.171' identified BY 'password';
   grant all privileges on zabbix.* to 'zabbix'@'10.16.16.171';
   flush privileges;
   ```

6. Quit MySQL and then make sure to run the secure `mysql` script with the following:

   ```
   mariadb_secure_installation
   ```

7. Log in to `lar-book-secure-zbx` and install the Zabbix server repo with the following command:

```
rpm -Uvh https://repo.zabbix.com/zabbix/7.0/rhel/9/x86_64/
zabbix-release-7.0-1.el8.noarch.rpm
dnf clean all
```

8. Let's add the MariaDB repository on our server:

```
wget https://downloads.mariadb.com/MariaDB/mariadb_repo_setup
chmod +x mariadb_repo_setup
./mariadb_repo_setup
```

9. Then, install the Zabbix server and its required components.

   Use the following RHEL-based command:

```
dnf install zabbix-server-mysql zabbix-web-mysql zabbix-apache-
conf zabbix-agent2 zabbix-sql-scripts mariadb-client
```

   Use the following Ubuntu command:

```
apt install zabbix-server-mysql zabbix-frontend-php zabbix-
apache-conf zabbix-agent2 mariadb-client
```

10. From the Zabbix server, connect to the remote database server and import the database schema and default data with the following command:

```
zcat /usr/share/doc/zabbix-sql-scripts/mysql/server.sql.gz |
mysql -h 10.16.16.170 -uzabbix -p zabbix
```

11. Now we are going to open the file called `openssl.cnf` and edit it by issuing the following command:

```
vim /etc/pki/tls/openssl.cnf
```

12. In this file, we need to edit the following lines:

```
countryName_default = XX
stateOrProvinceName_default = Default Province
localityName_default = Default City
0.organizationName_default = Default Company Ltd
organizationalUnitName_default.=
```

13. It will look like this filled out completely:

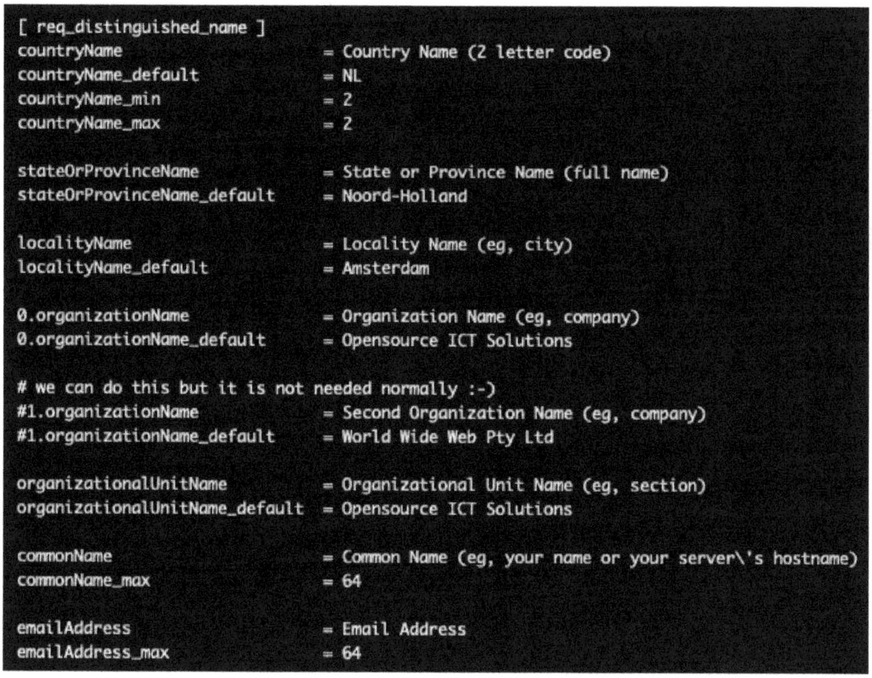

```
[ req_distinguished_name ]
countryName                     = Country Name (2 letter code)
countryName_default             = NL
countryName_min                 = 2
countryName_max                 = 2

stateOrProvinceName             = State or Province Name (full name)
stateOrProvinceName_default     = Noord-Holland

localityName                    = Locality Name (eg, city)
localityName_default            = Amsterdam

0.organizationName              = Organization Name (eg, company)
0.organizationName_default      = Opensource ICT Solutions

# we can do this but it is not needed normally :-)
#1.organizationName             = Second Organization Name (eg, company)
#1.organizationName_default     = World Wide Web Pty Ltd

organizationalUnitName          = Organizational Unit Name (eg, section)
organizationalUnitName_default  = Opensource ICT Solutions

commonName                      = Common Name (eg, your name or your server\'s hostname)
commonName_max                  = 64

emailAddress                    = Email Address
emailAddress_max                = 64
```

Figure 12.11 – OpenSSL config file with our personal defaults

14. We can also see this line:

```
dir = /etc/pki/CA    # Where everything is kept
```

15. This means the default directory is /etc/pki/CA; if yours is different, act accordingly. Close the file by saving, and continue.

16. Let's create a new folder for our private certificates using the following command:

```
mkdir -p /etc/pki/CA/private
```

17. Now, let's create our key pair in the new folder. Issue the following command:

```
openssl req -new -x509 -keyout /etc/pki/CA/private/cakey.pem
-out /etc/pki/CA/cacert.pem -days 3650 -newkey rsa:4096
```

18. You will be prompted for a password now:

```
Generating a RSA private key
.......................................................+++
.........................................................................................................+++
writing new private key to '/etc/pki/CA/private/cakey.pem'
Enter PEM pass phrase:
```

Figure 12.12 – Certificate generation response asking for a password

19. You might also be prompted to enter some information about your company. It will use the default we filled in earlier, so you can just press *Enter* up until Common Name.

20. Fill in Root CA for Common Name and add your email address like this:

```
Country Name (2 letter code) [NL]:
State or Province Name (full name) [Noord-Holland]:
Locality Name (eg, city) [Amsterdam]:
Organization Name (eg, company) [Opensource ICT Solutions]:
Organizational Unit Name (eg, section) [Opensource ICT Solutions]:
Common Name (eg, your name or your server's hostname) □:Root CA
Email Address □:nathan@oicts.nl
```

Figure 12.13 – Certificate generation response asking for information, Root CA

21. Next up is creating the actual signed certificates that our Zabbix server will use. Let's make sure that OpenSSL has the right files to keep track of signed certificates:

```
touch /etc/pki/CA/index.txt
echo 01 > /etc/pki/CA/serial
```

22. Then, create the folders to keep our certificates in:

```
mkdir /etc/pki/CA/unsigned
mkdir /etc/pki/CA/newcerts
mkdir /etc/pki/CA/certs
```

23. Now, let's create our certificate signing request for the lar-book-secure-zbx Zabbix server with the following command:

```
openssl req -nodes -new -keyout /etc/pki/CA/private/zbx-srv_key.
pem -out /etc/pki/CA/unsigned/zbx-srv_req.pem -newkey rsa:2048
```

24. You will be prompted to add a password and your company information again. Use the default up until Common Name. We will fill out our Common Name, which will be the server hostname, and we'll add our email address like this:

```
Country Name (2 letter code) [NL]:
State or Province Name (full name) [Noord-Holland]:
Locality Name (eg, city) [Amsterdam]:
Organization Name (eg, company) [Opensource ICT Solutions]:
Organizational Unit Name (eg, section) [Opensource ICT Solutions]:
Common Name (eg, your name or your server's hostname) □:lar-book-secure-db
Email Address □:nathan@oicts.nl
```

Figure 12.14 – Certificate generation response asking for information, lar-book-secure-zbx

25. Let's do the same for our `lar-book-secure-db` server:

```
openssl req -nodes -new -keyout /etc/pki/CA/private/mysql-
srv_key.pem -out /etc/pki/CA/unsigned/mysql-srv_req.pem -newkey
rsa:2048
```

The response will look like this:

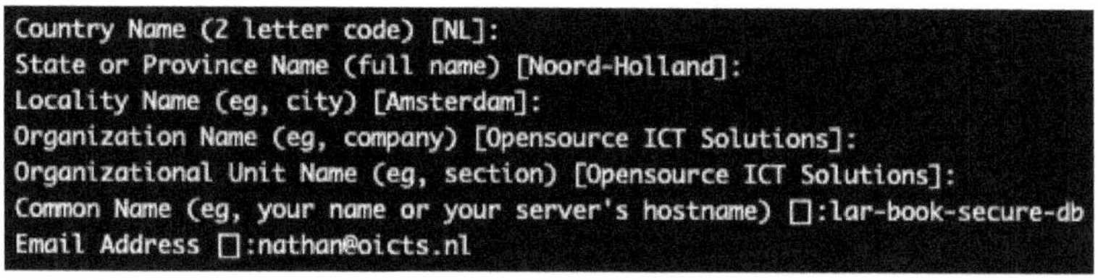

```
Country Name (2 letter code) [NL]:
State or Province Name (full name) [Noord-Holland]:
Locality Name (eg, city) [Amsterdam]:
Organization Name (eg, company) [Opensource ICT Solutions]:
Organizational Unit Name (eg, section) [Opensource ICT Solutions]:
Common Name (eg, your name or your server's hostname) []:lar-book-secure-db
Email Address []:nathan@oicts.nl
```

Figure 12.15 – Certificate generation response asking for information, lar-book-secure-db

> **Important note**
>
> Our certificates need to be created without a password; otherwise, our MariaDB and Zabbix applications won't be able to use them. Make sure to specify the `-nodes` option.

26. Now, sign the certificate for `lar-book-secure-zbx` with the following command:

```
openssl ca -policy policy_anything -days 365 -out /etc/pki/CA/
certs/zbx-srv_crt.pem -infiles /etc/pki/CA/unsigned/zbx-srv_req.
pem
```

27. You will be prompted with the question `Sign the certificate? [y/n]`. Answer this and all the following questions with `Y`.

28. Now, let's do the same thing for the `lar-book-secure-db` certificate:

```
openssl ca -policy policy_anything -days 365 -out/etc/ pki/CA/
certs/mysql-srv_crt.pem -infiles/etc/pki/CA/ unsigned/mysql-
srv_req.pem
```

29. Let's log in to the `lar-book-secure-db` MySQL server and create a directory for our newly created certificates:

```
mkdir /etc/my.cnf.d/certificates/
```

30. Add the right permissions to the folder:

```
chown -R mysql. /etc/my.cnf.d/certificates/
```

31. Now, back on the new `lar-book-secure-zbx` Zabbix server, copy over the files to the database server with the following commands:

```
scp /etc/pki/CA/private/mysql-srv_key.pem root@10.16.16.170:/
etc/my.cnf.d/certificates/mysql-srv.key
scp /etc/pki/CA/certs/mysql-srv_crt.pem root@10.16.16.170:/etc/
my.cnf.d/certificates/mysql-srv.crt
scp /etc/pki/CA/cacert.pem root@10.16.16.170:/etc/my.cnf.d/
certificates/cacert.crt
```

32. Now, back on the `lar-book-secure-db` MySQL server, add the right permissions to the files:

```
chown -R mysql:mysql /etc/my.cnf.d/certificates/
chmod 400 /etc/my.cnf.d/certificates/mysql-srv.key
chmod 444 /etc/my.cnf.d/certificates/mysql-srv.crt
chmod 444 /etc/my.cnf.d/certificates/cacert.crt
```

33. Edit the MariaDB configuration file with the following command:

```
vim /etc/my.cnf.d/server.cnf
```

34. Add the following lines to the configuration file under the [`mysqld`] block:

```
bind-address=lar-book-secure-db
ssl-ca=/etc/my.cnf.d/certificates/cacert.crt
ssl-cert=/etc/my.cnf.d/certificates/mysql-srv.crt
ssl-key=/etc/my.cnf.d/certificates/mysql-srv.key
```

35. Log in to MySQL with the following command:

```
mysql -u root -p
```

36. Make sure our Zabbix MySQL user requires SSL encryption with the following:

```
alter user 'zabbix'@'10.16.16.152' require ssl;
flush privileges;
```

Make sure the IP matches the IP from the Zabbix server (and one for the Zabbix frontend, if they are separated), just like we did in *step 2*.

37. Quit out of the MariaDB CLI and then restart MariaDB with the following command:

```
systemctl restart mariadb
```

38. Now, back on the `lar-book-secure-zbx` Zabbix server, create a new folder for our certificates:

```
mkdir -p /var/lib/zabbix/ssl/
```

39. Copy the certificates over to this folder with the following:

```
cp /etc/pki/CA/cacert.pem /var/lib/zabbix/ssl/
cp /etc/pki/CA/certs/zbx-srv_crt.pem/var/lib/zabbix/ssl/zbx-srv.
crt
cp /etc/pki/CA/private/zbx-srv_key.pem/var/lib/zabbix/ssl/
zbx-srv.key
```

40. Edit the Zabbix server configuration file to use these certificates:

```
vim /etc/zabbix/zabbix_server.conf
```

41. Make sure the following lines match our `lar-book-secure-db` database server's setup:

```
DBHost=lar-book-secure-db
DBName=zabbix
DBUser=zabbix
DBPassword=password
```

42. Now, make sure our SSL-related configuration matches our new files:

```
DBTLSConnect=verify_full
DBTLSCAFile=/var/lib/zabbix/ssl/cacert.pem
DBTLSCertFile=/var/lib/zabbix/ssl/zbx-srv.crt
DBTLSKeyFile=/var/lib/zabbix/ssl/zbx-srv.key
```

43. Also, make sure to add the right permissions to the SSL-related files:

```
chown -R zabbix:zabbix /var/lib/zabbix/ssl/
chmod 400 /var/lib/zabbix/ssl/zbx-srv.key
chmod 444 /var/lib/zabbix/ssl/zbx-srv.crt
chmod 444 /var/lib/zabbix/ssl/cacert.pem
```

44. Start and enable the Zabbix server with the following commands:

- RHEL-based systems:

- `systemctl restart zabbix-server zabbix-agent2 httpd php-fpm`

- `systemctl enable zabbix-server zabbix-agent2 httpd php-fpm`

- Ubuntu systems:

- `systemctl restart zabbix-server zabbix-agent2 apache2 php-fpm`

- `systemctl enable zabbix-server zabbix-agent2 apache2 php-fpm`

45. Then, navigate to the Zabbix frontend and fill in the right information, as shown in the following screenshot:

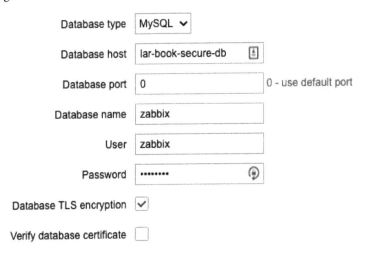

Figure 12.16 – Zabbix frontend configuration, database step

46. When we click **Next step**, we need to fill out some more information:

Zabbix server details

Please enter the host name or host IP address and port number of the Zabbix server, as well as the name of the installation (optional).

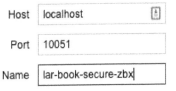

Figure 12.17 – Zabbix frontend configuration, server details step

47. Then, after clicking **Next step**, **Next step**, and **Finish**, the frontend should now be configured and working.

How it works...

This was quite a long recipe, so let's break it down quickly:

- In *steps 1* through *9*, we prepared our servers
- In *steps 10* through *37*, we executed everything needed to create our certificates
- In *steps 38* through *47*, we set up our Zabbix frontend for encryption

Going through all these steps, setting up your Zabbix database securely can seem like quite a daunting task, and it can be. Certificates, login procedures, loads of settings, and more can all add up to become very complicated, which is why I'd always recommend diving deeper into encryption methods before trying to set this up yourself.

If your setup requires encryption, though, this recipe is a solid starting point for your first-time setup. It works very well in an internal setting, as we are using private certificates.

> **Important note**
>
> Make sure to renew your SSL certificates, as they are only valid for however long we defined. In this case, it's 365 days, so we will renew them every year. It's also a good plan to monitor the expiry date of the certificate and create an alert in Zabbix for it.

All Zabbix components, except for communication between the Zabbix server and Zabbix frontend, can be encrypted, as shown in the following diagram:

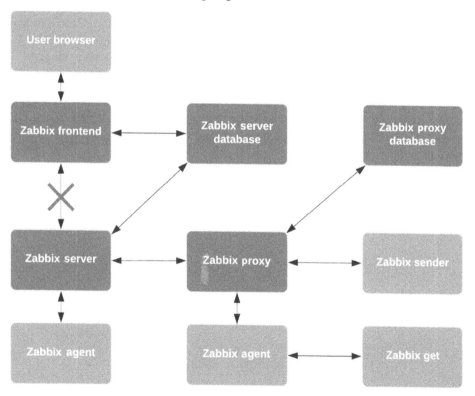

Figure 12.18 – Zabbix encryption scheme possibilities

We've set up encryption between the following:

- The Zabbix server and MariaDB

- The Zabbix frontend and MariaDB

This means that when our Zabbix server or frontend requests or writes data to our database, it will be encrypted. Because our Zabbix applications are running on a different server than our Zabbix database, this might be important. For example, our setup might look like this:

Figure 12.19 – Zabbix setup with an external network diagram

Let's say the cloud is called **Some company** in a network that isn't managed by us. There are several switches and routers in this network that are used for numerous clients with their own VLANs. If one of these devices gets compromised somehow, all of our Zabbix data could be seen by others.

Even if the network equipment is ours, there might still be a compromised device in the network and our data can be seen. This is why you might want to add encryption, to add that extra layer of security. Whether it's breaches in other companies and their network that you want to secure against or whether it's against your own breaches, securing your database as we did in this recipe might just save you from leaking all that data.

13

Bringing Zabbix to the Cloud with Zabbix Cloud Integration

For the last chapter, we have prepared something special. As a long-time Zabbix user, the importance of cloud integration for tools such as Zabbix has not gone unnoticed. For some people, the cloud can be daunting, and thus with this chapter, I want to show you just how easy it can be to start working with the most popular cloud providers and Zabbix.

We are going to start by talking about monitoring the **Amazon Web Services** (**AWS**) cloud with Zabbix. Then we will also see how the same things are done using Microsoft Azure so we can clearly see the differences.

After going through these cloud products, we'll also check out container monitoring with Docker, a very popular product that can also benefit greatly from setting up Zabbix monitoring. Follow these recipes closely and you will be able to monitor all of these products easily and work to extend the products using Zabbix. This chapter comprises the following recipes:

- Setting up AWS monitoring
- Setting up Microsoft Azure monitoring
- Building your Zabbix Docker monitoring

Technical requirements

As this chapter focuses on AWS, Microsoft Azure, and Docker monitoring, we are going to need a working AWS, Microsoft Azure, or Docker setup. The recipe does not cover how to set these up, so make sure to have your own infrastructure at the ready.

Furthermore, we are going to need our Zabbix server running Zabbix 7. We will call this server `lar-book-rocky` in this chapter.

You can download the code files for this chapter from the following GitHub link: `https://github.com/PacktPublishing/Zabbix-7-IT-Infrastructure-Monitoring-Cookbook/tree/main/chapter13`.

Setting up AWS monitoring

A lot of infrastructure is moving toward the cloud these days, and it's important to keep an eye on this infrastructure as much as you would if it were your own hardware. In this recipe, we are going to discover how to monitor EC2 instances, **Relational Database Service (RDS)** instances, and **S3 buckets** with our Zabbix setup.

Getting ready

For this recipe, we are going to need our AWS cloud with at least one of the following three resources:

- EC2 instances
- RDS instances
- S3 buckets

Of course, we will also need our Zabbix server, which we'll call `lar-book-rocky` in this recipe.

> **Important note**
> Using Amazon CloudWatch is not free, so you will incur costs. Make sure you check out the Amazon pricing for AWS CloudWatch before proceeding: `https://aws.amazon.com/cloudwatch/pricing/`.

How to do it...

Setting up AWS monitoring might seem like a daunting task at first, but once we get the hang of the technique, it's not that difficult. Let's waste no more time and check out one of the methods we could use:

1. Let's start by logging in to our Zabbix server with the hostname `lar-book-rocky`.
2. Log in to your AWS account by navigating to the following URL in your browser: `https://aws.amazon.com/`.
3. On this page, click on **Sign In to the Console**.

4. Once logged in, we can navigate to **My Security Credentials**, which should be listed in your user profile in the top-right corner:

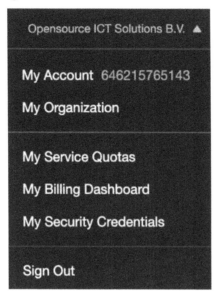

Figure 13.1 – AWS web frontend user profile

5. On the next page, on the left-hand side, click on **Users** under **Access management**.

6. Let's create a new dedicated Zabbix monitoring user by clicking on the **Add users** button. Add the user as follows.

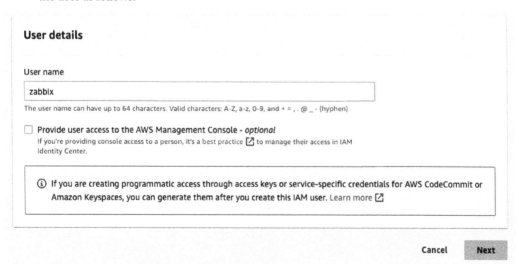

Figure 13.2 – AWS new user

7. Click on **Next**, and at the second step, if you'd like, you can add the user to a group to inherit some permissions, copy them, or set up a custom policy. I'll skip this step for now by clicking **Next** again.

8. Now click on **Create** to finish setting up this user.

9. Select the user from the list to edit it:

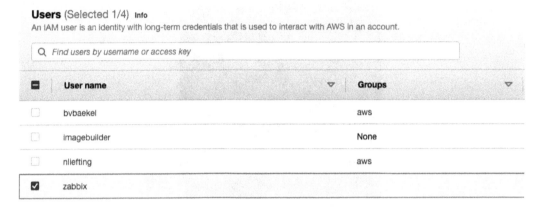

Figure 13.3 – AWS – edit new Zabbix user

10. In the list, we can see there are no policies assigned to this user yet, so let's create a new policy just for Zabbix monitoring.

11. Click on **Create inline policy** from the drop-down list:

Figure 13.4 – AWS – edit new Zabbix user policies

12. Then click on **JSON** to define the new policy in the JSON format. It should look like the following screenshot:

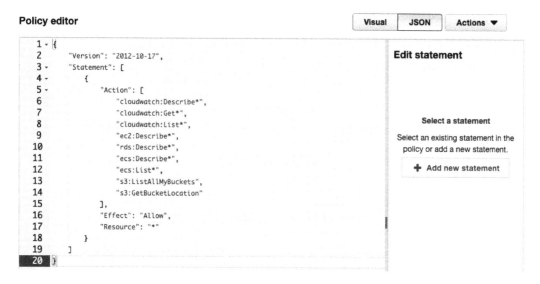

Figure 13.5 – AWS – edit new Zabbix user – add new policy

> **Tip**
> Check out Zabbix's integrations page for the latest required permissions for the AWS template you'll be using. Different templates need different permissions, and new permissions might be added later to incorporate new features or changes on the AWS side: `https://www.zabbix.com/integrations/aws`.

13. You can now click on **Next** and name your policy.

Policy details

Policy name
Enter a meaningful name to identify this policy.

    zabbix-monitoring

Maximum 128 characters. Use alphanumeric and '+=,.@-_' characters.

Figure 13.6 – AWS – edit new Zabbix user – add new policy name

14. Then, click on **Create policy** at the bottom of the page.

With the permissions out of the way, let's make sure we will be able to authenticate with this user account.

15. Still on the same page after creating the new policy, scroll down to **Access keys (access key ID and secret access key)** for this new user. This will show you the following:

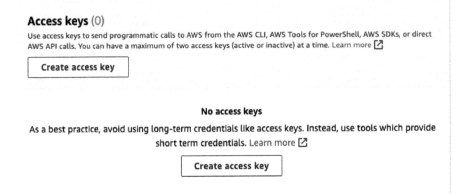

Figure 13.7 – AWS access keys page

16. Click on **Create access key** to create a new access key. You should see the following:

Figure 13.8 – AWS access key creation

17. Select a reason and click on **Next**. Make sure you understand the possible security implications.

18. Name your new access key:

Figure 13.9 – AWS access key creation naming

19. Lastly, click on **Create access key** and store the access key and secret access key somewhere safe (such as a password vault). After you've done that, click on **Done**:

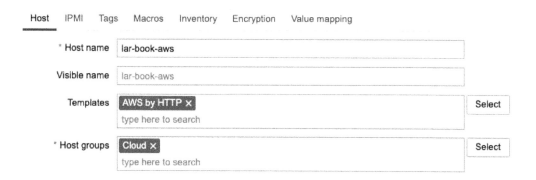

Figure 13.10 – AWS access key creation – copy keys

20. Now, let's finally move on to Zabbix. Log in to your Zabbix GUI and navigate to **Data collection | Hosts**.

21. Create a new host by clicking on **Create host** in the top-right corner. We'll create a new host called lar-book-aws and add the **AWS by HTTP** template and a host group such as **Cloud**.

New host

Host	IPMI	Tags	Macros	Inventory	Encryption	Value mapping

* Host name   lar-book-aws

Visible name   lar-book-aws

Templates   AWS by HTTP ×   Select
             type here to search

* Host groups   Cloud ×   Select
                 type here to search

Figure 13.11 – New AWS host

22. Before adding the host, make sure to go to **Macros**. We have to fill in a few macros to make this new template work. Make sure to fill out the keys you saved in *step 19* and fill them in as in the following screenshot. Also, make sure to add the region in which you want to discover your information.

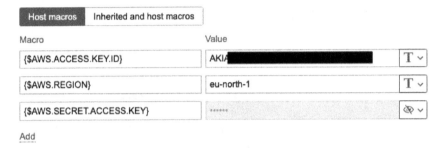

Figure 13.12 – New AWS host macros

> **Important note**
> In my case, all of the AWS resources I am running are within the same AWS regions. In a lot of production environments, this isn't the case. For those environments, you might want to create a Zabbix host per region to be able to discover all of your resources. All you have to do is define the {**$AWS.REGION**} macro uniquely per host.

23. Now click on **Add** to add this new host.

24. If done correctly, your AWS resources will be added once the discovery rule has been executed, as we can see in the following screenshot for some of my EC2 instances:

Name ▲	Items	Triggers	Graphs	Discovery	Web
☐ EC2 instances discovery: graveerhoekje	Items 29	Triggers 10	Graphs 4	Discovery 2	Web
☐ EC2 instances discovery: Nextcloud	Items 46	Triggers 12	Graphs 7	Discovery 2	Web

Figure 13.13 – Discovered EC2 instances

How it works...

Now that we've done all the setup work, let's have a look at what we have actually done. Zabbix 7.0 contains out-of-the-box cloud monitoring templates, which we've utilized to monitor some of the most common AWS resources.

The templates provided by Zabbix use a fairly extensive piece of JavaScript code to execute API calls toward AWS, parse through the received data, and then put that into a JSON array that Zabbix's low-level discovery understands.

Looking at the template at **Data collection | Templates** and then opening **Discovery** for the **AWS by HTTP** template, we can see three discovery rules:

	Template	Name ▲
☐	**AWS by HTTP**	EC2 instances discovery
☐	**AWS by HTTP**	RDS instances discovery
☐	**AWS by HTTP**	S3 buckets discovery

Figure 13.14 – AWS by HTTP discovery rules

These three rules discover the EC2 instances, RDS instances, and S3 buckets in AWS and use **Host prototypes** to create a new host for each instance or bucket found. Those created hosts will then in turn use their own templates to get the actual statistics from those instances or buckets, as we can see in the template list:

☐	AWS EC2 by HTTP	Hosts 2
☐	AWS RDS instance by HTTP	Hosts
☐	AWS S3 bucket by HTTP	Hosts

Figure 13.15 – The other three AWS templates in Zabbix 7.0

In my case, only two EC2 instances were discovered, and as such, those two hosts were added with the **AWS EC2 by HTTP** template, as seen in *Figure 13.13*.

All of the information is then collected by **Script** item types with their own unique JavaScript code. We can see a piece of the code in the following screenshot, where we make a call to AWS to a specific URL (underlined) with headers for things such as authentication:

```
var request = new HttpRequest(),
    url = 'https://' + host + canonical_uri + '?' + params;

if (typeof AwsEbs.params.proxy !== 'undefined' && AwsEbs.params.proxy !== '') {
    request.setProxy(AwsEbs.params.proxy);
}
request.addHeader('x-amz-date: ' + amzdate);
request.addHeader('Accept: application/json');
request.addHeader('Content-Type: application/json');
request.addHeader('Content-Encoding: amz-1.0');
request.addHeader('Authorization: ' + 'AWS4-HMAC-SHA256 Credential=' + AwsEbs.params.access_key + '/' +
```

Figure 13.16 – AWS template call

It is also possible to edit this JavaScript code to create entirely new calls to retrieve your own data and create different types of monitoring, as well as simply extending the out-of-the-box templates.

There's more...

It takes time to start monitoring with AWS CloudWatch as we need a good understanding of the AWS CLI commands with the use of CloudWatch. When you use the templates provided by Zabbix as a basis, you have a solid foundation on which to build.

Make sure to check out the AWS documentation for more information on the commands that we can use, using the following link:

`https://docs.aws.amazon.com/cli/latest/reference/#available-services`

Setting up Microsoft Azure monitoring

The Microsoft Azure cloud is a big player in the cloud market these days and it's important to keep an eye on this infrastructure as much as you would your own hardware. In this recipe, we are going to discover how to monitor Azure instances with our Zabbix setup.

Getting ready

For this recipe, we are going to need our Azure cloud with at least one of the following resources in it already.

- Cosmos DB for MongoDB databases
- Microsoft SQL databases
- MySQL servers
- PostgreSQL servers
- Virtual machines

The recipe does not cover how to set up any of these resources, so make sure to do this in advance. We will also need our Zabbix server, which we'll call `lar-book-rocky` in this recipe.

How to do it...

For Azure monitoring, we face some of the same techniques as we do for AWS monitoring. It can become a bit daunting if we dive into customization, but setting up the initial monitoring is a lot easier than it looks. Let's check it out:

1. With Azure monitoring, first we are going to need to set up our authentication correctly. To do so, navigate to `portal.azure.com` and log in.

2. In the search bar, search for `Enterprise applications` and select it from the list. Click on **New application**:

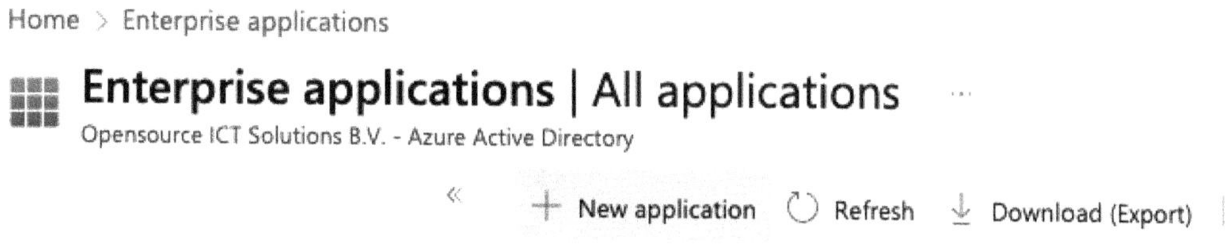

Home > Enterprise applications

Enterprise applications | All applications ...
Opensource ICT Solutions B.V. - Azure Active Directory

«   + **New application** ○ Refresh ↓ Download (Export)

Figure 13.17 – Azure enterprise application creation

3. Then click on **Create your own application**:

Browse Azure AD Gallery ...

+ Create your own application   ዱ⅃ Got feedback?

Figure 13.18 – Azure enterprise application creation – creating your own application

4. This is where we have to name our application. Name it something appropriate, as seen in the following screenshot:

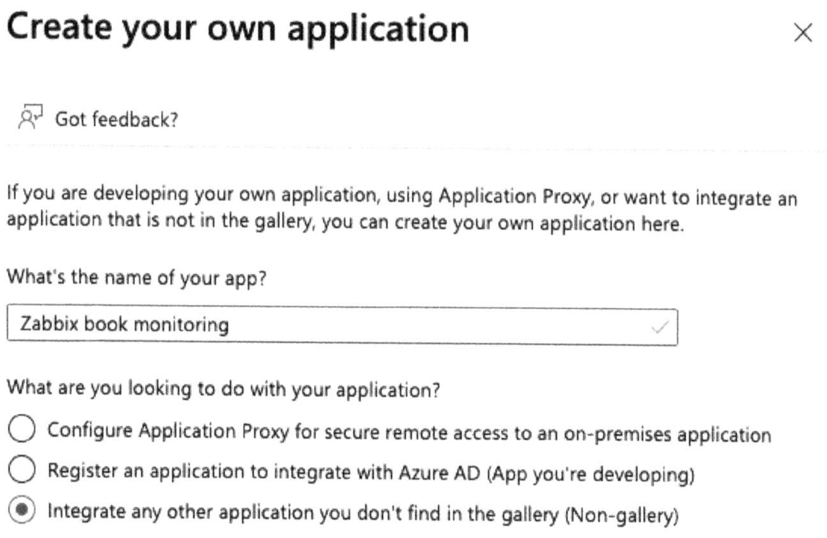

Create your own application   ✕

ዱ⅃ Got feedback?

If you are developing your own application, using Application Proxy, or want to integrate an application that is not in the gallery, you can create your own application here.

What's the name of your app?

Zabbix book monitoring   ✓

What are you looking to do with your application?

○ Configure Application Proxy for secure remote access to an on-premises application

○ Register an application to integrate with Azure AD (App you're developing)

◉ Integrate any other application you don't find in the gallery (Non-gallery)

Figure 13.19 – Azure enterprise application creation – setting the name of your application

5.   Then click **Create** at the bottom of the page to finish creating a new empty application. It will show you the application ID on this page. Make sure to write it down as we will need it later:

Figure 13.20 – Azure enterprise application overview page

6.   With the application created, let's immediately dive into setting up the credentials for it. To do so, use the Azure search bar at the top and search for `Azure Active Directory`, then select it from the list.

7.   In the left-hand sidebar, you should see **App registrations**. We are going to create a new registration, so click on **New registration**:

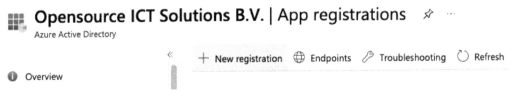

Figure 13.21 – Azure enterprise application – App registrations

8.   Simply give your registration a new name and keep the rest of the settings as the default:

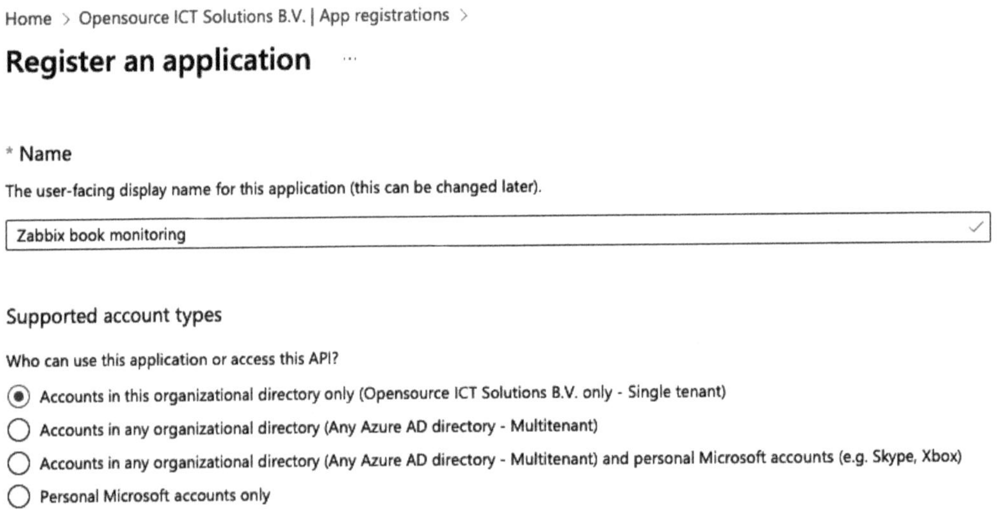

Home > Opensource ICT Solutions B.V. | App registrations >

Register an application   ⋯

* Name

The user-facing display name for this application (this can be changed later).

Zabbix book monitoring   ✓

Supported account types

Who can use this application or access this API?

◉ Accounts in this organizational directory only (Opensource ICT Solutions B.V. only - Single tenant)

○ Accounts in any organizational directory (Any Azure AD directory - Multitenant)

○ Accounts in any organizational directory (Any Azure AD directory - Multitenant) and personal Microsoft accounts (e.g. Skype, Xbox)

○ Personal Microsoft accounts only

Figure 13.22 – Azure enterprise application – new app registration

9. Click **Register** to finish this registration. This will redirect you to your newly created registration.

10. Now let's add the authentication. Go to **Certificates & secrets** in the left-hand sidebar.

11. We'll create a new client secret here. To do so, click on **New client secret**:

Certificates (0)   **Client secrets (0)**   Federated credentials (0)

A secret string that the application uses to prove its identity when requesting a token. Also can be referred to as application password.

+ New client secret

Description	Expires	Value ⓘ	Secret ID

No client secrets have been created for this application.

Figure 13.23 – Azure enterprise application – app registration secrets

12. All we have to do now is name the secret and set an expiry time period. Keep in mind that a shorter expiry means more administrative overhead. Faster expiry could mean better security as there is less time to potentially leak (or use once leaked) the secrets:

Figure 13.24 – Azure enterprise application – app registration secret creation

13. Now click on **Add** to finish setting up the new secret. It will show you the values once. Make sure to store them somewhere safe, such as in a password vault:

Figure 13.25 – Azure enterprise application secrets

14. With the authentication out of the way, there is only one thing left to do. We need to provide the correct permissions to this new enterprise application. To do so, search for `Subscriptions` in the Azure search bar at the top of the page.

15. For things such as Azure virtual machine and database instance monitoring, you will need to assign (read) permissions to your entire subscription. Find the subscription where your resources are located. Mine is called **OICTS Azure**:

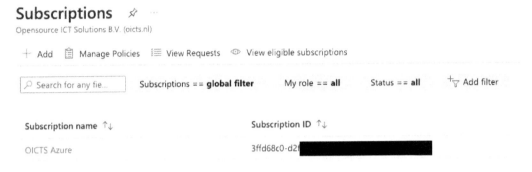

Figure 13.26 – Azure subscriptions

16. Now is also a great time to write down the subscription ID, as we will need it in a later step!

17. Select your subscription, and then from the list, select **Access control (IAM)**. Then, click on **Add role assignment**.

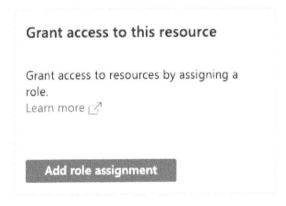

Figure 13.27 – Azure subscription – role assignment

18. On the next page, select the **Reader** role from the list and then press **Next**.

19. At the **Members** part of the creation process, click on **+ Select members**. We'll add the **Zabbix book monitoring** member. It will look as follows:

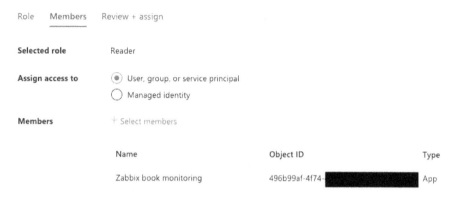

Figure 13.28 – Azure subscription – role assignment members

20. Now click on **Review + assign** and the permissions will be added.

21. There's one more thing to do in the Azure portal. In the search bar at the top of the page, type in `Tenant properties` and select it from the list. On this page, make sure to note down the tenant ID as we will need it shortly.

22. With the application set up, the authentication created, and the permissions assigned, let's move on to the Zabbix frontend. Navigate to **Data collection | Hosts** and create a new host by clicking on **Create host** in the top-right corner:

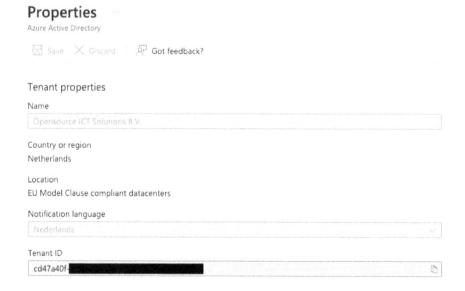

Figure 13.29 – Azure tenant properties

23. Create the following host, with the name `lar-book-azure`, the **Azure by HTTP** template, and a host group such as **Cloud**:

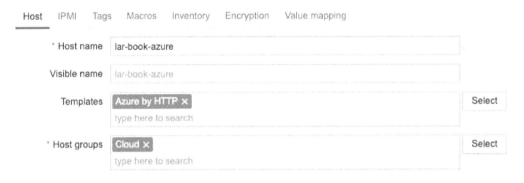

Figure 13.30 – New Azure monitoring host in Zabbix

24. Before adding the host, switch to the **Macros** tab:

Figure 13.31– New Azure monitoring host macros in Zabbix

We will have to add at least the following macros here:

- For {$AZURE.APP.ID}, fill in the application ID from *step 5*.

- For {$AZURE.PASSWORD}, fill in the value under the **Value** column from *step 13*.

- For {$AZURE.SUBSCRIPTION.ID}, fill in the subscription ID from *step 15*.

- For {$AZURE.TENANT.ID}, fill in the tenant ID from *step 22*.

25. That's it; you can now add the new host by clicking on the **Add** button.

26. After the discovery rule runs for the first time, your discovered instances will be added as new hosts, as you can see in the following screenshot:

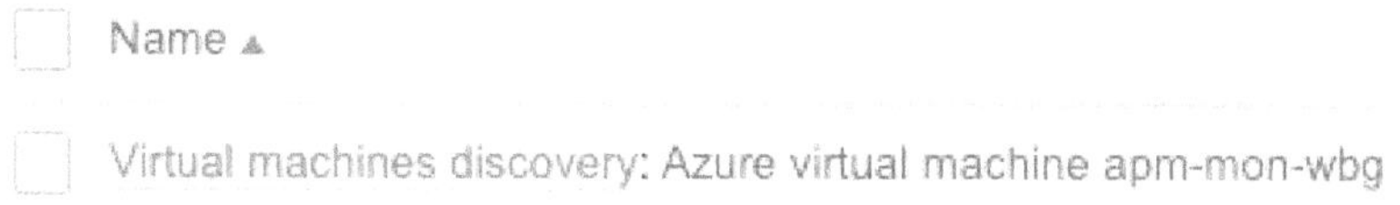

☐ Name ▲

☐ Virtual machines discovery: Azure virtual machine apm-mon-wbg

Figure 13.32 – New Azure-discovered virtual machine

That's it, your automated Azure monitoring is now working as expected. Let's have a look at how it works.

How it works...

If you've followed the recipe on AWS monitoring, you might think that Azure monitoring works in the exact same way. To an extent, that is true; the monitoring is completely based on API calls made from Zabbix toward the Azure API.

What is different between AWS and Azure is of course going to be the JavaScript scripts used in the Zabbix items on the templates.

The templates provided by Zabbix 7.0 out of the box use a fairly extensive piece of JavaScript code to execute API calls toward Azure, parse through the received data, and then put that into a JSON array that Zabbix low-level discovery understands.

Looking at the template found at **Data collection | Templates** and then opening **Discovery** for the **Azure by HTTP** template, we can see six discovery rules.

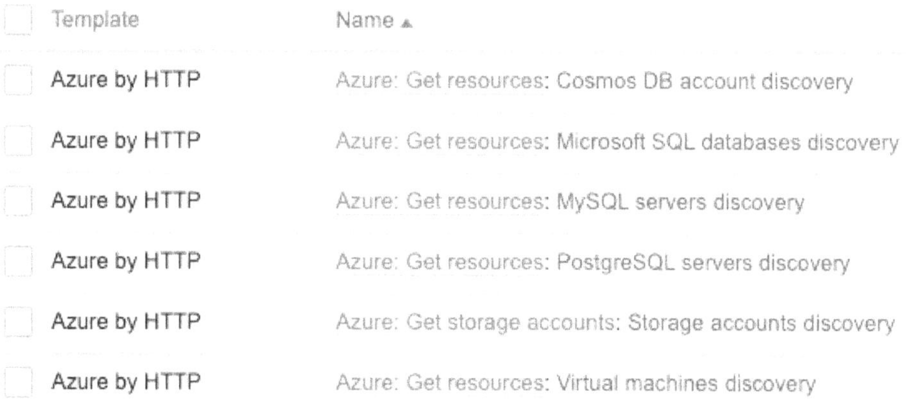

Template	Name ▲
Azure by HTTP	Azure: Get resources: Cosmos DB account discovery
Azure by HTTP	Azure: Get resources: Microsoft SQL databases discovery
Azure by HTTP	Azure: Get resources: MySQL servers discovery
Azure by HTTP	Azure: Get resources: PostgreSQL servers discovery
Azure by HTTP	Azure: Get storage accounts: Storage accounts discovery
Azure by HTTP	Azure: Get resources: Virtual machines discovery

Figure 13.33 – Azure by HTTP discovery rules

These six rules discover the different types of Azure database instances and virtual machines and use host prototypes to create a new host for each instance found. The only difference here is that storage accounts won't use host prototypes but item prototypes to supply you with information. The hosts created by host prototypes will then in turn use their own templates to get the actual statistics from those instances or buckets, as we can see in the template list:

Azure Cosmos DB for MongoDB by HTTP

Azure Microsoft SQL Database by HTTP

Azure Microsoft SQL Serverless Database by HTTP

Azure MySQL Flexible Server by HTTP

Azure MySQL Single Server by HTTP

Azure PostgreSQL Flexible Server by HTTP

Azure PostgreSQL Single Server by HTTP

Azure Virtual Machine by HTTP

Figure 13.34 – The other Azure templates in Zabbix 7.0

In my case, only one virtual machine was discovered, and as such, that host was added with the **Azure Virtual Machine by HTTP** template, as seen in *Figure 13.32*.

All of the information is then collected by **Script** item types with their own unique JavaScript code. We can see a piece of the code in the following screenshot, where we make a call to Azure to a specific URL (underlined):

```
if (!('auth' in data.errors)) {
  try{
    health = AzureVM.request('https://management.azure.com' + AzureVM.params.resource_id + '/providers/Microsoft.ResourceHealth/availabilityStatuses?api-version=2020-05-01');
    if ('value' in health && Array.isArray(health.value) && health.value.length > 0 && 'properties' in health.value[0] && typeof health.value[0].properties === 'object') {
    data.health = health.value[0].properties;
    }
  }
  catch (error) {
    data.errors.health = error.toString();
  }
```

Figure 13.35 – Azure template call

It is also possible to edit this JavaScript code to create entirely new calls to retrieve your own data and create different types of monitoring, as well as simply extending the out-of-the-box templates.

There's more...

We can discover way more from Azure using the method applied in this recipe. The JavaScript we employ is used to get metrics from Azure, which can be edited to gather almost any metric from the Azure API.

Check out the Azure API documentation for more information on the metrics retrieved using JavaScript:

```
https://learn.microsoft.com/en-us/rest/api/azure/
```

Building your Zabbix Docker monitoring

Ever since the release of Zabbix 5, monitoring our Docker containers became a lot easier with the introduction of Zabbix agent 2 and plugins. Using Zabbix agent 2 and Zabbix 7, we are able to monitor our Docker containers out of the box.

In this recipe, we are going to see how to set this up and how it works.

Getting ready

For this recipe, we require some Docker containers. We won't go over the setup of Docker containers, so make sure to do this yourself. Furthermore, we are going to need Zabbix agent 2 installed on the host running these Docker containers. Zabbix agent does not work in relation to this recipe; Zabbix agent 2 is required.

We also need our Zabbix server to actually monitor the Docker containers. We will call our Zabbix server zbx-home.

How to do it...

Let's waste no more time and dive right into the process of monitoring your Docker setup with Zabbix:

1. First things first, log in to the Linux CLI of the host running your Docker container(s).
2. Add the repository for installing Zabbix components.

   For RHEL-based systems, use the following:

   ```
   rpm -Uvh https://repo.zabbix.com/zabbix/7.0/rhel/8x86_64/zabbix-
   release-7.0-1.el8.noarch.rpm
   dnf clean all
   ```

   For Ubuntu systems, use the following:

   ```
   wget https://repo.zabbix.com/zabbix/7.0/ubuntu/pool/main/z/
   zabbix-release/zabbix-release_7.0-1+ubuntu22.04_all.deb
   dpkg -i zabbix-release_7.0-1+ubuntu22.04_all.deb
   apt update
   ```

3. Now, install Zabbix agent 2 with the following command.

    For RHEL-based systems, use the following:

    ```
    dnf install zabbix-agent2
    ```

    For Ubuntu systems, use the following:

    ```
    apt install zabbix-agent2
    ```

4. Following installation, make sure to edit the configuration file of the newly installed Zabbix agent 2 with the help of the following command:

    ```
    vim /etc/zabbix/zabbix_agent2.conf
    ```

5. Find the line that says `Server` and add your Zabbix server IP address to the file, as follows:

    ```
    Server=10.16.16.102
    ```

6. Now, we need to add the `zabbix` user to the Docker group by executing the following command:

    ```
    gpasswd -a zabbix docker
    ```

7. Make sure to save the file and then restart Zabbix agent 2 with the following command:

    ```
    systemctl restart zabbix-agent2
    ```

8. Now, navigate to your Zabbix server frontend. Go to **Data collection | Hosts** and click on the blue **Create host** button.

9. Let's create a new host called `Docker containers` and make sure to link the **Docker by Zabbix agent 2** template to the host.

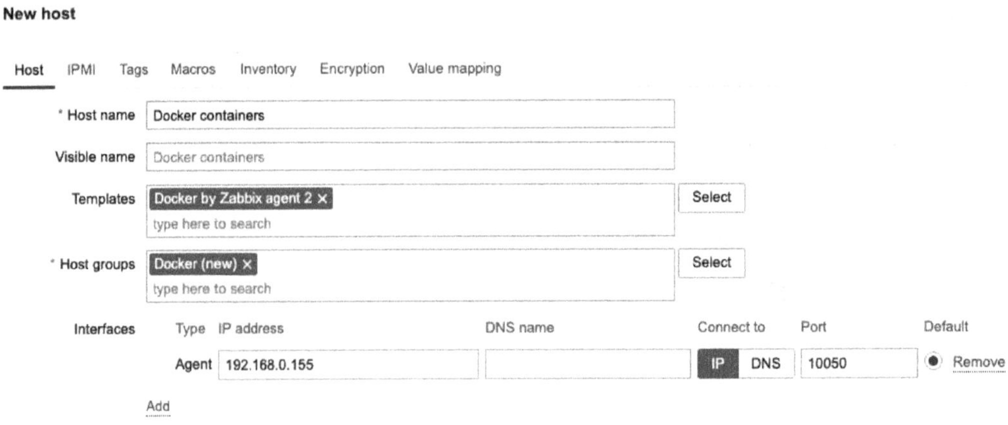

Figure 13.36 – New Docker host configuration

That's all there is to monitoring Docker containers with the Zabbix server. Let's now see how it works.

How it works...

Docker monitoring in Zabbix these days is easy, due to the new Zabbix agent 2 support and default templates. On occasion, though, a default template does not cut it, so let's break down the items used.

Almost all the items we can see on our host are dependent items, most of which are dependent on the master item, `Docker: Get info`. This master item is the most important item on our Docker template. It executes the `docker.info` item key, which is built into the new Zabbix agent 2. This item retrieves a list with all kinds of information from our Docker setup. We use the dependent items and preprocessing to get the values we want from this master item.

The Docker template also contains two Zabbix discovery rules, one to discover Docker images and one to discover Docker containers. If we check out the discovery rule for Docker containers called `Containers discovery`, we can see what happens. Our Zabbix Docker host will use the `docker.containers.discovery` item key to find every container and put this in the `{#NAME}` LLD macro. In the item prototypes, we then use this `{#NAME}` LLD macro to discover statistics with another master item, such as `docker.container_info`. From this master item, we then use the dependent items and preprocessing again to include this information in other item prototypes as well. We are now monitoring a bunch of statistics straight from our Docker setup.

If you want to get values from Docker that aren't in the default template, check out the information collected with the master items on the template. Use a new dependent item (prototype) and then use preprocessing to get the correct data from the master item.

There's more...

If you want to learn more about the Zabbix agent 2 Docker item keys, check out the supported item key list for Zabbix agent 2 in the Zabbix documentation:

```
https://www.zabbix.com/documentation/current/en/manual/config/items/
itemtypes/zabbix_agent/zabbix_agent2?s[]=docker.
```

Index

packtpub.com

Subscribe to our online digital library for full access to over 7,000 books and videos, as well as industry leading tools to help you plan your personal development and advance your career. For more information, please visit our website.

Why subscribe?

- Spend less time learning and more time coding with practical eBooks and Videos from over 4,000 industry professionals

- Improve your learning with Skill Plans built especially for you

- Get a free eBook or video every month

- Fully searchable for easy access to vital information

- Copy and paste, print, and bookmark content

Did you know that Packt offers eBook versions of every book published, with PDF and ePub files available? You can upgrade to the eBook version at packtpub.com and as a print book customer, you are entitled to a discount on the eBook copy. Get in touch with us at customercare@packtpub.com for more details.

At www.packtpub.com, you can also read a collection of free technical articles, sign up for a range of free newsletters, and receive exclusive discounts and offers on Packt books and eBooks.

Other Books You May Enjoy

If you enjoyed this book, you may be interested in these other books by Packt:

Network Automation with Nautobot

Jason Edelman, Glenn Matthews, Josh VanDeraa, Ken Celenza, Christian Adell, Brad Haas, Bryan Culver, John Anderson, Gary Snider

ISBN: 978-1-83763-786-7

- Understand network sources of truth and the role they play in network automation architecture
- Gain an understanding of Nautobot as a network source and a network automation platform
- Convert Python scripts to enable self-service Nautobot Jobs
- Understand how YAML files in Git can be easily integrated into Nautobot
- Get to grips with the NetDevOps ecosystem around Nautobot and its app ecosystem
- Delve into popular Nautobot Apps including Single Source of Truth and Golden Config

Security Monitoring with Wazuh

Rajneesh Gupta

ISBN: 978-1-83763-215-2

- Find out how to set up an intrusion detection system with Wazuh
- Get to grips with setting up a file integrity monitoring system
- Deploy Malware Information Sharing Platform (MISP) for threat intelligence automation to detect indicators of compromise (IOCs)
- Explore ways to integrate Shuffle, TheHive, and Cortex to set up security automation
- Apply Wazuh and other open source tools to address your organization's specific needs
- Integrate Osquery with Wazuh to conduct threat hunting

Packt is searching for authors like you

If you're interested in becoming an author for Packt, please visit `authors.packtpub.com` and apply today. We have worked with thousands of developers and tech professionals, just like you, to help them share their insight with the global tech community. You can make a general application, apply for a specific hot topic that we are recruiting an author for, or submit your own idea.

Share Your Thoughts

Now you've finished *Zabbix 7 IT Infrastructure Monitoring Cookbook*, we'd love to hear your thoughts! Scan the QR code below to go straight to the Amazon review page for this book and share your feedback or leave a review on the site that you purchased it from.

`https://packt.link/r/1801078327`

Your review is important to us and the tech community and will help us make sure we're delivering excellent quality content.

Download a free PDF copy of this book

Thanks for purchasing this book!

Do you like to read on the go but are unable to carry your print books everywhere?

Is your eBook purchase not compatible with the device of your choice?

Don't worry, now with every Packt book you get a DRM-free PDF version of that book at no cost.

Read anywhere, any place, on any device. Search, copy, and paste code from your favorite technical books directly into your application.

The perks don't stop there, you can get exclusive access to discounts, newsletters, and great free content in your inbox daily

Follow these simple steps to get the benefits:

1. Scan the QR code or visit the link below

https://packt.link/free-ebook/978-1-80107-832-0

2. Submit your proof of purchase

3. That's it! We'll send your free PDF and other benefits to your email directly